GEOLOGY AND MINERALOGY RESEARCH DEVELOPMENTS

APATITE

SYNTHESIS, STRUCTURAL CHARACTERIZATION AND BIOMEDICAL APPLICATIONS

GEOLOGY AND MINERALOGY RESEARCH DEVELOPMENTS

Additional books in this series can be found on Nova's website under the Series tab.

Additional e-books in this series can be found on Nova's website under the e-book tab.

GEOLOGY AND MINERALOGY RESEARCH DEVELOPMENTS

APATITE

SYNTHESIS, STRUCTURAL CHARACTERIZATION AND BIOMEDICAL APPLICATIONS

MICHELE IAFISCO, PH.D.
AND
JOSÉ MANUEL DELGADO-LÓPEZ, PH.D.
EDITORS

nova publishers
New York

Library of Congress Cataloging-in-Publication Data

Apatite : synthesis, structural characterization, and biomedical applications / editors, Michele Iafisco and Josi Manuel Delgado-Lspez (Laboratory of Bioceramics and Bio-Hybrid Composites, CNR-Institute of Science and Technology for Ceramics, Italy and others).
pages cm. -- (Geology and mineralogy research developments)
Includes bibliographical references and index.
ISBN: 978-1-63321-500-9 (hardcover)
1. Apatite. 2. Apatite in the body. 3. Phosphates. I. Iafisco, Michele, editor. II. Delgado-Lspez, Josi Manuel, editor.
QE391.A6A45 2014
615.2'39324--dc23

2014026711

Published by Nova Science Publishers, Inc. † New York

CONTENTS

PREFACE

Mimicking Nature and designing bioinspired materials represents a promising way to reach smart technological innovations in many interdisciplinary scientific fields, including medicine. Biological materials exhibit a high degree of hierarchical organisation, hybridisation, efficiency, resistance and adaptability. These properties, which biogenic materials have achieved through specific building principles selected by evolution, can only be partially replicate in man-made materials by the present synthetic and technological processes. For this reason, nature is the main school for material scientists and a growing focus is devoted to prepare synthetic analogues, so-called "biomimetic", able to precisely mimic the morphological and physical-chemical features of biological compounds.

Calcium phosphates are the principal inorganic constituent of normal (bones, teeth, fish enameloid, deer antlers and some species of shells) and pathological (dental and urinary calculus and stones, atherosclerotic lesions) calcifications. Except for small portions of the inner ear, all hard tissues of the human body are made of calcium phosphates. The mineral phase of bone and dentine is commonly assimilated to synthetic hydroxyapatite ($Ca_{10}(PO_4)_6(OH)_2$), however, biological apatites differ from the stoichiometric ones in several features, including non-stoichiometry, crystals with dimensions in the nano-scale and low degree of structural order. Moreover, they are Ca and OH deficient and they contain significant amounts of foreign ions (such as CO_3, Na, K, Mg, etc.), either fully incorporated in the crystal lattice or just inserted in the non-apatitic sites of the surface. The low degree of crystallinity, the small crystal sizes, the existence of a surface hydrated layer non-apatitic in nature, and the Ca and OH deficiency together with the foreign ions substitution are responsible for the relatively high solubility and specific biological activity of bone apatite with respect to stoichiometric hydroxyapatite.

Materials based on calcium phosphates are well known for their excellent biocompatibility, particularly in establishing tight bone bonding that yield effective and rapid osteointegration. Thus, calcium phosphate based biomaterials are now used in many different applications throughout the body, covering all areas of the skeleton. Among the calcium phosphate compounds used as bone grafts, the most common is stoichiometric hydroxyapatite processed as dense or porous ceramics, coatings and composites. However, it is reasonable to hypothesize that by virtue of their high similarity with the mineral phase of bone, nano-sized and ionic substituted apatites can display better biological performances than coarser crystals. For this reason, the study of their crystallization mechanisms and the possibility to tailor the structural and physical-chemical properties of apatites open several possibilities to prepare optimal devices in medical sciences and also to investigate the biomineralization processes of living organisms.

This book is devoted to highlight the recent progresses on the preparation and characterization of multifunctional nanocrystalline apatites with tailored properties to be potentially used as healthcare materials (*i.e.* nanoparticles for drug or gene delivery and materials for bone substitution and regeneration).

According to the title, the book is divided in 3 main parts. The first part (chapters 1-3) is intended to be a comprehensive review of the importance and the occurrence of calcium phosphates in nature (geological and biological occurrence) and to underline the peculiar characteristics of nanocrystalline apatites. In this section new strategies to synthesize biomimetic and bioinspired apatites as well as the latest findings about their physical-chemical and thermodynamic properties are also analysed and described.

The aim of the second part (chapters 4-7) is to show the most important analytical techniques for the characterization of apatite features and to describe the new theories behind the apatite crystallization and dissolution processes. This part covers the use of vibrational spectroscopies as ideal technique to characterize nano-apatites: e.g., to study in detail the distribution of carbonate in biological apatites and to diagnose apatites deposits in skin calcified tissues. Finally, new progresses on high-temperature *in-situ* X-Ray investigations of apatites are also shown.

The third part (chapters 8-12) describes the traditional and the most exciting and innovative applications of pure, ionic-substituted, and organic-functionalized apatite as biomaterial. In particular, the use of apatite based materials as versatile tools in bone tissue engineering (as bioactive coatings for bioceramics, and reinforcement and bioactive phase for polymeric hydrogels) and for the delivery of genes, siRNAs, drug and proteins in nanomedicine are emphasised and highlighted.

This collection of contributions was only possible thank to the tough work and effort of all authors, who have gently shared their knowledge with us and devoted their valuable time to this project. The chapters of this book written by eminent scientists, everyone specialized in their own field, will serve as an important and useful reference source for a broad audience that consists of both expert (materials scientists, biomedical engineers and biologists) and non-expert readers wanting to enter in the field of apatite based biomaterials.

Michele Iafisco, Ph.D.
Laboratory of Bioceramics and Bio-hybrid Composites
CNR - Institute of Science and Technology for Ceramics
Via Granarolo,
Faenza (Ra) Italy
Tel: +39 0546 699730
Fax: +39 0546 46381
E-mail: michele.iafisco@istec.cnr.it

José Manuel Delgado-López, Ph.D.
Laboratorio de Estudios Cristalogr´aficos
Instituto Andaluz de Ciencias de la Tierra (CSIC-UGR)
Armilla, Granada, Spain
Tel: 958 230 000 (Ext. 190109)
jmdl@lec.csic.es

In: Apatite
Editors: M. Iafisco and J. M. Delgado-López ISBN: 978-1-63321-500-9

Chapter 1

CALCIUM ORTHOPHOSPHATES: OCCURRENCE, PROPERTIES AND MAJOR APPLICATIONS

Sergey V. Dorozhkin*
Moscow, Russia

ABSTRACT

The present overview is intended to point the readers' attention to the important subject of calcium orthophosphates. They are of the special significance for the human beings because they represent the inorganic part of major normal (bones, teeth and antlers) and pathological (i.e., those appearing due to various diseases) calcified tissues of mammals. Therefore, the majority of the artificially prepared calcium orthophosphates of high purity appear to be well tolerated by human tissues *in vivo* and possess the excellent biocompatibility, osteoconductivity and bioresorbability. These biomedical properties of calcium orthophosphates are widely used to construct bone grafts. In addition, natural calcium orthophosphates are the major source of phosphorus, which are used to produce agricultural fertilizers, detergents and various phosphorus-containing chemicals. Thus, there is a great significance of calcium orthophosphates for the humankind and, here, an overview on the current knowledge on this subject is provided.

Keywords: Calcium orthophosphates, hydroxyapatite, fluorapatite, occurrence, properties, applications

INTRODUCTION

Due to the abundance in nature (as phosphate ores) and presence in living organisms (as bones, teeth, deer antlers and the majority of various pathological calcifications), calcium orthophosphates are the inorganic compounds of a special interest for human being. They were discovered in 1769 and have been investigated since then [1, 2]. According to the databases of scientific literature (Web of knowledge, Scopus, Medline, *etc.*), the total amount

* E-mail: sedorozhkin@yandex.ru.

of currently available publications on the subject exceeds 40,000 with the annual increase for, at least, 2000 papers. This is a clear confirmation of the importance.

Table 1. Existing calcium orthophosphates and their major properties [7, 8]

Ca/P molar ratio	Compound	Formula	Solubility at 25 °C, -log(K_s)	Solubility at 25 °C, g/L	pH stability range in aqueous solutions at 25°C
0.5	Monocalcium phosphate monohydrate (MCPM)	$Ca(H_2PO_4)_2{\cdot}H_2O$	1.14	~ 18	0.0 – 2.0
0.5	Monocalcium phosphate anhydrous (MCPA or MCP)	$Ca(H_2PO_4)_2$	1.14	~ 17	[c]
1.0	Dicalcium phosphate dihydrate (DCPD), mineral brushite	$CaHPO_4{\cdot}2H_2O$	6.59	~ 0.088	2.0 – 6.0
1.0	Dicalcium phosphate anhydrous (DCPA or DCP), mineral monetite	$CaHPO_4$	6.90	~ 0.048	[c]
1.33	Octacalcium phosphate (OCP)	$Ca_8(HPO_4)_2(PO_4)_4{\cdot}5H_2O$	96.6	~ 0.0081	5.5 – 7.0
1.5	α-Tricalcium phosphate (α-TCP)	α-$Ca_3(PO_4)_2$	25.5	~ 0.0025	[a]
1.5	β-Tricalcium phosphate (β-TCP)	β-$Ca_3(PO_4)_2$	28.9	~ 0.0005	[a]
1.2 – 2.2	Amorphous calcium phosphates (ACP)	$Ca_xH_y(PO_4)_z{\cdot}nH_2O$, n = 3 – 4.5; 15 – 20% H_2O	[b]	[b]	~ 5 – 12 [d]
1.5 – 1.67	Calcium-deficient hydroxyapatite (CDHA or Ca-def HA)[e]	$Ca_{10-x}(HPO_4)_x(PO_4)_{6-x}(OH)_{2-x}$ $(0<x<1)$	~ 85	~ 0.0094	6.5 – 9.5
1.67	Hydroxyapatite (HA, HAp or OHAp)	$Ca_{10}(PO_4)_6(OH)_2$	116.8	~ 0.0003	9.5 – 12
1.67	Fluorapatite (FA or FAp)	$Ca_{10}(PO_4)_6F_2$	120.0	~ 0.0002	7 – 12
1.67	Oxyapatite (OA, OAp or OXA)[f], mineral voelckerite	$Ca_{10}(PO_4)_6O$	~ 69	~ 0.087	[a]
2.0	Tetracalcium phosphate (TTCP or TetCP), mineral hilgenstockite	$Ca_4(PO_4)_2O$	38 – 44	~ 0.0007	[a]

[a] These compounds cannot be precipitated from aqueous solutions.

[b] Cannot be measured precisely. However, the following values were found: 25.7±0.1 (pH = 7.40), 29.9±0.1 (pH = 6.00), 32.7±0.1 (pH = 5.28) [236]. The comparative extent of dissolution in acidic buffer is: ACP >> α-TCP >> β-TCP > CDHA >> HA > FA [107].

[c] Stable at temperatures above 100°C.

[d] Always metastable.

[e] Occasionally, it is called "precipitated HA (PHA)".

[f] Existence of OA remains questionable.

Briefly, by definition, all calcium orthophosphates consist of three major chemical elements: calcium (oxidation state +2), phosphorus (oxidation state +5) and oxygen (reduction state –2), as a part of the orthophosphate anions. These three chemical elements are present in abundance on the surface of our planet: oxygen is the most widespread chemical element of the earth's surface (~ 47 mass %), calcium occupies the fifth place (~ 3.3 – 3.4 mass %) and phosphorus (~ 0.08 – 0.12 mass %) is among the first twenty of the chemical elements most widespread on our planet [3]. In addition, the chemical composition of many calcium orthophosphates includes hydrogen, as an acidic orthophosphate anion (for example, HPO_4^{2-} or $H_2PO_4^-$), hydroxide (for example, $Ca_{10}(PO_4)_6(OH)_2$) and/or incorporated water (for example, $CaHPO_4 \cdot 2H_2O$). Regarding their chemical composition, diverse combinations of CaO and P_2O_5 oxides (both in the presence of water and without it) provide a large variety of calcium phosphates, which are differentiated by the type of the phosphate anion. Namely, ortho- (PO_4^{3-}), meta- (PO_3^-), pyro- ($P_2O_7^{4-}$) and poly- ($(PO_3)_n^{n-}$) phosphates are known. Furthermore, in the case of multi-charged anions (valid for orthophosphates and pyrophosphates), the calcium phosphates are also differentiated by the number of hydrogen ions attached to the anion. Examples include mono- ($Ca(H_2PO_4)_2$), di- ($CaHPO_4$), tri- ($Ca_3(PO_4)_2$) and tetra- ($Ca_2P_2O_7$) calcium phosphates. Here, one must stress that prefixes "mono", "di", "tri" and "tetra" are related to the amount of hydrogen ions replaced by calcium [4-6]. However, to narrow the subject, calcium orthophosphates will be considered and discussed only. Their names, the standard abbreviations, chemical formulae and solubility values are listed in Table 1 [7, 8]. Since all of them belong to calcium *ortho*phosphates, strictly speaking, all abbreviations in Table 1 are incorrect; however, they have been extensively used in literature for decades and, to avoid any confusion, there is no need to modify them.

In general, the atomic arrangement of all calcium orthophosphates is built up around a network of orthophosphate (PO_4) groups, which stabilize the entire structure. Therefore, the majority of calcium orthophosphates are sparingly soluble in water (Table 1); however, all of them are easily soluble in acids but insoluble in alkaline solutions. In addition, all chemically pure calcium orthophosphates are colorless transparent crystals of moderate hardness but, as powders, they are of white color. Nevertheless, natural minerals of calcium orthophosphates are always colored due the presence of impurities and dopants, such as ions of Fe, Mn and rare earth elements [9, 10]. Biologically formed calcium orthophosphates are the major component of all mammalian calcified tissues [11], while the geologically formed ones are the major raw material to produce phosphorus-containing agricultural fertilizers, chemicals and detergents [12-14].

Geological and Biological Occurrences

Geologically, natural calcium orthophosphates are found in different regions mostly as deposits of apatites, mainly as ion-substituted FA (igneous rocks), and phosphorites (sedimentary rocks) [13-15]. In addition, natural ion-substituted CDHA was also found [16] but it is a very rare mineral. Some types of sedimentary rocks can be formed by weathering of igneous rocks into smaller particles [17]. Other types of sedimentary rocks can be composed of minerals precipitated from the dissolution products of igneous rocks or minerals produced by biomineralization (Figure 1) [18]. Thus, due to a sedimentary origin, both a general appearance and a chemical composition of natural phosphorites vary a lot [19, 20]. It is a common practice to consider francolite (or carbonate-hydroxyfluorapatite regarded as its synonym) as the basic phosphorite mineral [15, 21-24]. According to Henry [25], the name francolite was given by Mr. Brooke and Mr. Nuttall to a mineral from Wheal Franco, Tavistock, Devon, some years prior to 1850. A cryptocrystalline (almost amorphous) variety of francolite (partly of a biological origin) is called collophane (synonyms: collophanit, collophanita, collophanite, grodnolite, kollophan) [26-28]. It occurs in natural phosphorites predominantly as fossil bones and phosphatized microbial pseudomorphs: phosphatic crusts of chasmolithic biofilms (or microstromatolites) and globular clusters with intra-particular porosities [29-32]. Natural phosphorites (therefore, francolite and collophane as well) occur in various forms, such as nodules, crystals or masses. Occasionally, other types of natural calcium orthophosphates are found as minerals, for example clinohydroxylapatite [33], staffelite (synonyms: staffelit, staffelita) belonging to carbonate-rich fluorapatites (chemical formula: $Ca_5[(F,O)(PO_4,CO_3)_3]$) [34, 35] and DCPD [36]. Furthermore, calcium orthophosphates were found in meteoric stones [37, 38]. The world deposits of natural calcium orthophosphates are estimated to exceed 150 billion tons; from which approximately 85 % belong to phosphorites and the remaining ~ 15 % belong to apatites.

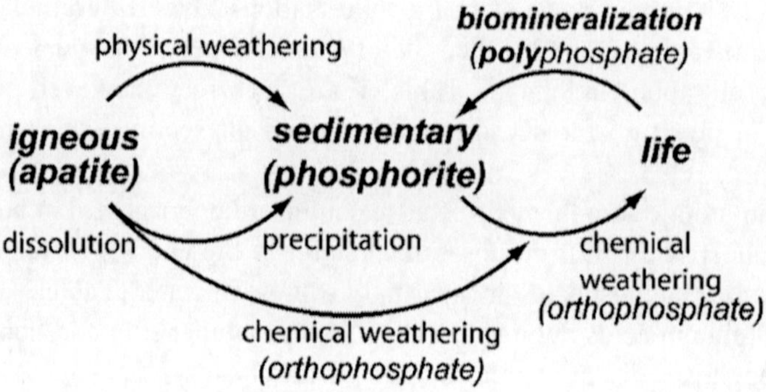

Figure 1. Simplified schematic of the phosphorus cycle from apatitic igneous rock to phosphorite sedimentary rock through chemical or physical weathering. Life forms accumulate soluble phosphorus species and can produce apatite through biomineralization. Reprinted from Ref. [18] with permission.

As minor constituents (< ~ 5 %), natural calcium orthophosphates (both apatites and phosphorites) occur in many geological environments. Concentrations sufficient for economic use (> 15 %) are also available. Namely, the largest world deposits of natural apatites are located in Russia (the Khibiny and Kovdor massifs, Kola peninsula [39, 40]), Brazil and Zambia, while the largest world deposits of natural phosphorites are located in Morocco, Russia, Kazakhstan, USA (Florida, Tennessee), China and Australia [12-15]. In addition, they are found at seabed and ocean floor [41]. The majority of natural calcium orthophosphates occur as small polycrystalline structures (spherulitic clusters). Larger crystals are rare [42]. They usually have the crystal structure of apatites (hexagonal system, space group $P6_3/m$). Giant crystals including "a solid but irregular mass of green crystalline apatite, 15 feet long and 9 feet wide" [43] and a single euhedral crystal from the Aetna mine measuring 2.1 × 1.2 m with an estimated weight of 6 tons [44] were found. None of them is a pure compound; they always contain dopants of other elements. For example, ions of calcium might be partially replaced by Sr, Ba, Mg, Mn, K, Na, Fe; ions of orthophosphate may be partly replaced by AsO_4^{3-}, CO_3^{2-} and VO_4^{3-} [45]; ions of hydroxide, chloride, bromide, carbonate and oxide may to a certain extent substitute fluoride in the crystal lattice of natural apatites

[23]. Furthermore, organic radicals have been found in natural apatites [46, 47]. In principle, the crystal structure of apatites can incorporate half the Periodic Chart in its atomic arrangement. In medicine, this property might be used as an antidote for heavy metal intoxication [48]. Ease of atomic substitution for apatite leaves this mineral open to a wide array of compositions. This might be related to the fact that the apatite structure type displays porous properties [49]. The substitutions in apatites are usually in trace concentrations but for certain dopants large concentrations and even complete solid solutions exist (e.g., F^- and OH^-). To make things even more complicated, some ions in the crystal structure may be missing, leaving the crystallographic defects, which leads to formation of non-stoichiometric compounds, such as CDHA. Figure 2 shows examples of polycrystalline and single-crystalline samples of natural FA.

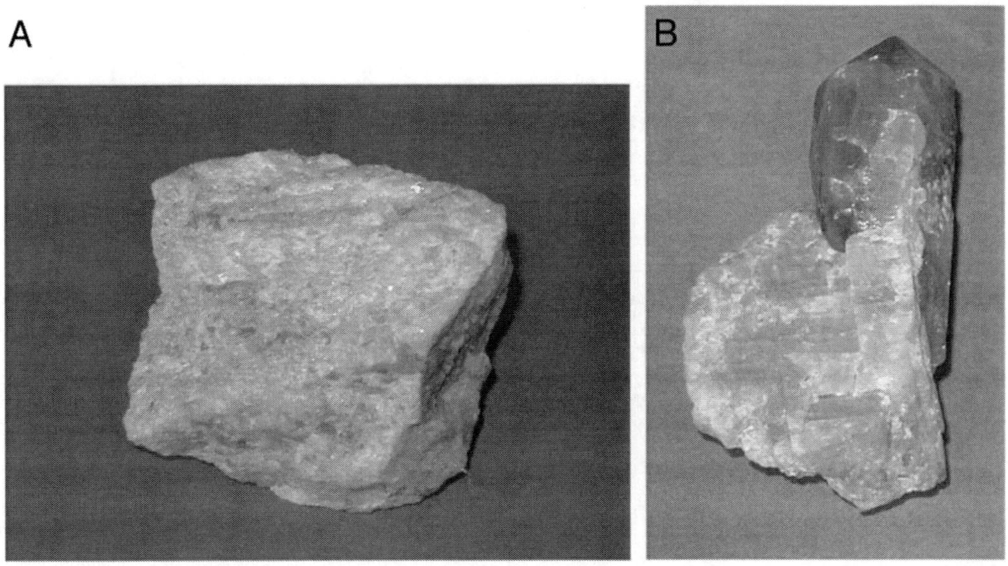

Figure 2. Polycrystalline (a) and single-crystalline (b) FA of a geological origin. The single crystal has a grey-green color due to incorporated ions of transition metals.

Manufacturing of elementary phosphorus (white and red) [50, 51], phosphoric acids [13, 52-55], various phosphorus-containing chemicals, agricultural fertilizers (namely, superphosphate [56-58], ammonium orthophosphates [59]) and detergents (principally sodium tripolyphosphate [60]) are the major industrial applications of natural calcium orthophosphates. The annual consumption of a phosphate rock has approached ~ 150 million tons and about 95 percent of this production is utilized in the fertilizer industry [61, 62].

In biological systems, many organisms, ranging from bacteria and isolated cells to invertebrates and vertebrates, synthesize calcium orthophosphates [18]. Formation of calcium orthophosphates in primitive organisms is believed to enable the storage and regulation of essential elements such as calcium, phosphorus and, possibly, magnesium. The morphology of precipitates in these organisms (small intracellular nodules of ACP often located in mitochondria) complies with the necessities for rapid mobilization and intracellular control of the concentration of these elements [63]. In vertebrates calcium orthophosphates occur as the principal inorganic constituent of normal (bones, teeth, fish enameloid, deer antlers and some species of shells) and pathological (dental and urinary calculus and stones, atherosclerotic lesions, *etc.*) calcifications [11, 64-69]. Except for small portions of the inner ear, all hard tissue of the human body is formed of calcium orthophosphates. Structurally, they occur mainly in the form of poorly crystalline, non-stoichiometric, Ca-deficient, Na-, Mg- and carbonate-containing HA. It is often called "biological apatite" [70-72] (which might be abbreviated as BAp [73, 74]), bioapatite [75-78e] or dahllite. The latter was named in 1888 by Broeger and Bäckstroem [79] after the Swedish mineralogist brothers Tellef and Johan Martin Dahll.

The main constituents of human bones are calcium orthophosphates (~ 60 – 70 wt.%), collagen (~ 20 – 30 wt.%) and water (up to 10 wt.%) [68, 75, 80, 81]. Detailed information on the chemical composition of the most important human normal calcified tissues is comprised in Table 2. One should note that the values mentioned in Table 2 are approximate; the main constituents can vary by a percent or more [82].

Table 2. Comparative composition and structural parameters of inorganic phases of adult human calcified tissues. Due to the considerable variation found in biological samples, typical values are given in these cases [4, 81]

Composition, wt%	Enamel	**Dentine**	**Cementum**	**Bone**	**HA**
Calcium[a]	36.5	35.1	~ 35	34.8	39.6
Phosphorus (as P)[a]	17.7	16.9	~ 16	15.2	18.5
Ca/P (molar ratio)[a]	1.63	1.61	~ 1.65	1.71	1.67
Sodium[a]	0.5	0.6	[c]	0.9	–
Magnesium[a]	0.44	1.23	0.5 – 0.9	0.72	–
Potassium[a]	0.08	0.05	[c]	0.03	–
Carbonate (as CO_3^{2-})[b]	3.5	5.6	[c]	7.4	–
Fluoride[a]	0.01	0.06	up to 0.9	0.03	–
Chloride[a]	0.30	0.01	[c]	0.13	–
Pyrophosphate (as $P_2O_7^{4-}$)[b]	0.022	0.10	[c]	0.07	–
Total inorganic[b]	97	70	60	65	100
Total organic[b]	1.5	20	25	25	–
Water[b]	1.5	10	15	10	–
Crystallographic properties: Lattice parameters (± 0.003 Å)					
a-axis, Å	9.441	9.421	[c]	9.41	9.430
c-axis, Å	6.880	6.887	[c]	6.89	6.891
Crystallinity index (HA = 100)	70 – 75	33 – 37	~ 30	33 – 37	100
Typical crystal sizes (nm) [11]	100 µm ×50×50	35×25×4	[c]	50×25×4	200 – 600
Ignition products (800 ºC)	β-TCP + HA	β-TCP+ HA	β-TCP+ HA	HA + CaO	HA
Elastic modulus (GPa)	80	23.8 ± 3.7	15.0 ± 3.6	0.34 – 13.8	10
Tensile strength (MPa)	10	100	[c]	150	100

[a] Ashed samples.

[b] Unashed samples.

[c] Numerical values were not found in the literature but they should be similar to those for dentine.

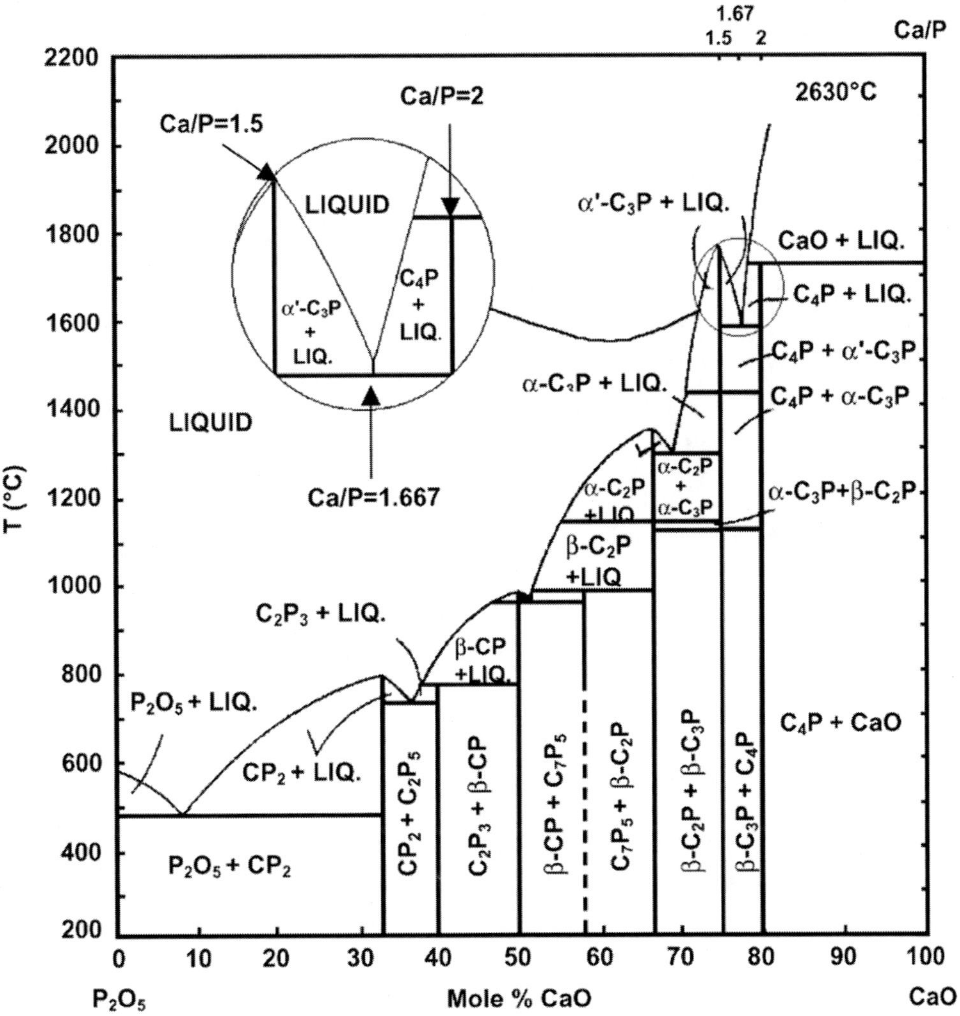

Figure 3. Phase diagram of the system CaO – P_2O_5 (C = CaO, P = P_2O_5) at elevated temperatures. Here: C_7P_5 means $7CaO \cdot 5P_2O_5$; other abbreviations should be written out in the same manner. Reprinted from Refs. [86, 87] with permission.

THE MEMBERS OF CALCIUM ORTHOPHOSPHATE FAMILY

In the ternary aqueous system $Ca(OH)_2$ – H_3PO_4 – H_2O (or CaO – P_2O_5 – H_2O) [83-85] there are twelve known non-ion-substituted calcium orthophosphates with the Ca/P molar ratio ranging between 0.5 and 2.0 (Table 1). An anhydrous phase diagram CaO – P_2O_5 at temperatures within 200 – 2200 °C is shown in Figure 3 [86, 87]. Table 3 comprises crystallographic data of the existing calcium orthophosphates [5, 88, 89]. The most important parameters of calcium orthophosphates are the ionic Ca/P ratio, basicity / acidity and solubility. All these parameters strongly correlate with the solution pH. The lower the Ca/P molar ratio is, the more acidic and water-soluble the calcium orthophosphate is [4-6]. One can see that the solubility ranges from high values for acidic compounds, such as MCPM, to very low values for basic compounds, such as apatites, which allow calcium orthophosphates to be dissolved, transported from one place to another and precipitated, when necessary. Crystallization, dissolution and phase transformation processes of different calcium orthophosphates under various experimental conditions have been reviewed [90]. Regarding applications, some of them might be used in food industry and, according to the European classification of food additives, calcium orthophosphates of food grade quality are known as E341 additive.

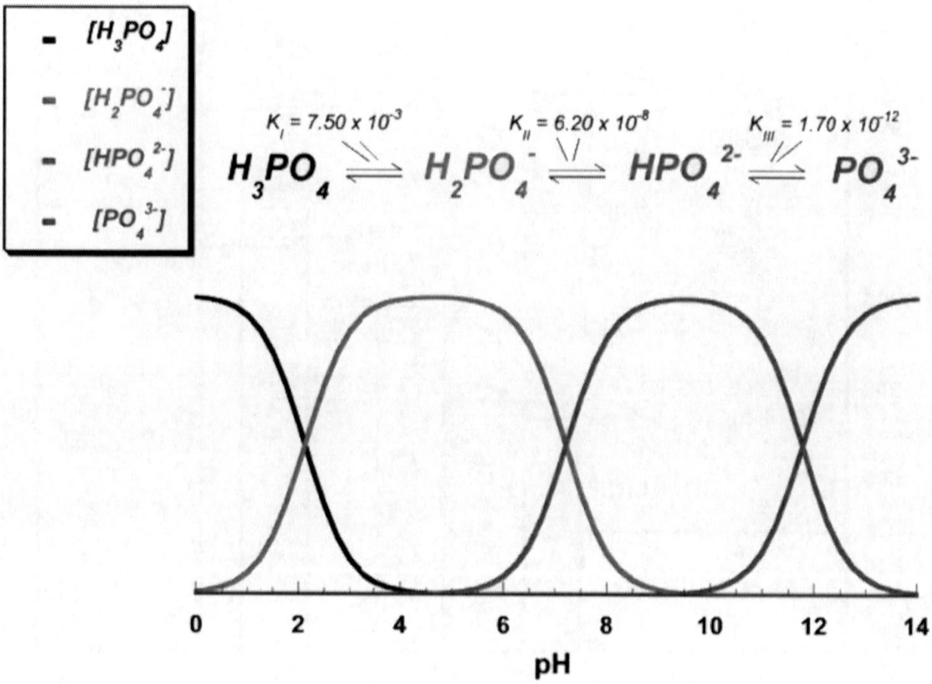

Figure 4. pH variation of ionic concentrations in triprotic equilibrium for phosphoric acid solutions. Reprinted from Ref. [91] with permission.

Due to the triprotic equilibrium that exists within orthophosphate-containing solutions, variations in pH alter the relative concentrations of the four polymorphs of orthophosphoric acid (Figure 4) [91] and thus both the chemical composition (Figure 5) [92] and the amount of the calcium orthophosphates that are formed by a direct precipitation. The solubility isotherms of different calcium orthophosphates are shown in Figure 6 [5, 6, 84, 85, 93, 94]. However, in 2009, the classic solubility data of calcium orthophosphates [5, 6, 84, 85, 93, 94] were mentioned to be inappropriate [95]. According to the authors of the latter study, all previous solubility calculations were based on simplifications, which were only crudely approximate. The problem lies in incongruent dissolution, leading to phase transformations and lack of the detailed solution equilibria. Using an absolute solid-titration approach, the true solubility isotherm of HA was found to lie substantially lower than previously reported. In addition, contrary to a wide belief, DCPD appeared not to be the most stable phase below pH ~ 4.2, where CDHA was less soluble [95].

Table 3. Crystallographic data of calcium orthophosphates [5, 88, 89]

Compound	Space group	Unit cell parameters	Z[a]	Density, g cm^{-3}
MCPM	triclinic $P\bar{1}$	a = 5.6261(5), b = 11.889(2), c = 6.4731(8) Å, α = 98.633(6)°, β = 118.262(6)°, γ = 83.344(6)°	2	2.23
MCPA	triclinic $P\bar{1}$	a = 7.5577(5), b = 8.2531(6), c = 5.5504(3) Å, α = 109.87(1)°, β = 93.68(1)°, γ = 109.15(1)°	2	2.58
DCPD	monoclinic *I*a	a = 5.812(2), b = 15.180(3), c = 6.239(2) Å, β = 116.42(3)°	4	2.32
DCPA	triclinic $P\bar{1}$	a = 6.910(1), b = 6.627(2), c = 6.998(2) Å, α = 96.34(2)°, β = 103.82(2)°, γ = 88.33(2)°	4	2.89
OCP	triclinic $P\bar{1}$	a = 19.692(4), b = 9.523(2), c = 6.835(2) Å, α = 90.15(2)°, β = 92.54(2)°, γ = 108.65(1)°	1	2.61
α-TCP	monoclinic $P2_1/a$	a = 12.887(2), b = 27.280(4), c = 15.219(2) Å, β = 126.20(1)°	24	2.86
β-TCP	rhombohedral *R*3cH	$a = b$ = 10.4183(5), c = 37.3464(23) Å, γ = 120°	21[b]	3.08
HA	monoclinic $P2_1/b$ or hexagonal $P6_3/m$	a = 9.84214(8), b = 2a, c = 6.8814(7) Å, γ = 120° (monoclinic) $a = b$ = 9.4302(5), c = 6.8911(2) Å, γ = 120° (hexagonal)	4 2	3.16
FA	hexagonal $P6_3/m$	$a = b$ = 9.367, c = 6.884 Å, γ = 120°	2	3.20
OA	hexagonal $P\bar{6}$	$a = b$ = 9.432, c = 6.881 Å, α = 90.3°, β = 90.0°, γ = 119.9°	1	~ 3.2
TTCP	monoclinic $P2_1$	a = 7.023(1), b = 11.986(4), c = 9.473(2) Å, β = 90.90(1)°	4	3.05

[a] Number of formula units per unit cell.
[b] Per the hexagonal unit cell.

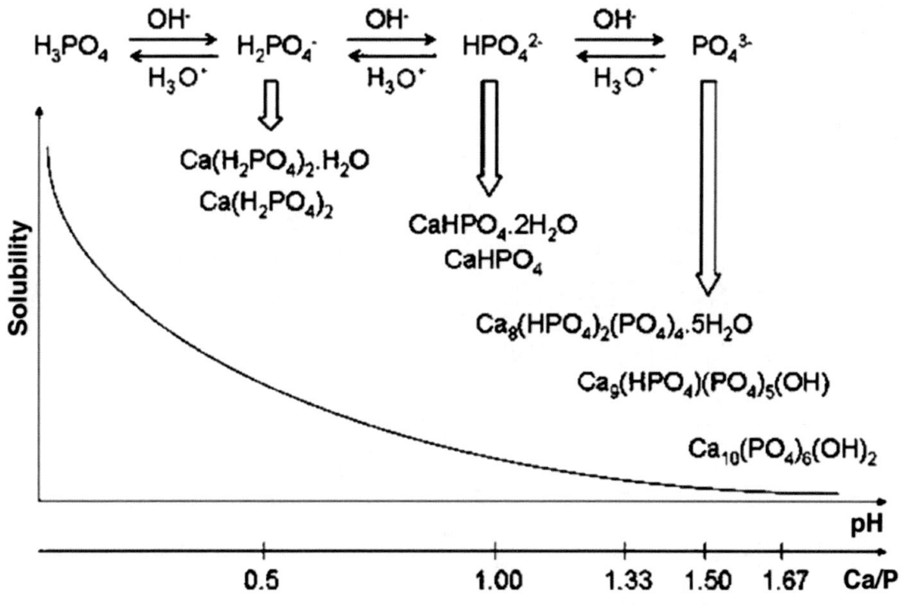

Figure 5. Various calcium orthophosphates obtained by neutralizing of orthophosphoric acid. Ca/P are reported in the figure. The solubility of calcium orthophosphates in water decreases drastically from left to right, HA being the most insoluble and stable phase. Reprinted from Ref. [91] with permission.

A brief description of all known calcium orthophosphates (Table 1) is given below.

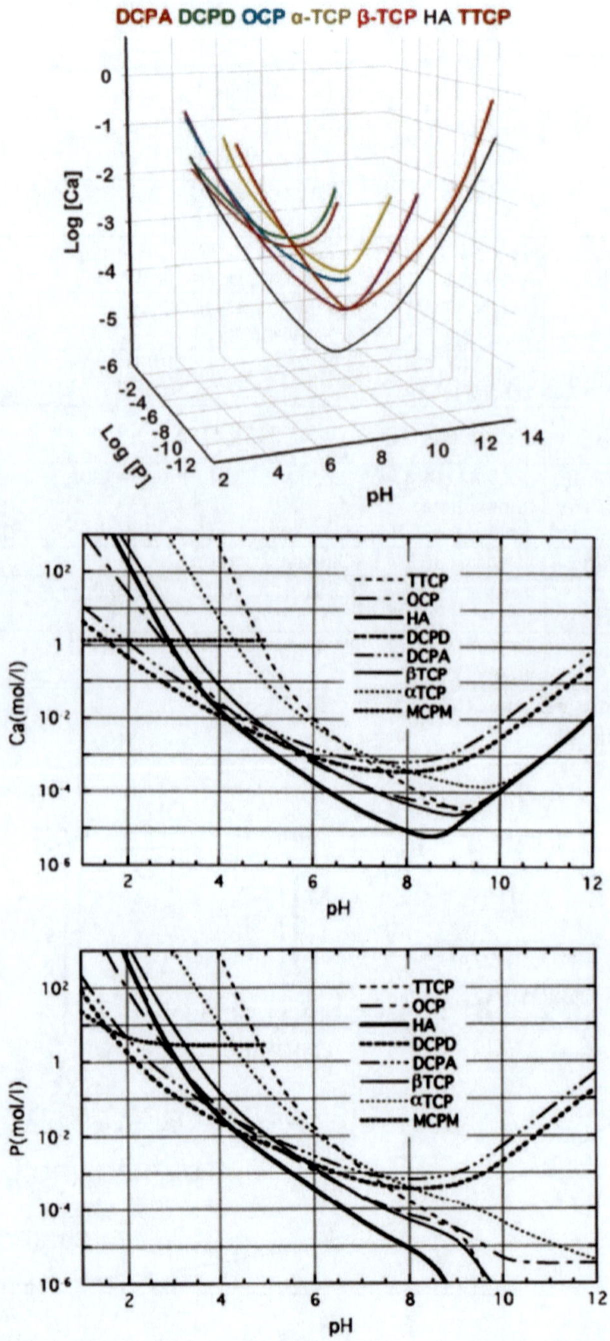

Figure 6. Top: a 3D version of the classical solubility phase diagrams for the ternary system $Ca(OH)_2$ – H_3PO_4 – H_2O. Reprinted from Ref. [94] with permission. Middle and bottom: solubility phase diagrams in two-dimensional graphs, showing two logarithms of the concentrations of (a) calcium and (b) orthophosphate ions as a function of the pH in solutions saturated with various salts. Reprinted from Ref. [384] with permission.

MCPM

Monocalcium phosphate monohydrate ($Ca(H_2PO_4)_2{\cdot}H_2O$; the IUPAC name is calcium dihydrogen orthophosphate monohydrate) is both the most acidic and water-soluble calcium orthophosphate. Although acidic calcium orthophosphates in general were known by 1795 as "super-phosphate of lime" [96], their differentiation started in 1800-s. Namely, by 1807, researchers first prepared a calcium phosphate, which could be attributed to MCPM [97].

MCPM crystallizes from aqueous solutions containing dissolved ions of $H_2PO_4^-$ and Ca^{2+} at the Ca/P ratio ~ 0.5 and solution pH below ~ 2.0. Besides, MCPM might be precipitated from aqueous solutions containing organic solvents [98, 99]. At temperatures above ~ 100 °C, MCPM releases a molecule of water and transforms into MCPA but at temperatures > ~ 500 °C MCPA further transforms into $Ca(PO_3)_2$ [100]. The results of the spectroscopic investigations of MCPM are available in literature [101].

Due to high acidity and solubility, MCPM is never found in biological calcifications. Moreover, pure MCPM is not biocompatible with bones [102]. However, in medicine MCPM is used as a component of several self-hardening calcium orthophosphate formulations [103, 104]. In addition, MCPM is used as a nutrient, acidulant and mineral supplement for dry baking powders, food, feed and some beverages [105, 106]. Coupled with $NaHCO_3$, MCPM is used as a leavening agent for both dry baking powders and bakery dough. MCPM might be added to salt-curing preserves, pickled and marinated foods. In addition, MCPM might be added to tooth pastes and chewing gums [107]. Besides, MCPM might be added to ceramics and glasses, while agriculture is the main consumer of a technical grade MCPM, where it is used as a fertilizer, triple superphosphate [108].

MCPA (or MCP)

Monocalcium phosphate anhydrous ($Ca(H_2PO_4)_2$; the IUPAC name is calcium dihydrogen orthophosphate anhydrous) is the anhydrous form of MCPM. Although MCPM has been known since 1807 [1, 2], MCPA was differentiated as "tetra-hydrogen calcium phosphate, $H_4Ca(PO_4)_2$" by 1879 [109]. It crystallizes under the same conditions as MCPM but at temperatures above ~ 100 °C (e.g., from concentrated hot mother liquors during fertilizer production). In addition, MCPA might be prepared from MCPM by dehydration. Furthermore, it might be also prepared at ambient temperatures by crystallization in water-restricted or non-aqueous systems. Like MCPM, MCPA never appears in calcified tissues and is not biocompatible due to its acidity. There is no current application of MCPA in medicine. Due to the similarity with MCPM, in many cases, MCPA might be used instead of MCPM; however, highly hydroscopic properties of MCPA reduce its commercial applications [13, 105].

DCPD

Dicalcium phosphate dihydrate ($CaHPO_4{\cdot}2H_2O$; the IUPAC name is calcium hydrogen orthophosphate dihydrate; the mineral brushite) has been known since, at least, 1804 [96]. As a mineral, brushite was first discovered in phosphatic guano from Avis Island (Caribbean) in

1865 [110] and named to honor an American mineralogist Prof. George Jarvis Brush (1831 – 1912), Yale University, New Haven, Connecticut, USA.

DCPD can be easily crystallized from aqueous solutions containing dissolved ions of HPO_4^{2-} and Ca^{2+} at the Ca/P ratio ~ 1 and solution pH within ~ 2.0 < pH < ~ 6.5. Other preparation techniques such as neutralization of H_3PO_4 and/or MCPM solutions by CaO, $CaCO_3$ or more basic calcium orthophosphates (α- or β-TCP, CDHA, HA, TTCP) are also known. Interestingly, that precipitation of DCPD by mixing a $Ca(OH)_2$ suspension and a H_3PO_4 solution in the equimolar quantities was found to occur in five stages, being HA the first precipitated phase [111, 112]. Besides, DCPD might be prepared in gels [113, 114]. DCPD transforms into DCPA at temperatures above ~ 80 ºC and this transformation is accompanied by ~ 11% decrease in volume [115] and structural changes [116]. The value for $\Delta_r G^0$ for DCPD → DCPA transformation is –1.032 kJ/mol [116]. Briefly, DCPD crystals consist of $CaPO_4$ chains arranged parallel to each other, while lattice water molecules are interlayered between them [117]. The atomic structure of the {010} interface of DCPD with water was determined [118]. Recently, data on DCPD solubility have been updated by solid titration technique [119]. The optical properties of DCPD were described [120], while many additional data on DCPD including a good drawing of its atomic structure might be found in Ref. [121].

DCPD is of biological importance because it is often found in pathological calcifications (dental calculi, crystalluria, chondrocalcinosis and urinary stones) and some carious lesions [4, 64-66]. It was proposed as an intermediate in both bone mineralization and dissolution of enamel in acids (dental erosion) [4, 64, 65]. In medicine, DCPD is used in self-setting calcium orthophosphate formulations [104] and as an intermediate for tooth remineralization. DCPD is added to toothpaste both for caries protection (in this case, it is often coupled with F-containing compounds such as NaF and/or Na_2PO_3F) and as a gentle polishing agent [107]. Other applications include a flame retardant [122], a slow release fertilizer, using in glass production, as well as calcium supplement in food, feed and cereals. In food industry, it serves as a texturizer, bakery improver and water retention additive. In diary industry, DCPD is used as a mineral supplement. In addition, plate-like crystals of DCPD might be used as a non-toxic, anticorrosive and passivating pigment for some ground coat paints [105].

DCPA (or DCP)

Dicalcium phosphate anhydrous ($CaHPO_4$; the IUPAC name is calcium hydrogen orthophosphate anhydrate; the mineral monetite) is the anhydrous form of DCPD. Although DCPD has been known since, at least, 1804 [1, 2], DCPA was differentiated as "mono-hydrogen calcium orthophosphate, $HCaPO_4$" by 1879 [107]. As a mineral, monetite was first described in 1882 in rock-phosphate deposits from the Moneta (now Monito) Island (archipelago of Puerto Rico), which contains a notable occurrence [123].

Due to the absence of water inclusions, DCPA is less soluble than DCPD (Table 1). Like DCPD, DCPA can be crystallized from aqueous solutions containing Ca/P ratio ~ 1 at solution pH within ~ 2.0 < pH < ~ 6.5 but at temperatures > ~ 90 ºC. In addition, DCPA might be prepared by dehydration of DCPD. Furthermore, it might be also prepared at ambient temperatures in water-restricted or non-aqueous systems, such as gels [113], ethanol [124], as well as in the oil-in-water and water-in-oil systems [125]. DCPA is physically stable and resisted hydration even when dispersed in water for over 7 months in the temperature range of 4 – 50°C [126]. A calcium-deficient DCPA was also prepared. It might be sintered at ~ 300 °C [127]. Unlike DCPD, DCPA occurs in neither normal nor pathological calcifications. It is used in self-setting calcium orthophosphate formulations [104]. Besides, DCPA might be implanted as bioceramics [128]. Other applications include using as a polishing agent, a source of calcium and phosphate in nutritional supplements (e.g., in prepared breakfast cereals, enriched flour and noodle products), a tabletting aid [129] and a toothpaste component [107]. In addition, it is used as a dough conditioner in food industry [105].

OCP

Octacalcium phosphate ($Ca_8(HPO_4)_2(PO_4)_4 \cdot 5H_2O$; the IUPAC name is tetracalcium hydrogen orthophosphate diorthophosphate pentahydrate, another name is octacalcium bis(hydrogenphosphate) tetrakis(phosphate) pentahydrate) is often found as an unstable transient intermediate during the precipitation of the thermodynamically more stable calcium orthophosphates (e.g., CDHA) in aqueous solutions. To the best of my findings [1, 2], OCP has been known since, at least, 1843, when Percy published a paper [130], in which he described formation of "a new hydrated phosphate of lime" with a chemical formula 2CaO + PO5 + 6HO, in which "1 equiv. water being basic and 5 constitutional". However, according to Bjerrum [131], a calcium orthophosphate with the OCP composition was first described by Berzelius in 1836.

The preparation techniques of OCP are available in literature [132-135]. Briefly, to prepare OCP, Ca- and PO_4-containing chemicals must be mixed to get the supersaturated aqueous solutions with the Ca/P ratio equal to 1.33. However, OCP might be non-stoichiometric and be either Ca-deficient (Ca/P = 1.26) or include excessive calcium (up to Ca/P = 1.48) in the structure [135]. Furthermore, a partially hydrolyzed form of OCP with Ca/P molar ratio of 1.37 might be prepared [135, 136]. The full hydrolysis of OCP into CDHA occurs within ~ 6 hours [134]. Ion-substituted OCP might be prepared as well [137, 138]. Crystals of OCP are typically small, extremely platy and almost invariably twinned.

The triclinic structure of OCP displays apatitic layers (with atomic arrangements of calcium and orthophosphate ions similar to those of HA) separated by hydrated layers (with atomic arrangements of calcium and orthophosphate ions similar to those in DCPD) [4-6, 135]. A similarity in crystal structure between OCP and HA [139, 140] is one reason that the epitaxial growth of these phases is observed. It is generally assumed that, in solutions, the hydrated layer of the (100) face is the layer most likely exposed to solution. The water content of OCP crystals is about 20% that of DCPD and this is partly responsible for its lower solubility. New data on OCP solubility have been published recently [141].

OCP is of a great biological importance because it is one of the stable components of human dental and urinary calculi [142, 143]. OCP was first proposed by W. E. Brown to participate as the initial phase in enamel mineral formation and bone formation through subsequent precipitation and stepwise hydrolysis of OCP [139, 140, 144]. It plays an important role in formation of apatitic biominerals *in vivo* [145]. A "central OCP inclusion" (also known as "central dark line") is seen by transmission electron microscopy in many biological apatites and in synthetically precipitated CDHA [146-148]. Although OCP has not been observed in vascular calcifications, it has been strongly suggested as a precursor phase to biological apatite found in natural and prosthetic heart valves [149, 150]. In surgery, OCP is used for implantation into bone defects [151-154]. For the comprehensive information on OCP, the readers are referred to other reviews [135, 142].

β-TCP

β-tricalcium phosphate (β-$Ca_3(PO_4)_2$; the IUPAC name is tricalcium diorthophosphate beta, other names are calcium orthophosphate tribasic beta or tricalcium bis(orthophosphate) beta) is one of the polymorphs of TCP. Although calcium orthophosphates with the composition close to that of TCP, CDHA and HA were known in 1770-s [1, 2], α- and β-polymorphs of TCP were differentiated only by 1932 [155, 156].

β-TCP cannot be precipitated from aqueous solutions. It is a high temperature phase, which can be prepared at temperatures above ~ 800 °C by thermal decomposition of CDHA or by solid-state interaction of acidic calcium orthophosphates, e.g., DCPA, with a base, e.g., CaO. In all cases, the chemicals must be mixed in the proportions to get the Ca/P ratio equal to 1.50. However, β-TCP can also be prepared at relatively low temperatures (~ 150 °C) by precipitation in water-free mediums, such as ethylene glycol [157, 158]. Apart from the chemical preparation routes, ion-substituted β-TCP can be prepared by calcining of bones [159, 160]: such type of calcium orthophosphates is occasionally called "bone ash" [161]. At temperatures above ~ 1125 °C, β-TCP is transformed into a high-temperature phase α-TCP. Being the stable phase at room temperature, β-TCP is less soluble in water than α-TCP (Table 1). Both ion-substituted [162-165] and organically modified [166-168] forms of β-TCP can be synthesized, as well. The modern structural data on β-TCP are available in Refs. [169-171], those on Vicker's and Knoop microhardness studies might be found if Ref. [172], while solubility data – in Ref. [220173 Furthermore, an ability of β-TCP to store an electrical charge by electrical polarization was studied and this material was found to have a suitable composition and structure for both ion conduction and charge storage [174].

Pure β-TCP never occurs in biological calcifications. Only the Mg-substituted form (β-TCMP – β-tricalcium magnesium phosphate, β-$(Ca,Mg)_3(PO_4)_2$), which is often called whitlockite to honor Mr. Herbert Percy Whitlock (1868 – 1948), an American mineralogist, the curator of the American Museum of Natural History, New York City, New York, USA [175]. Since β-TCMP is less soluble than β-TCP [176], it is formed instead of β-TCP in dental calculi and urinary stones, dentineal caries, salivary stones, arthritic cartilage, as well as in some soft-tissue deposits [4, 64-66, 177-180]. However, it has not been observed in enamel, dentine or bone. In medicine, β-TCP is used in the self-setting calcium orthophosphate formulations [104] and other types of bone grafts [159, 181-185]. Dental applications of β-TCP are also known. For example, β-TCP is added to some brands of toothpaste as a gentle polishing agent [107]. Multivitamin complexes with calcium orthophosphate are widely available in the market and β-TCP is used as the calcium phosphate there. In addition, β-TCP serves as a texturizer, bakery improver and anti-clumping agent for dry powdered food (flour, milk powder, dried cream, cocoa powder). Besides, β-TCP is added as a dietary or mineral supplement to food and feed [186]. Occasionally, β-TCP might be used as inert filler in pelleted drugs. Other applications comprise porcelains, pottery, enamel, using as a component for mordants and ackey, as well as a polymer stabilizer [105].

β-TCP of a technical grade (as either calcined natural phosphorites or bone dust) is used as a slow release fertilizer for acidic soils [13].

α-TCP

α-tricalcium phosphate (α-$Ca_3(PO_4)_2$; the IUPAC name is tricalcium diorthophosphate alpha, other names are calcium orthophosphate tribasic alpha or tricalcium bis(orthophosphate) alpha) is another polymorph of TCP, which was differentiated by 1932 [155, 156]. α-TCP is also a high temperature phase; therefore, it cannot be precipitated from aqueous solutions either. Thus, α-TCP is usually prepared by the same techniques as β-TCP (see the previous section) but, since the β-TCP → α-TCP transition temperature is ~ 1125 °C [187], calcining is performed at temperatures above ~ 1200 °C [188]. Consequently, α-TCP is often considered as a high temperature polymorph of β-TCP. However, data are available that α-TCP might be prepared at lower temperatures. Namely, at the turn of the millennium, the previously forgotten data that the presence of silicates stabilized α-TCP at temperatures of 800 – 1000 °C [189] were rediscovered again. Such type of α-TCP is called "silica stabilized α-TCP" [190-192]. Furthermore, sometimes, α-TCP might be prepared at even lower temperatures (~ 700 °C) by a thermal decomposition of low temperature ACPs [193].

Although α-TCP and β-TCP have exactly the same chemical composition, they differ by the crystal structure (Table 3) and solubility (Table 1). In the absence of humidity, both polymorphs of TCP are stable at room temperatures; however, according to a density functional study, stability of β-TCP crystal lattice exceeds that of α-TCP [170]. Therefore, of them, α-TCP is more reactive in aqueous systems, has a higher specific energy and in aqueous solutions it can be hydrolyzed to CDHA [194-196]. Milling was found to increase the α-TCP reactivity even more [197]. Although, α-TCP never occurs in biological calcifications, in medicine, it is used as a component of self-setting calcium orthophosphate formulations [104]. On the other hand, the chemically pure α-TCP has received not much interest in the biomedical field [183]. The disadvantage for using α-TCP is its quick resorption rate (faster than formation of a new bone), which limits its application in this area. However, the silicon stabilized α-TCP (more precisely as a biphasic composite with HA) has been commercialized as a starting material to produce bioresorbable porous ceramic scaffolds to be used as artificial bone grafts [190-192]. Upon implantation, α-TCP tends to convert to CDHA, which drastically reduces further degradation rate. Theoretical insights into bone grafting properties of the silicon-stabilized α-TCP might be found in Ref. [198]. The structure of α-TCP is well described in literature [170, 171], while the surface and adsorption properties are available in Ref. [199]. Similar to β-TCP, α-TCP of a technical grade might be used slow release fertilizer for acidic soils [105].

To conclude, one should briefly mention on an existence of α′-TCP polymorph, which was discovered in 1959 [189]. However, this TCP polymorph lacks of any practical interest because it only exists at temperatures between ~ 1450 °C and its melting point (~ 1756 °C). It reverts to α-TCP polymorph by cooling below the transition temperature.

ACP

Amorphous calcium phosphates (ACPs) represent a special class of calcium orthophosphate salts, having variable chemical but rather identical glass-like physical properties, in which there are neither translational nor orientation long-range orders of the atomic positions. To the best of my findings [1, 2], ACP was first prepared in 1845 [200]. Nevertheless, until recently [201], ACP has often been considered as an individual calcium orthophosphate with a variable chemical composition, while, in reality, ACP is just an amorphous state of other calcium orthophosphates. Therefore, in principle, all compounds mentioned in Table 1 might be somehow fabricated in an amorphous state but, currently, only few of them (e.g., an amorphous TCP) are known [201]. Thus, strictly speaking, ACP should be excluded from Table 1. Furthermore, since ACPs do not have the definite chemical composition, the IUPAC nomenclature is not applicable to describe them.

Depending on the production temperatures, all types of ACP are divided into two major groups: low-temperature ACPs (prepared in solutions, usually aqueous ones) and high-temperature ACPs [201]. Low-temperature ACPs (described by the chemical formula $Ca_xH_y(PO_4)_z{\cdot}nH_2O$, $n = 3 - 4.5$; 15 – 20% H_2O) are often encountered as a transient precursor phase during precipitation of other calcium orthophosphates in aqueous systems. Usually, an ACP is the first phase precipitated from supersaturated solutions (the higher supersaturation, the better) prepared by rapid mixing of solutions containing ions of calcium and orthophosphate [5, 201]. Such ACP precipitates usually look like spherical particles with diameters in the range 200 to 1200 Å without a definite structure. Generally, the ACP particles are smaller if prepared under conditions of high supersaturation and/or high pH, while for a given pH, higher temperatures give larger particles [202]. The freshly precipitated ACPs contain 10 – 20% by weight of tightly bound water, which is removed by vacuum drying at elevated temperature [204]. The amorphization degree of ACPs increases with the concentration increasing of Ca- and PO_4-containing solutions, as well as at a high solution pH and a low crystallization temperature. A continuous gentle agitation of as precipitated ACPs in the mother solution, especially at elevated temperatures, results in a slow recrystallization and formation of better crystalline calcium orthophosphates, such as CDHA [4, 5]. In addition, other production techniques of ACPs are known [201].

The lifetime of ACPs in aqueous solutions was reported to be a function of the presence of additive molecules and ions, pH, ionic strength and temperature. Thus, ACPs may persist for appreciable periods and retain the amorphous state under some specific experimental conditions [203]. The chemical composition of ACPs strongly depends on the solution pH and the concentrations of mixing solutions. For example, ACPs with Ca/P ratios in the range of 1.18 (precipitated at solution pH = 6.6) to 1.53 (precipitated at solution pH = 11.7) [5, 205] and even to 2.5 [4, 64, 65] were described. It should be noted that unsubstituted ACPs are unstable in aqueous solutions and even when stored dry they tend to transform into more crystalline calcium orthophosphates, such as poorly crystalline CDHA. The presence of poly(ethylene glycol) [206], ions of pyrophosphate, carbonate and/or magnesium in solutions during the crystallization promotes formation of ACPs and slows down their further transformation, while the presence of fluoride has the opposite effect [4-6, 81, 207]. In general, low temperatures ACPs heated to ~ 550 °C (so that all volatiles have already escaped) remain amorphous, but further heating above ~ 650 °C causes their transformation into crystalline calcium orthophosphates, such as α- or β-TCP, HA, mixtures thereof, depending on the Ca/P ratio of the ACP heated.

High-temperature ACPs might be prepared using high energy processing at elevated temperatures [201]. This method is based on a rapid quenching of melted calcium orthophosphates occurring, e.g., during plasma spraying of HA [87, 208, 209]. A plasma jet, possessing very high temperatures (~ 5000 – ~ 20000 °C), partly decomposes HA. That results in formation of a complicated mixture of products, some of which would be ACPs. Obviously, all types of high-temperature ACPs are definitively anhydrous contrary to the precipitated ACPs. Unfortunately, no adequate chemical formula is available to describe the high-temperature ACPs.

In general, as all amorphous compounds are characterized by a lack of long-range order, it is problematic to discuss the structure of ACPs (they are X-ray amorphous). Concerning a short-range order (SRO) in ACPs, it exists, just due to the nature of chemical bonds. Unfortunately, in many cases, the SRO in ACPs is uncertain either, because it depends on many variables, such as Ca/P ratio, preparation conditions, storage, admixtures, *etc.* Infrared spectra of ACPs show broad featureless phosphate absorption bands. Electron microscopy of freshly precipitated ACPs usually shows featureless nearly spherical particles with diameters in the range of 20 to 200 nm. However, there is a questionable opinion that ACPs might have an apatitic structure but with a crystal size so small, that they are X-ray amorphous. This is supported by X-ray absorption spectroscopic data (EXAFS) on biogenic and synthetic samples [210-213]. On the other hand, it was proposed that the basic structural unit of the precipitated ACPs is a 9.5 Å diameter, roughly spherical cluster of ions with the composition of $Ca_9(PO_4)_6$ (Figure 7) [5, 205, 214, 215]. These clusters were found experimentally as first nuclei during the crystallization of CDHA and a model was developed to describe the crystallization of HA as a step-wise assembly of these units [216] (see section *3.10. HA (or HAp, or OHAp)* below). Biologically, ion-substituted ACPs (always containing ions of Na, Mg, carbonate and pyrophosphate) are found in soft-tissue pathological calcifications (e.g., heart valve calcifications of uremic patients) [4, 64-66].

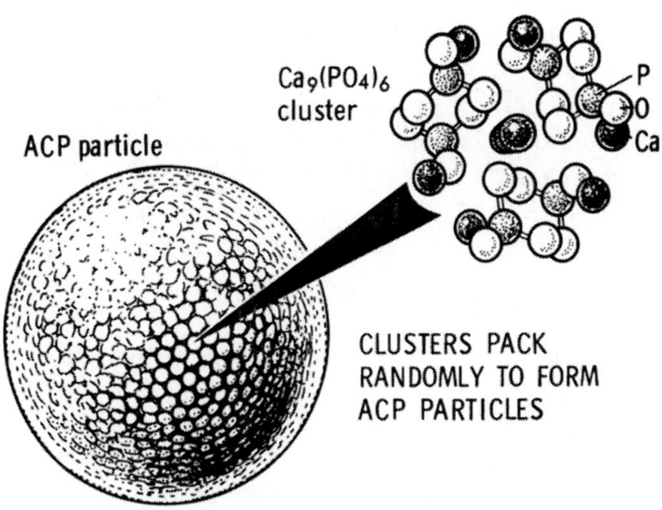

Figure 7. A model of ACP structure. Reprinted from Ref. [214] with permission.

In medicine, ACPs are used in self-setting calcium orthophosphate formulations [104]. Bioactive composites of ACPs with polymers have properties suitable for use in dentistry [107, 201] and surgery [201, 217]. Due to a reasonable solubility and physiological pH of aqueous solutions, ACPs appeared to be consumable by some microorganisms and, due to this reason, it might be added as a mineral supplement to culture media. Non-biomedical applications of ACPs comprise their using as a component for mordants and ackey. In food industry, ACPs are used for syrup clearing. Occasionally, they might be used as inert filler in pelleted drugs. In addition, ACPs are used in glass and pottery production and as a raw material for production of some organic phosphates. To get further details on ACPs, the readers are referred to special reviews [201, 215, 218].

CDHA (or Ca-def HA)

Calcium-deficient hydroxyapatite ($Ca_{10-x}(HPO_4)_x(PO_4)_{6-x}(OH)_{2-x}$ ($0 < x < 1$)) became known since the earliest experiments on establishing the chemical composition of bones performed in 1770-s [1, 2]. However, the initial appropriate term "sub-phosphate of lime" appeared by 1819 [219]. Other chemical formulae such as $Ca_{10-x}(HPO_4)_{2x}(PO_4)_{6-2x}(OH)_2$ ($0 < x < 2$), $Ca_{10-x-y}(HPO_4)_x(PO_4)_{6-x}(OH)_{2-x-2y}$ ($0 < x < 2$ and $y < x/2$), $Ca_{10-x}(HPO_4)_x(PO_4)_{6-x}(OH)_{2-x}(H_2O)_x$ ($0 < x < 1$), $Ca_{9-x}(HPO_4)_{1+2x}(PO_4)_{5-2x}(OH)$, *etc.* were also proposed to describe its variable composition [5]. As seen from these formulae, Ca-deficiency is always coupled with both OH-deficiency and protonation of some PO_4 groups with simultaneous formation of the ionic vacations in the crystal structure [220]. In addition, CDHA often contains tightly bound water molecules, which might occupy some of these ionic vacations. For example, there is an approach describing a lack of the hydroxide vacancies in CDHA: to perform the necessary charge compensation of the missing Ca^{2+} ions, a portion of OH^- anions is substituted by neutral water molecules [221]. This water is removed by vacuum drying at elevated temperature. Concerning possible vacancies of orthophosphate ions, nothing is known about their presence in CDHA. It is just considered that a portion of PO_4^{3-} ions is either protonated (as HPO_4^{2-}) or substituted by other ions (e.g., CO_3^{2-}) [222]. Since CDHA does not have any definite chemical composition, the IUPAC nomenclature is not applicable to describe it.

CDHA can be easily prepared by simultaneous addition of Ca- and PO_4-containing solutions in the proportions to get Ca/P ratio within 1.50 – 1.67 into boiling water followed by boiling the suspension for several hours (an ageing stage). That is why, in literature it might be called as "precipitated HA (PHA)" [223, 224]. Besides, it might be prepared by hydrolysis of α-TCP [194-196]. Other preparation techniques of CDHA are known as well [225-228]. During ageing, initially precipitated ACPa are restructured and transformed into CDHA. Therefore, there are many similarities in the structure, properties and application between the precipitated in alkaline solutions (pH > 8) ACPs and CDHA. Some data indicated on a presence of intermediate phases during further hydrolysis of CDHA to a more stable HA-like phase [229]. In general, CDHA crystals are poorly crystalline and of submicron dimensions. They have a very large specific surface area, typically 25 – 100 m^2/g. On heating above ~ 700 °C, CDHA with Ca/P = 1.5 converts to β-TCP and that with 1.5 < Ca/P < 1.67 converts into a biphasic composite of HA and β-TCP (see section *3.14. Biphasic, triphasic and multiphasic calcium orthophosphate formulations* below) [230]. A solid-state transformation mechanism of CDHA into HA + β-TCP biocomposite was proposed [231, 232].

The variability in Ca/P molar ratio of CDHA has been explained through different models: surface adsorption, lattice substitution and intercrystalline mixtures of HA and OCP [233]. Due to a lack of stoichiometry, CDHA is usually doped by other ions [63]. The doping extent depends on the counter-ions of the chemicals used for CDHA preparation. Direct determinations of the CDHA structures are still missing and the unit cell parameters remain uncertain. However, unlike that in ACPs (see section *3.8. ACP* above), a long-range order exists in CDHA. Namely, the following lattice parameters were reported for CDHA with Ca/P = 1.5: $a = 9.4418(20)$ Å and $c = 6.8745(17)$ Å [228].

A systematic study of defect constellations in CDHA is available in literature [221]. As a first approximation, CDHA may be considered as HA with some ions missing (ionic vacancies) [234]. The more amount of Ca is deficient, the more disorder, imperfections and vacancies are in the CDHA structure [235]. Furthermore, a direct correlation between the Ca-deficiency and the mechanical properties of the crystals was found: calcium deficiency lead to an 80% reduction in the hardness and elastic modulus and at least a 75% reduction in toughness in plate-shaped HA crystals [236]. Theoretical investigations of the defect formation mechanism relevant to non-stoichiometry in CDHA are available elsewhere [237].

Unsubstituted CDHA (i.e., that containing ions of Ca^{2+}, PO_4^{3-}, HPO_4^{2-} and OH^- only) does not exist in biological systems. However, the ion substituted CDHA: Na^+, K^+, Mg^{2+}, Sr^{2+} for Ca^{2+}; CO_3^{2-} for PO_4^{3-} or HPO_4^{2-}; F^-, Cl^-, CO_3^{2-} for OH^-, plus some water forms biological apatite – the main inorganic part of animal and human normal and pathological calcifications [4, 63, 64]. Therefore, CDHA is a very promising compound for industrial manufacturing of artificial bone substitutes [238], including drug delivery applications [239]. Non-biomedical applications of CDHA are similar to those of ACP and HA. Interestingly that CDHA was found to possess a catalytic activity to produce biogasoline [240].

HA (or HAp, or OHAp)

Hydroxyapatite ($Ca_5(PO_4)_3(OH)$, but is usually written as $Ca_{10}(PO_4)_6(OH)_2$ to denote that the crystal unit cell comprises two molecules; the IUPAC name is pentacalcium hydroxide tris(orthophosphate)) is the second most stable and least soluble calcium orthophosphate after FA. Apatites were recognized as calcium phosphates by, at least, 1789 [1, 2]. Here, it is worth noting that *hydroxylapatite* would be a more accurate abbreviation expansion of HA (perhaps, *hydroxideapatite* would be even better because it relates to calcium hydroxide) while by both the medical and material communities HA is usually expanded as *hydroxyapatite*.

Chemically pure HA crystallizes in the monoclinic space group $P2_1/b$ [241]. However, at temperatures above ~ 250 °C, there is a monoclinic to hexagonal phase transition in HA (space group $P6_3/m$) [5, 89, 205, 242, 243]. The detailed description of the HA structure was first reported in 1964 [244] and its interpretation in terms of aggregation of $Ca_9(PO_4)_6$ clusters, the so-called Posner's clusters, has been widely used since publication of the article by Posner and Betts [204]. In hexagonal HA, the hydroxide ions are more disordered within each row, when compared with the monoclinic form, pointing either upward or downward in the structure. This induces strains that are compensated for by substitutions or ion vacancies. Some impurities, like partial substitution of hydroxide by fluoride or chloride, stabilize the hexagonal structure of HA at ambient temperature. Due to this reason, hexagonal HA is seldom the stoichiometric phase and very rare single crystals of natural HA always exhibit the hexagonal space group. The crystal structure of HA is well described elsewhere [5, 88, 89], the detailed analysis of the electronic structure, bonding, charge transfer, optical and elastic properties are also available [245-248], while the readers interested in Posner's clusters are referred to other papers [249-251]. A shell model was developed to study the lattice dynamics of HA [252], while a cluster growth model was created to illustrate its growth [216]. Polarization characteristics [253, 254] and pyroelectrical properties [255] of HA bioceramics have been investigated. First-principles calculations for the elastic properties of doped HA [256] and vacancy formation in HA [257] were performed. Computer simulations of the structures and properties of HA are well described in feature articles [258, 259].

Several techniques might be utilized for HA preparation; they can be divided into solid-state reactions and wet methods [260], which include precipitation, hydrothermal synthesis and hydrolysis of other calcium orthophosphates. However, in all cases, Ca- and PO_4-containing chemicals must be mixed to get the Ca/P ratio strictly equal to 1.67. Nevertheless, even under the ideal stoichiometric conditions, the precipitates are generally non-stoichiometric, suggesting intermediate formation of precursor phases, such as ACP and CDHA. Usually unsintered HA is poorly crystalline and often non-stoichiometric, resembling the aforementioned CDHA. However, well crystalline HA can be prepared from an aqueous solution [261]. HA with the Ca/P ratio > 1.67 (Ca-rich HA) might be prepared as well [262]. The detailed information on HA synthesis is available elsewhere [263-267]. In addition, there are good reviews on HA solubility, crystal growth and intermediate phases of HA crystallization [268], as well as on HA dissolution [269].

Pure HA never occurs in biological systems. However, due to the chemical similarities to bone and teeth mineral (Table 2), HA is widely used as coatings on orthopedic (e.g., hip joint prosthesis) and dental implants [107, 270-274]. HA bioceramics is very popular as well [275-277]. Due to a great similarity to biological apatite, over a long time HA has been used in liquid chromatography of nucleic acids, proteins and other biological compounds [278-282] and for drug delivery purposes [283-287]. Also, HA is added to some brands of toothpaste as a gentle polishing agent instead of calcium carbonate [288, 289]. Non-biomedical applications of HA include its using as an environmentally friendly filler for elastomers [290], a sorbent of poisonous chemical elements [291, 292] and a carrier for various catalysts [293-295]. Furthermore, HA by itself might act as a catalyst for formaldehyde combustion at room temperature [296].

FA (or FAp)

Fluorapatite ($Ca_5(PO_4)_3F$, but is usually written as $Ca_{10}(PO_4)_6F_2$ to denote that the crystal unit cell comprises two molecules; the IUPAC name is pentacalcium fluoride tris(orthophosphate)) is the only ion-substituted calcium orthophosphate, considered in this review. Since the presence of 2.5% of fluorides in natural apatites was established by 1798 [297], this date might be accepted as the earliest hearing of FA.

FA is the hardest (5 according to the Mohs' scale of mineral hardness), most stable and least soluble compound among all calcium orthophosphates (Table 1). Perhaps, such "extreme" properties of FA are related to the specific position of F^- ions in the center of Ca(2) triangles of the crystal structure [5, 88]. Due to its properties, FA is the only calcium orthophosphate that naturally forms large deposits suitable for the commercial use [12-14] (see also Figure 2). Preparation techniques of the chemically pure FA are similar to the aforementioned ones for HA but the synthesis must be performed in presence of the necessary amount of F^- ions (usually, NaF or NH_4F is added). Under some special crystallization conditions (e.g., in presence of gelatin or citric acid), FA might form unusual dumbbell-like fractal morphology that finally are closed to spheres (Figure 8) [298, 299]. In addition, FA is the only calcium orthophosphate, which melts without decomposition (the melting point is ~ 1650 °C); therefore, big (up to 30 cm long and, for shorter lengths, up to 1.9 cm wide) single FA crystals might be grown from FA melts [300, 301]. Similar to that for HA (see CDHA), an existence of CaF_2-deficient FA was also detected but for the crystals grown from the FA melt only [300, 302]. In addition, FA with an excess of CaF_2 was prepared [303]. A hierarchical structure for FA was proposed [304]. The crystal structure of FA for the first time was studied in 1930 [305, 306] and is well described elsewhere [5, 88, 89]. The detailed analysis of the electronic structure, bonding, charge transfer and optical properties is available as well [246]. In addition, there are reviews on FA solubility [268] and the dissolution mechanism [269].

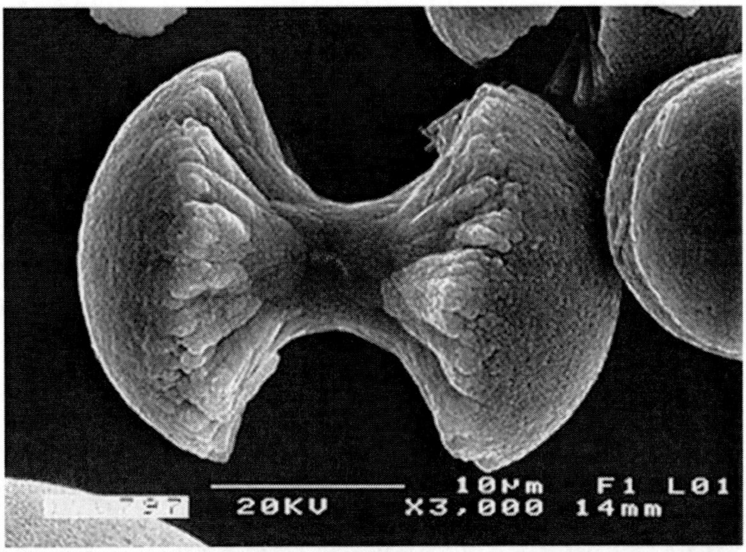

Figure 8. A biomimetically grown aggregate of FA that was crystallized in a gelatin matrix. Its shape can be explained and simulated by a fractal growth mechanism. Scale bar: 10 μm. Reprinted from Ref. [298] with permission.

FA easily forms solid solutions with HA with any desired F/OH molar ratio. Such compounds are called fluorhydroxyapatites (FHA) [307-309] or hydroxyfluorapatites (HFA) [310, 311] and described with a chemical formula $Ca_{10}(PO_4)_6(OH)_{2-x}F_x$, where $0 < x < 2$. If the F/OH ratio is either uncertain or not important, the chemical formula of FHA and HFA is often written as $Ca_{10}(PO_4)_6(F,OH)_2$. The lattice parameters, crystal structure, solubility and other properties of FHA and HFA lay in between of those for the chemically pure FA and HA. Namely, the substitution of F for OH results in a contraction in the *a*-axis with no significant change in the *c*-axis dimensions and greater resolution of the IR absorption spectra.

Similar to pure HA, pure FA never occurs in biological systems. Obviously, a lack of the necessary amount of toxic fluorides (the acute toxic dose of fluoride is ~ 5 mg/kg of body weight) in living organisms is the main reason of this fact (pure FA contains 3.7 % mass. F). Enameloid of shark teeth [11, 81, 312-316] and some exoskeletons of mollusks [317] seem to be the only exclusions because they contain substantial amounts of fluoride, with is presented there as ion-substituted, non-stoichiometric FHA or HFA. Among all normal calcified tissues of humans, the highest concentration of fluorides is found in dentine and cementum, while the lowest – in dental enamel (Table 2). Nevertheless, one should stress that the amount of fluorides on the very surface of dental enamel might be subtantially increased by using fluoride-containing toothpastes and mouthwashes [318, 319]. However, in no case, the total amount of fluorides is enough to form FA.

Contrary to the initial expectations [320], chemically pure FA is not used for grafting purposes. Presumably, this is due to the lowest solubility, good chemical stability of FA and toxicity of high amounts of fluorides. However, attempts to test FA-containing formulations [321-325], ion-substituted FA [326, 327], FHA [328, 329] and porous FA bioceramics [330] are kept performing. The effect of fluoride contents in FHA on both osteoblast behavior [331, 332] and leukemia cells proliferation [333] has been described. Non-biomedical applications of FA include fluorescent light tubes [334], laser materials [335, 336] (in both cases dopants are necessary), as well as catalysts [337].

OA (or OAp, or OXA)

Oxyapatite ($Ca_{10}(PO_4)_6O$; the IUPAC name is decacalcium oxide hexakis(phosphate), mineral voelckerite) is the least stable and, therefore, the least known calcium orthophosphate, which, probably, does not exist at all. Nevertheless, a name “voelckerite” was introduced in 1912 by A.F. Rogers (1887 – 1957) for a hypothetical mineral with the chemical composition of $3Ca_3(PO_4)_2 + CaO$ [338, 339], to honor an English agricultural chemist John Christopher Augustus Voelcker (1822 – 1884), who, in 1883, first showed an apparent halogen deficiency in some natural apatites [340]. Therefore, 1883 might be accepted as the earliest hearing of OA.

To the best of my findings, pure OA has never been prepared; therefore, its properties are not well established. Furthermore, still there are serious doubts that pure OA can exist. However, a mixture (or a solid solution?) of OA and HA (oxy-HA) might be prepared by a partial dehydroxylation of HA at temperatures exceeding ~ 900 °C (e.g., during plasma-spray of HA) only in the absence of water vapor [341-345]. It also might be crystallized in glass-ceramics [346]. OA is very unstable and has no stability field in aqueous conditions [347]. Namely, data are available, that oxy-HA containing less than 25% HA (i.e., almost OA) during further dehydration decomposes to a mixture of α-TCP and TTCP. In addition, OA is very reactive and transforms to HA in contact with water vapor [341]. Computer modeling techniques have been employed to qualitatively and quantitatively investigate the dehydration of HA to OA [348]. OA has the hexagonal space group symmetry $P\bar{6}$ (174) of cesanite type [88], while the space group symmetry for partially dehydrated HA was found to change from hexagonal P63/m to triclinic $P\overline{1}$ when more than *ca.* 35% of the structurally bound water had been removed [343]. On the *c*-axis, pure OA should have a divalent ion O^{2-} coupled with a vacancy instead of two neighboring monovalent OH^- ions.

Due to the aforementioned problems with OA preparation, it cannot be found in biological systems. In addition, no information on the biomedical applications of OA is available either. Plasma-sprayed coatings of calcium orthophosphates, in which OA might be present as an admixture phase, seem to be the only exception [274].

TTCP (or TetCP)

Tetracalcium phosphate or tetracalcium diorthophosphate monoxide ($Ca_4(PO_4)_2O$; the IUPAC name is tetracalcium oxide bis(orthophosphate); the mineral hilgenstockite) is the most basic calcium orthophosphate, however, its solubility in water is higher than that of HA (Table 1). TTCP has been known since 1883 [349], while the mineral hilgenstockite was named to honor a German metallurgist Gustav Hilgenstock (1844 – 1913), who first discovered it in Thomas slag from blast furnaces [349, 350]. Its major industrial importance stems from the fact that it is formed by the reactions between orthophosphates and lime in the manufacture of iron, and through these reactions, TTCP has a significant role in controlling the properties of the metal.

TTCP cannot be precipitated from aqueous solutions. It can be prepared only under the anhydrous conditions by solid-state reactions at temperatures above ~ 1300 °C, e.g., by heating homogenized equimolar quantities of DCPA and $CaCO_3$ in dry air, or in a flow of dry nitrogen [5, 205, 351]. These reactions should be carried out in a dry atmosphere, in vacuum or with rapid cooling (to prevent uptake of water and formation of HA). Easily DCPA might be replaced by ammonium orthophosphates [352, 353], while calcium carbonate might be replaced by calcium acetate [353]; however, in all cases, Ca/P ratio must be equal to 2.00. Furthermore, TTCP often appears as an unwanted by-product in plasma-sprayed HA coatings, where it is formed as a result of the thermal decomposition of HA to a mixture of high-temperature phases of α-TCP, TTCP and CaO [354]. TTCP is metastable: in both wet environment and aqueous solutions it slowly hydrolyses to HA and calcium hydroxide [5, 205, 355]. Consequently, TTCP is never found in biological calcifications. In medicine, TTCP is widely used for preparation of various self-setting calcium orthophosphate formulations [104, 354]; however, to the best of my knowledge, there is no commercial bone-substituting product consisting solely of TTCP. For the comprehensive information on TTCP, the readers are referred to a special review [354], while the spectra [356] and solubility [173] of TTCP are well described elsewhere.

To conclude the description of the individual calcium orthophosphates, one should mention on an interesting opinion [89, 142], that all calcium orthophosphates listed in Table 1 might be classified into three major structural types. They comprise: (i) the apatite type, $Ca_{10}(PO_4)_6X_2$, which includes HA, FA, OA, CDHA, OCP and TTCP; (ii) the glaserite type, named after the mineral glaserite, $K_3Na(SO_4)_2$, which includes all polymorphs of TCP and, perhaps, ACP; (iii) the Ca–PO_4 sheet-containing compounds, which include DCPD, DCPA, MCPM and MCPA. According to the authors, a closer examination of the structures revealed that all available calcium orthophosphates could be included into distorted glaserite type structures, but with varying degrees of distortion [89, 142].

Biphasic, Triphasic and Multiphasic Calcium Orthophosphate Formulations

Calcium orthophosphates might form biphasic, triphasic and multiphasic (polyphasic) formulations, in which the individual components cannot be separated from each other. Presumably, the individual phases of such compositions are homogeneously "mixed" at a far submicron level (< 0.1 μm) and strongly integrated with each other. Nevertheless, the presence of all individual phases is easily seen by X-ray diffraction technique [230].

The usual way to prepare multiphasic formulations consists of sintering non-stoichiometric compounds, such as ACP and CDHA, at temperatures above ~ 700 °C. Furthermore, a thermal decomposition of the stoichiometric calcium orthophosphates at temperatures above ~ 1300 °C might be used as well; however, this approach often results in formation of complicated mixtures of various products including admixtures of CaO, calcium pyrophosphates, *etc.* [230, 499]. Namely, transformation of HA into polyphasic calcium orthophosphates by annealing in a vacuum occurs as this: the outer part of HA is transformed into α-TCP and TTCP, while the α-TCP phase of the surface further transforms into CaO. Besides, in the boundary phase, HA is transformed into TTCP [357].

Historically, Nery and Lynch with co-workers first used the term biphasic calcium phosphate (BCP) in 1986 to describe a bioceramic, that consisted of a mixture of HA and β-TCP [358]. Based on the results of X-ray diffraction analysis, these authors found that the "tricalcium phosphate" preparation material used in their early publication [359] was in fact a mixture of ~ 20% HA and ~ 80% β-TCP. Currently, only biphasic and triphasic calcium orthophosphate formulations are known; perhaps, more complicated formulations will be manufactured in future. Furthermore, nowadays, multiphasic and/or polyphasic compositions consisting of high-temperature phases of calcium orthophosphates, such as α-TCP, β-TCP, HA and, perhaps, high-temperature ACP, OA and TTCP, are known only. No precise information on multiphasic compositions, containing MCPM, MCPA, DCPD, DCPA, low-temperature ACP, OCP and CDHA has been found in literature [230]. Perhaps, such formulations will be produced in future.

All BCP formulations might be subdivided into two major groups: those consisting of calcium orthophosphates having either the same (e.g., α-TCP and β-TCP) or different (e.g., β-TCP and HA) molar Ca/P ratios. Among all known BCP formulations, BCP consisting of HA and β-TCP is both the most known and the best investigated [230]. In 1986, LeGeros in USA and Daculsi in France initiated the basic studies on preparation of this type of BCP and its *in vitro* properties. This material is soluble and gradually dissolves in the body, seeding new bone formation as it releases calcium and orthophosphate ions into the biological medium. Presently, commercial BCP products of different or similar HA/β-TCP ratios are manufactured in many parts of the world as bone-graft or bone substitute materials for orthopaedic and dental applications under various trademarks and several manufacturers [230]. A similar combination of α-TCP with HA forms BCP as well [191, 360-362].

Recently the concept of BCP has been extended by preparation and characterization of biphasic TCP (BTCP), consisting of α-TCP and β-TCP phases [363-365]. It is usually prepared by heating ACP precursors [363-365], in which the α-TCP/β-TCP ratio can be controlled by aging time and pH value during synthesis of the amorphous precursor [364]. Furthermore, triphasic formulations, consisting of HA, α-TCP and β-TCP [366] or HA, α-TCP and TTCP [357] have been prepared [230].

It is important to recognize, that the major biomedical properties (such as bioactivity, bioresorbability, osteoconductivity and osteoinductivity) of the multiphasic formulations might be adjusted by changing the ratios among the phases. When compared to both α- and β-TCP, HA is a more stable phase under the physiological conditions, as it has a lower solubility (Table 1) and, thus, slower resorption kinetics. Therefore, due to a higher biodegradability of the α- or β-TCP component, the reactivity of BCP increases with the TCP/HA ratio increasing. Thus, *in vivo* bioresorbability of BCP can be adjusted through the phase composition. Similar conclusions are also valid for both the biphasic TCP (in which α-TCP is a more soluble phase) and the triphasic (HA, α-TCP and β-TCP) formulations. Further details on this subject might be found in a topical review [230].

Ion-substituted Calcium Orthophosphates

Finally, one should very briefly mention on existence of carbonate-apatite [367-369], chlorapatite [370-372], as well as on a great number of calcium orthophosphates with various ionic substitutions [23, 63, 373, 374]. In principle, any ion in calcium orthophosphates might be substituted by other ion(s). Usually, the ion-substituted calcium orthophosphates are of a non-stoichiometric nature with just a partial ionic substitution and there are too many of them to be described here. Currently, this is a hot investigation topic; therefore, the readers are referred to the special literature [23, 63, 373, 374]. In addition, there is a very good review, in which the structures of more than 75 chemically different apatites have been discussed [88].

To conclude this topic, it is interesting to note, that chemical elements not found in natural bones can be intentionally incorporated into calcium orthophosphate biomaterials to get special properties. For example, addition of Ag^+ [375-377], Zn^{2+} [377, 378] and Cu^{2+} [377-379] has been used for imparting antimicrobial effect, while radioactive isotopes of ^{90}Y [380], ^{153}Sm [381-383] and ^{186}Re [381] have been incorporated into HA bioceramics and injected into knee joints to treat rheumatoid joint synovitis [380, 381, 383]. More to the point, apatites were found to incorporate individual molecules, such as water, oxygen and carbon dioxide [63].

References

[1] Dorozhkin, S.V. Calcium orthophosphates and human beings. A historical perspective from the 1770s until 1940. *Biomatter* 2012, 2, 53-70.

[2] Dorozhkin, S.V. A detailed history of calcium orthophosphates from 1770s till 1950. *Mater. Sci. Eng.* C 2013, 33, 3085-3110.

[3] Lide, D.R. *The CRC handbook of chemistry and physics.* 86th Ed., CRC Press, Boca Raton, Florida, 2005, 2544 pp.

[4] LeGeros, R.Z. Calcium phosphates in oral biology and medicine. *Monographs in Oral Science.* Vol. 15. Karger, Basel, 1991, 201 pp.

[5] Elliott, J.C. Structure and chemistry of the apatites and other calcium orthophosphates. *Studies in inorganic chemistry,* Vol. 18; Elsevier: Amsterdam, Netherlands, 1994; 389 pp.
[6] Amjad, Z., Ed. *Calcium phosphates in biological and industrial systems.* Kluwer Academic Publishers: Boston, MA, USA, 1997, 529 pp.
[7] Dorozhkin, S.V. Calcium orthophosphates: occurrence, properties, biomineralization, pathological calcification and biomimetic applications. *Biomatter* 2011, 1, 121-164.
[8] Dorozhkin, S.V. *Calcium orthophosphates: applications in nature, biology, and medicine.* Pan Stanford: Singapore, 2012, 854 pp.
[9] Cantelar, E., Lifante, G., Calderón, T., Meléndrez, R., Millán, A., Alvarez, M.A., Barboza-Flores, M. Optical characterisation of rare earths in natural fluorapatite. *J. Alloy Compd.* 2001, 323-324, 851-854.
[10] Ribeiro, H.B., Guedes, K.J., Pinheiro, M.V.B., Greulich-Weber, S., Krambrock, K. About the blue and green colours in natural fluorapatite. *Physica Status Solidi C* 2005, 2, 720-723.
[11] Lowenstam, H.A., Weiner, S. *On biomineralization.* Oxford University Press, 1989, 324 pp.
[12] McConnell, D. Apatite: its crystal chemistry, mineralogy, utilization, and geologic and biologic occurrences. *Applied Mineralogy,* Vol. 5. Springer-Verlag: Vienna and New York, USA, 1973; 111 pp.
[13] Becker, P. *Phosphates and phosphoric acid: raw materials technology and economics of the wet process.* 2nd Ed. Fertilizer science and technology series. Marcel Dekker: New York, USA, 1989; 760 pp.
[14] Smith, D.K. *Calcium phosphate apatites in nature. In: Hydroxyapatite and related materials.* Brown, P.W., Constantz, B. Eds. CRC Press Inc.: Boca Raton, FL, USA, 1994; pp. 29-44.
[15] Cook, P.J., Shergold, J.H., Davidson, D.F., Eds., *Phosphate deposits of the world: phosphate rock resources.* Cambridge University Press: Cambridge, MA, USA, 2005; Vol. 2, 600 pp.
[16] Mitchell, L., Faust, G.T., Hendricks, S.B., Reynolds, D.S. The mineralogy and genesis of hydroxylapatite. *Am. Miner.* 1943, 28, 356-371.
[17] Zhang, J.Z., Guo, L., Fischer, C.J. Abundance and chemical speciation of phosphorus in sediments of the Mackenzie river delta, the Chukchi sea and the Bering sea: importance of detrital apatite. *Aquatic Geochem.* 2010, 16, 353-371.
[18] Omelon, S.J., Grynpas, M.D. Relationships between polyphosphate chemistry, biochemistry and apatite biomineralization. *Chem. Rev.* 2008, 108, 4694-4715.
[19] Jarvis, I. Phosphorite geochemistry: state-of-the-art and environmental concerns. *Eclogae Geologicae Helvetiae* 1994, 87, 643-700.
[20] Glenn, C.R. Phosphorus and phosphorites: sedimentology and environments of formation. *Eclogae Geologicae Helvetiae* 1994, 87, 747-788.
[21] McClellan, G.H. Mineralogy of carbonate fluorapatites (Francolites). *J. Geol. Soc.* 1980, 137, 675-681.
[22] Mcarthur, J.M. Francolite geochemistry-compositional controls during formation, diagenesis, metamorphism and weathering. *Geochim. Cosmochim. Acta* 1985, 49, 23-35.

[23] Pan, Y., Fleet, M.E. Compositions of the apatite-group minerals: substitution mechanisms and controlling factors. In: Phosphates: geochemical, geobiological and materials importance. Series: Reviews in Mineralogy and Geochemistry. Vol. 48. Hughes, J.M., Kohn, M., Rakovan, J., Eds. *Mineralogical Society of America:* Washington, D.C., USA, 2002; pp. 13-49.

[24] Zanin, Y.N. The classification of calcium phosphates of phosphorites. *Lithol. Miner. Resour.* 2004, 39, 281-282.

[25] Henry, T.H. On francolite, a supposed new mineral. *Phil. Mag.* 1850, 36, 134-135.

[26] Rogers, A.F. Collophane, a much neglected mineral. *Am. J. Sci.* 1922, 3, 269-276.

[27] Cao, Q., Wen, S., Li, C., Bai, S., Liu, D. Investigation on beneficiation strategy for collophane. *Adv. Mater. Res.* 2013, 634-638, 3404-3411.

[28] http://www.mindat.org/min-10072.html (accessed in February 2014).

[29] Elorza, J., Astibia, H., Murelaga, X., Pereda-Suberbiola, X. Francolite as a diagenetic mineral in dinosaur and other upper cretaceous reptile bones (Lano, Iberian peninsula): microstructural, petrological and geochemical features. *Cretaceous Res.* 1999, 20, 169-187.

[30] Hubert, B., Álvaro, J.J., Chen, J.Y. Microbially mediated phosphatization in the Neoproterozoic Doushantuo Lagerstätte, South China. *Bull. Soc. Géol. Fr.* 2005, 176, 355-361.

[31] Xiao, S., Zhang, Y., Knoll, A.H. Three-dimensional preservation of algae and animal embryos in a neoproterozoic phosphorite. *Nature* 1998, 391, 553-558.

[32] Xiao, S., Yuan, X., Knoll, A.H. Eumetazoan fossils in terminal Proterozoic phosphorites? *Proc. Natl. Acad. Sci. USA* 2000, 97, 13684-13689.

[33] Chakhmouradian, A.R., Medici, L. Clinohydroxylapatite: a new apatite-group mineral from northwestern Ontario (Canada), and new data on the extent of Na-S substitution in natural apatites. *Eur. J. Mineral.* 2006, 18, 105-112.

[34] Mason, H.E., Mccubbin, F.M., Smirnov, A.E., Phillips, B.L. Solid-state NMR and IR spectroscopic investigation of the role of structural water and F in carbonate-rich fuorapatite. *Am. Miner.* 2009, 94, 507-516.

[35] http://www.mindat.org/gallery.php?min = 9293 (accessed in February 2014).

[36] Klein, C. *Brushite from the island of Mona* (between Haiti and Puerto Rico). Sitzber. K. Preuss. Aka. 1901, 720-725.

[37] Merrill, G.P. On the calcium phosphate in meteoric stones. *Am. J. Sci.* 1917, 43, 322-324.

[38] McCubbin, F.M., Nekvasil, H. Maskelynite-hosted apatite in the Chassigny meteorite: insights into late-stage magmatic volatile evolution in martian magmas. *Am. Miner.* 2008, 93, 676-684.

[39] Tyrrell, G.W. Apatite, nepheline, and rare-earth mining in the Kola Peninsula. *Nature* 1938, 141, 354-355.

[40] Kogarko, L.N. Problems of the genesis of giant apatite and rare metal deposits of the Kola Peninsula, Russia. *Geology of Ore Deposits* 1999, 41, 351-366.

[41] Baturin, G.N. Phosphorites of the Sea of Japan. *Oceanology* 2012, 52, 666-676.

[42] Ford, A.K. A remarkable crystal of apatite from Mt. Apatite, Auburn, Maine. *Am. J. Sci.* 1917, 44, 245-246.

[43] Hogarth, D.D. The discovery of apatite on the Lievre River, Quebec. *Mineral. Rec.* 1974, 5, 178-182.

[44] van Velthuizen, J. Giant fluorapatite crystals: a question of locality. *Mineral. Rec.* 1992, 23, 459-463.

[45] Trueman, N.A. Substitutions for phosphate ions in apatite. *Nature* 1966, 210, 937-938.

[46] Gilinskaya, L.G. Organic radicals in natural apatites according to EPR data: potential genetic and paleoclimatic indicators. *J. Struct. Chem.* 2010, 51, 471-481.

[47] Gilinskaya, L.G. A stable perinaphthenyl radical in natural apatites. *J. Struct. Chem.* 2010, 51, 761-764.

[48] Sánchez-Salcedo, S., Vila, M., Izquierdo-Barba, I., Cicuéndez, M., Vallet-Regí, M. Biopolymer-coated hydroxyapatite foams: a new antidote for heavy metal intoxication. *J. Mater. Chem.* 2010, 20, 6956-6961.

[49] White, T., Ferraris, C., Kim, J., Madhavi, S. Apatite – an adaptive framework structure. In: Micro- and mesoporous mineral phases. Series: Reviews in Mineralogy and Geochemistry. Vol. 57. Ferraris, G., Merlino S., Eds. *Mineralogical Society of America:* Washington, D.C., USA, 2005, pp. 307-401.

[50] Jacob, K.D., Reynolds, D.S. Reduction of tricalcium phosphate by carbon. *Ind. Eng. Chem.* 1928, 20, 1204-1210.

[51] Emsley, J. *The shocking history of phosphorus: a biography of the devil's element.* Pan Books, Macmillan, UK. 2001, 336 pp.

[52] van der Sluis, S., Meszaros, Y., Marchee, W.G.J., Wesselingh, H.A., van Rosmalen, G.M. The digestion of phosphate ore in phosphoric acid. *Ind. Eng. Chem. Res.* 1987, 26, 2501-2505.

[53] Dorozhkin, S.V. Fundamentals of the wet-process phosphoric acid production. 1. Kinetics and mechanism of the phosphate rock dissolution. *Ind. Eng. Chem. Res.* 1996, 35, 4328-4335.

[54] Dorozhkin, S.V. Fundamentals of the wet-process phosphoric acid production. 2. kinetics and mechanism of $CaSO_4{\cdot}0.5H_2O$ surface crystallization and coating formation. *Ind. Eng. Chem. Res.* 1997, 36, 467-473.

[55] Dorozhkin, S.V. Ecological principles of wet-process phosphoric acid technology. *J. Chem. Technol. Biotechnol.* 1998, 71, 227-233.

[56] Copson, R.L., Newton, R.H., Lindsay, J.D. Superphosphate manufacture – mixing phosphate rook with concentrated phosphoric acid. *Ind. Eng. Chem.* 1936, 28, 923-927.

[57] Newton, R.H., Copson, R.L. Superphosphate manufacture – composition of superphosphate made from phosphate rock and concentrated phosphoric acid. *Ind. Eng. Chem.* 1936, 28, 1182-1186.

[58] Rossete, A.L.R.M., Carneiro, J.M.T., Bendassolli, J.A., Tavares, C.R.O., Sant'Ana Filho, C.R. Production of single superphosphate labeled with 34S. *Scientia Agricola* 2008, 65, 91-94.

[59] Magda, A., Pode, V., Niculescu, M., Muntean, C., Bandur, G., Iovi, A. Studies on process of obtaining the fertilizers based on ammonium phosphates with addition of boric acid. *Rev. Chim. Bucharest* 2008, 59, 1340-1344.

[60] Kijkowska, R., Kowalski, Z., Pawlowska-Kozinska, D., Wzorek, Z., Gorazda, K. Effect of purification from sulfates on phase composition of sodium tripolyphosphate obtained from wet-process phosphoric acid derived from Kola apatite. *Phosphorus Sulfur* 2007, 182, 2667-2683.

[61] Abouzeid, A.Z.M. Upgrading of phosphate ores – a review. *Powder Handling and Processing* 2007, 19, 92-109.

[62] Abouzeid, A.Z.M. Physical and thermal treatment of phosphate ores – an overview. *Int. J. Mineral Process.* 2008, 85, 59-84.

[63] Rey, C., Combes, C., Drouet, C., Sfihi, H. Chemical diversity of apatites. *Adv. Sci. Technol.* 2006, 49, 27-36.

[64] O'Neill, W.C. The fallacy of the calcium – phosphorus product. *Kidney Int.* 2007, 72, 792-796.

[65] LeGeros, R.Z. Formation and transformation of calcium phosphates: relevance to vascular calcification. *Z. Kardiol.* 2001, 90, Suppl. 3, III116-III125.

[66] Becker, A., Epple, M., Müller, K.M., Schmitz, I. A comparative study of clinically well-characterized human atherosclerotic plaques with histological, chemical, and ultrastructural methods. *J. Inorg. Biochem.* 2004, 98, 2032-2038.

[67] Wopenka, B., Pasteris, J.D. A mineralogical perspective on the apatite in bone. *Mater. Sci. Eng.* C 2005, 25, 131-143.

[68] Pasteris, J.D., Wopenka, B., Valsami-Jones, E. Bone and tooth mineralization: why apatite? *Elements* 2008, 4, 97-104.

[69] Sun, Y., Hanley, E.N., Jr. Calcium-containing crystals and osteoarthritis. *Curr. Opin. Orthoped.* 2007, 18, 472-478.

[70] Nakano, T., Kaibara, K., Tabata, Y., Nagata, N., Enomoto, S., Marukawa, E., Umakoshi, Y. Unique alignment and texture of biological apatite crystallites in typical calcified tissues analyzed by microbeam x-ray diffractometer system. *Bone* 2002, 31, 479-487.

[71] Grynpas, M.D., Omelon, S. Transient precursor strategy or very small biological apatite crystals? *Bone* 2007, 41, 162-164.

[72] Bazin, D., Dessombz, A., Nguyen, C., Ea, H.K., Lioté, F., Rehr, J., Chappard, C., Rouzière, S., Thiaudière, D., Reguer, S., Daudon, M. The status of strontium in biological apatites: an XANES/EXAFS investigation. *J. Synchrotron Radiat.* 2014, 21, 136-142.

[73] Lee, J.W., Sasaki, K., Ferrara, J.D., Akiyama, K., Sasaki, T., Nakano, T. Evaluation of preferential alignment of biological apatite (BAp) crystallites in bone using transmission X-ray diffraction optics. *J. Jpn. Ins. Metal.* 2009, 73, 786-793.

[74] Ishimoto, T., Sakamoto, T., Nakano, T. Orientation of biological apatite in rat calvaria analyzed by microbeam X-ray diffractometer. *Mater. Sci. Forum* 2010, 638-642, 576-581.

[75] Meneghini, C., Dalconi, M.C., Nuzzo, S., Mobilio, S., Wenk, R.H. Rietveld refinement on X-ray diffraction patterns of bioapatite in human fetal bones. *Biophys. J.* 2003, 84, 2021-2029.

[76] Eagle, R.A., Schauble, E.A., Tripati, A.K., Tütken, T., Hulbert, R.C., Eiler, J.M. Body temperatures of modern and extinct vertebrates from ^{13}C-^{18}O bond abundances in bioapatite. *Proc. Natl. Acad. Sci. USA* 2010, 107, 10377-10382.

[77] Cherkinsky, A., Dantas, M.A.T., Cozzuol, M.A. Bioapatite ^{14}C age of giant mammals from Brazil. *Radiocarbon* 2013, 55, 464-471.

[78] Šupová, M. Isolation and preparation of nanoscale bioapatites from natural sources: a review. *J. Nanosci. Nanotechnol.* 2014, 14, 546-563.

[79] Brögger, W.C., Bäckström, H. Über den Dahllit, ein neues Mineral von Ödegärden, Bamle, Norwegen. *Neues Jb. Miner. Geol. Paläont.* 1890, 223-224.

[80] Skinner, H.C.W. Biominerals. Mineral. Mag. 2005, 69, 621-641.

[81] Daculsi, G., Bouler, J.M., LeGeros, R.Z. Adaptive crystal formation in normal and pathological calcifications in synthetic calcium phosphate and related biomaterials. *Int. Rev. Cytology* 1997, 172, 129-191.

[82] Driessens, F.C.M., Verbeeck, R.M.H. *Biominerals.* CRC Press: Boca Raton, FL, USA, 1990; 440 pp.

[83] Clark, N.A. The system P_2O_5 – CaO – H_2O and the recrystallization of monocalcium phosphate. *J. Phys. Chem.* 1931, 35, 1232-1238.

[84] Brown, P.W. Phase relationships in the ternary system CaO – P_2O_5 – H_2O at 25°C. J. *Am. Ceram. Soc.* 1992, 75, 17-22.

[85] Martin, R.I., Brown, P.W. Phase equilibria among acid calcium phosphates. *J. Am. Ceram. Soc.* 1997, 80, 1263-1266.

[86] Kreidler, E.R., Hummel, F.A. Phase relationships in the system SrO-P_2O_5 and the influence of water vapor on the formation of $Sr_4P_2O_9$. *Inorg. Chem.* 1967, 6, 884-891.

[87] Carayon, M.T., Lacout, J.L. Study of the Ca/P atomic ratio of the amorphous phase in plasma-sprayed hydroxyapatite coatings. *J. Solid State Chem.* 2003, 172, 339-350.

[88] White, T.J., Dong, Z.L. Structural derivation and crystal chemistry of apatites. *Acta Crystallogr.* B 2003, B59, 1-16.

[89] Mathew, M., Takagi, S. Structures of biological minerals in dental research. *J. Res. Natl. Inst. Stand. Technol.* 2001, 106, 1035-1044.

[90] Wang, L., Nancollas, G.H. Calcium orthophosphates: crystallization and dissolution. *Chem. Rev.* 2008, 108, 4628-4669.

[91] Lynn, A.K., Bonfield, W. A novel method for the simultaneous, titrant-free control of pH and calcium phosphate mass yield. *Acc. Chem. Res.* 2005, 38, 202-207.

[92] León, B., Jansen, J.J., Eds. *Thin calcium phosphate coatings for medical implants.* Springer: New York, USA, 2009; 326 pp.

[93] McDowell, H., Gregory, T.M., Brown, W.E. Solubility of $Ca_5(PO_4)_3OH$ in the system $Ca(OH)_2$ – H_3PO_4 – H_2O at 5, 15, 25, and 37 degree C. *J. Res. Natl. Bur. Stand. Sect.* A Phys. Chem. 1977, 81A, 273-281.

[94] Chow, L.C. Next generation calcium phosphate-based biomaterials. *Dent. Mater. J.* 2009, 28, 1-10.

[95] Pan, H.B., Darvell, B.W. Calcium phosphate solubility: the need for re-evaluation. *Cryst. Growth Des.* 2009, 9, 639-645.

[96] Fourcroy, A.F. *A general system of chemical knowledge; and its application to the phenomena of nature and art. In eleven volumes.* Translated from the original French by William Nicholson. Vol. III. Printed for Cadell and Davies, Strand; Longman and Rees, G. and J. Robinson, and J. Walker, Paternoster-row; Vernor and Hood, Poultry; Clarke and sons, Portugal-street; Cuthell and Martin, and Ogilvy and son, Holborn; and S. Bagster, Strand. London, 1804, 472 pp.

[97] *A dictionary of chemistry and mineralogy, with an account of the processes employed in many of the most important chemical manufactures.* To which are added a description of chemical apparatus, and various useful tables of weights and measures, chemical instruments, &c. &c. Illustrated with fifteen engravings. By A. & C. R. Aikin. Vol. II. London: Printed for John and Arthur Arch, Couninll; and William Phillips, George Yard, Lombard Street. 1807, 176 pp.

[98] Boonchom, B. Parallelogram-like microparticles of calcium dihydrogen phosphate monohydrate ($Ca(H_2PO_4)_2{\cdot}H_2O$) obtained by a rapid precipitation route in aqueous and acetone media. *J. Alloy Compd.* 2009, 482, 199-202.

[99] Kongteweelert, S., Ruttanapun, C., Thongkam, M., Chaiyasith, P., Woramongkonchai, S., Boonchom, A.B. Facile, alternative synthesis of spherical-like $Ca(H_2PO_4)_2{\cdot}H_2O$ nanoparticle by aqueous-methanol media. *Adv. Mater. Res.* 2013, 717, 49-53.

[100] Tynsuaadu, K. Influence of silicic acid and glauconite on thermal dehydration of $Ca(H_2PO_4)_2{\cdot}H_2O$. *J. Therm. Anal. 1990*, 36, 1785-1793.

[101] Xu, J., Gilson, D.F.R., Butler, I.S. FT-Raman and high-pressure FT-infrared spectroscopic investigation of monocalcium phosphate monohydrate, $Ca(H_2PO_4)_2{\cdot}H_2O$. *Spectrochim. Acta* A 1998, 54, 1869-1878.

[102] Köster, K., Heide, H., König, R. Resorbierbare Calciumphosphatkeramik im Tierexperiment unter Belastung. *Langenbecks Arch. Chir.* 1977, 343, 173-181.

[103] Huan, Z., Chang, J. Novel bioactive composite bone cements based on the β-tricalcium phosphate – monocalcium phosphate monohydrate composite cement system. *Acta Biomater.* 2009, 5, 1253-1264.

[104] Dorozhkin. S.V. Self-setting calcium orthophosphate formulations. J. Funct. Biomater. 2013, 4 209-311.

[105] Budavari, S., O'Neil, M.J., Smith, A., Heckelman, P.E., Kinneary, J.F., Eds., *The Merck Index: an encyclopedia of chemicals, drugs, and biologicals.* 12th Ed. Chapman & Hall: USA, 1996; 1741 pp.

[106] Stein, H.H., Kadzere, C.T., Kim, S.W., Miller, P.S. Influence of dietary phosphorus concentration on the digestibility of phosphorus in monocalcium phosphate by growing pigs. *J. Animal Sci.* 2008, 86, 1861-1867.

[107] Dorozhkin, S.V. Calcium orthophosphates in dentistry. *J. Mater. Sci. Mater. Med.* 2013, 24, 1335-1363.

[108] Smalberger, S.A., Chien, S.H., Singh, U., Henao, J. Relative agronomic effectiveness of phosphate rock compared with triple superphosphate for initial canola, wheat, or ryegrass, and residual wheat in two acid soils. *Soil Sci.* 2010, 175, 36-43.

[109] Roscoe, H.E., Schorlemmer, C. *A treatise on chemistry.* Volume II: Metals. Part 1. Macmillan and Co., London, UK, 1879, 504 pp.

[110] Moore, G.E. On brushite, a new mineral occurring in phosphatic guano. *Am. J. Sci.* 1865, 39, 43-44.

[111] Ferreira, A., Oliveira, C., Rocha, F. The different phases in the precipitation of dicalcium phosphate dihydrate. *J. Cryst. Growth* 2003, 252, 599-611.

[112] Oliveira, C., Ferreira, A., Rocha, F. Dicalcium phosphate dihydrate precipitation: characterization and crystal growth. *Chem. Eng. Res. Design* 2007, 85, 1655-1661.

[113] Sivkumar, G.R., Girija, E.K., Kalkura, S.N., Subramanian, C. Crystallization and characterization of calcium phosphates: brushite and monetite. *Crystal Res. Technol.* 1997, 33, 197-205.

[114] Madhurambal, G., Subha, R., Mojumdar, S.C. Crystallization and thermal characterization of calcium hydrogen phosphate dihydrate crystals. *J. Therm. Anal. Calorim.* 2009, 96, 73-76.

[115] MacDowell, H., Brown W.E., Sutter, J.R. Solubility study of calcium hydrogenphosphate. Ion pair formation. *Inorg. Chem.* 1971, 10, 1638-1643.

[116] Landin, M., Rowe, R.C., York, P. Structural changes during the dehydration of dicalcium phosphate dihydrate. *Eur. J. Pharmac. Sci.* 1994, 2, 245-252.
[117] Curry, N.A., Jones, D.W. Crystal structure of brushite, calcium hydrogen orthophosphate dihydrate: a neutron-diffraction investigation. *J. Chem. Soc. A Inorg. Phys. Theoret. Chem.* 1971, 3725-3729.
[118] Arsic, J., Kaminski, D., Poodt, P., Vlieg, E. Liquid ordering at the brushite – {010} – water interface. *Phys. Rev.* B 2004, 69, 245406 (4 pages).
[119] Pan, H.B., Darvell, B.W. Solubility of dicalcium phosphate dihydrate by solid titration. *Caries Res.* 2009, 43, 254-260.
[120] Lundager-Madsen, H.E. Optical properties of synthetic crystals of brushite ($CaHPO_4 \cdot 2H_2O$). *J. Cryst. Growth* 2008, 310, 617-623.
[121] Qiu, S.R., Orme, C.A. Dynamics of biomineral formation at the near-molecular level. *Chem. Rev.* 2008, 108, 4784-4822.
[122] Mostashari, S.M., Haddadi, H., Hashempoor, Z. Effect of deposited calcium hydrogen phosphate dihydrate on the flame retardancy imparted to cotton fabric. *Asian J. Chem.* 2006, 18, 2388-2390.
[123] Shepard, C.U. On two new minerals, monetite and monite, with a notice of pyroclasite. *Am. J. Sci.* 1882, 23, 400-405.
[124] Tas, A.C. Monetite ($CaHPO_4$) synthesis in ethanol at room temperature. *J. Am. Ceram. Soc.* 2009, 92, 2907-2912.
[125] Chen, G.G., Luo, G.S., Yang, L.M., Xu, J.H., Sun, Y., Wang, J.D. Synthesis and size control of $CaHPO_4$ particles in a two-liquid phase micro-mixing process. *J. Cryst. Growth* 2005, 279, 501-507.
[126] Miyazaki, T., Sivaprakasam, K., Tantry, J., Suryanarayanan, R. Physical characterization of dibasic calcium phosphate dihydrate and anhydrate. *J. Pharmac. Sci.* 2009, 98, 905-916.
[127] Eshtiagh-Hosseini, H., Houssaindokht, M.R., Chahkandhi, M., Youssefi, A. Preparation of anhydrous dicalcium phosphate, DCPA, through sol-gel process, identification and phase transformation evaluation. *J. Non-Cryst. Solids* 2008, 354, 3854-3857.
[128] Tamimi, F., Torres, J., Bassett, D., Barralet, J., Cabarcos, E.L. Resorption of monetite granules in alveolar bone defects in human patients. *Biomaterials* 2010, 31, 2762-2769.
[129] Takami, K., Machimura, H., Takado, K., Inagaki, M., Kawashima, Y. Novel preparation of free flowing spherically granulated dibasic calcium phosphate anhydrous for direct tabletting. *Chem. Pharm. Bull.* 1996, 44, 868-870.
[130] Percy, J. Notice of a new hydrated phosphate of lime. *Mem. Proc. Chem. Soc.* 1843, 2, 222-223.
[131] Bjerrum, N. Calciumorthophosphate. I. Die Festen Calciumorthophosphate. II. Komplexbildung in Lösung von Calcium-und Phosphate-Ionen. *Math. Fys. Medd.* Dan. Vid. Selsk. 1958, 31, 1-79.
[132] LeGeros, R.Z. Preparation of octacalcium phosphate (OCP): a direct fast method. *Calcif. Tissue Int.* 1985, 37, 194-197.
[133] Nakahira, A., Aoki, S., Sakamoto, K., Yamaguchi, S. Synthesis and evaluation of various layered octacalcium phosphates by wet-chemical processing. *J. Mater. Sci. Mater. Med. 2001*, 12, 793-800.

[134] Arellano-Jiménez, M.J., García-García, R., Reyes-Gasga, J. Synthesis and hydrolysis of octacalcium phosphate and its characterization by electron microscopy and X-ray diffraction. *J. Phys. Chem. Solids* 2009, 70, 390-395.
[135] Suzuki, O. Octacalcium phosphate: osteoconductivity and crystal chemistry. *Acta Biomater.* 2010, 6, 3379-3387.
[136] Miyatake, N., Kishimoto, K.N., Anada, T., Imaizumi, H., Itoi, E., Suzuki, O. Effect of partial hydrolysis of octacalcium phosphate on its osteoconductive characteristics. *Biomaterials* 2009, 30, 1005-1014.
[137] Matsunaga, K. First-principles study of substitutional magnesium and zinc in hydroxyapatite and octacalcium phosphate. *J. Chem. Phys.* 2008, 128, 245101.
[138] Boanini, E., Gazzano, M., Rubini, K., Bigi, A. Collapsed octacalcium phosphate stabilized by ionic substitutions. *Cryst. Growth Des.* 2010, 10, 3612-3617.
[139] Brown, W.E. Octacalcium phosphate and hydroxyapatite: crystal structure of octacalcium phosphate. *Nature* 1962, 196, 1048-1050.
[140] Brown, W.E., Smith, J.P., Lehr, J.R., Frazier, A.W. Octacalcium phosphate and hydroxyapatite: crystallographic and chemical relations between octacalcium phosphate and hydroxyapatite. *Nature* 1962, 196, 1050-1055.
[141] Pan, H.B., Darvell, B.W. Solid titration of octacalcium phosphate. *Caries Res.* 2009, 43, 322-330.
[142] Chow, L.C., Eanes, E.D. (Eds.). *Octacalcium phosphate. Monographs in Oral Science. Vol.* 18. Karger: Basel, Switzerland, 2001, 167 pp.
[143] Kakei, M., Sakae, T., Yoshikawa, M. Electron microscopy of octacalcium phosphate in the dental calculus. *J. Electron Microsc.* 2009, 58, 393-398.
[144] Brown, W.E. Crystal growth of bone mineral. *Clin. Orthop. Relat. Res.* 1966, 44, 205-220.
[145] Suzuki, O. Biological role of synthetic octacalcium phosphate in bone formation and mineralization. *J. Oral Biosci.* 2010, 52, 6-14.
[146] Iijima, M., Nelson, D.G.A., Pan, Y., Kreinbrink, A.T., Adachi, M., Goto, T., Moriwaki, Y. Fluoride analysis of apatite crystals with a central planar OCP inclusion: concerning the role of F^- ions on apatite / OCP / apatite structure formation. *Calcif. Tissue Int.* 1996, 59, 377-384.
[147] Bodier-Houllé, P., Steuer, P., Voegel, J.C., Cuisinier, F.J.G. First experimental evidence for human dentine crystal formation involving conversion of octacalcium phosphate to hydroxyapatite. *Acta Crystallogr. D: Biol. Crystallogr.* 1998, 54, 1377-1381.
[148] Rodríguez-Hernández, A.G., Fernández, M.E., Carbajal-de-la-Torre, G., García-García, R., Reyes-Gasga, J. Electron microscopy analysis of the central dark line defect of the human tooth enamel. *Mater. Res. Soc. Symp. Proc.* 2005, 839, 157-162.
[149] Tomazic, B.B., Brown, W.E., Shoen, F.J. Physicochemical properties of calcific deposits isolated from porcine bioprosthetic heart valves removed from patients following 2 – 13 years function. *J. Biomed. Mater. Res.* 1994, 28, 35-47.
[150] Nancollas, G.H., Wu, W. Biomineralization mechanisms: a kinetics and interfacial energy approach. *J. Cryst. Growth* 2000, 211, 137-142.
[151] Suzuki, O., Kamakura, S., Katagiri, T., Nakamura, M., Zhao, B., Honda, Y., Kamijo, R. Bone formation enhanced by implanted octacalcium phosphate involving conversion into Ca-deficient hydroxyapatite. *Biomaterials* 2006, 27, 2671-2681.

[152] Kikawa, T., Kashimoto, O., Imaizumi, H., Kokubun, S., Suzuki, O. Intramembranous bone tissue response to biodegradable octacalcium phosphate implant. *Acta Biomater.* 2009, 5, 1756-1766.

[153] Murakami, Y., Honda, Y., Anada, T., Shimauchi, H., Suzuki, O. Comparative study on bone regeneration by synthetic octacalcium phosphate with various granule sizes. *Acta Biomater.* 2010, 6, 1542-1548.

[154] Suzuki, O. Octacalcium phosphate (OCP)-based bone substitute materials. *Jpn. Dent. Sci. Rev.* 2013, 49, 58-71.

[155] Bredig, M.A., Franck, H.H., Fülnder, H. Beiträge zur Kenntnis der Kalk-Phosphorsäure-Verbindungen. II. Z. *Elktrochem. Angew. P.* 1932, 38, 158-164.

[156] Trömel, G. Beiträge zur Kenntnis des Systems Kalziumoxyd-Phosphorpentoxyd. Mitt. Kaiser-Wilhelm-Inst. Eisenforsch. *Düsseldorf* 1932, 14, 25-34.

[157] Tao, J., Jiang, W., Zhai, H., Pan, H., Xu X., Tang, R. Structural components and anisotropic dissolution behaviors in one hexagonal single crystal of β-tricalcium phosphate. *Cryst. Growth Des.* 2008, 8, 2227-2234.

[158] Tao, J., Pan, H., Zhai, H., Wang, J., Li, L., Wu, J., Jiang, W., Xu, X., Tang, R., Controls of tricalcium phosphate single-crystal formation from its amorphous precursor by interfacial energy. *Cryst. Growth Des.* 2009, 9, 3154-3160.

[159] Hou, X.J., Mao, K.Y., Chen, D.F. Bone formation performance of beta-tricalcium phosphate sintered bone. *J. Clin. Rehabil. Tiss. Eng. Res. 2008*, 12, 9627-9630.

[160] Santos, C.F.L., Silva, A.P., Lopes, L., Pires, I., Correia, I.J. Design and production of sintered β-tricalcium phosphate 3D scaffolds for bone tissue regeneration. *Mater. Sci. Eng.* C 2012, 32, 1293-1298.

[161] Lee, N.H., Hwang, K.H., Lee, J.K. Fabrication of biphasic calcium phosphate bioceramics from the recycling of bone ash. *Adv. Mater. Res.* 2013, 610-613, 2328-2331.

[162] Ito, A., LeGeros, R.Z. Magnesium- and zinc-substituted beta-tricalcium phosphates as potential bone substitute biomaterials. *Key Eng. Mater.* 2008, 377, 85-98.

[163] 163 Kannan, S., Goetz-Neunhoeffer, F., Neubauer, J., Ferreira, J.M.F. Synthesis and structure refinement of zinc-doped β-tricalcium phosphate powders. *J. Am. Ceram. Soc.* 2009, 92, 1592-1595.

[164] Kannan, S., Goetz-Neunhoeffer, F., Neubauer, J., Pina, S., Torres, P.M.C., Ferreira, J.M.F. Synthesis and structural characterization of strontium- and magnesium-co-substituted β-tricalcium phosphate. *Acta Biomater.* 2010, 6, 571-576.

[165] Araújo, J.C., Sader, M.S., Moreira, E.L., Moraes, V.C.A., LeGeros, R.Z., Soares, G.A. Maximum substitution of magnesium for calcium sites in Mg-β-TCP structure determined by X-ray powder diffraction with the Rietveld refinement. *Mater. Chem. Phys.* 2009, 118, 337-340.

[166] Karlinsey, R.L., Mackey, A.C. Solid-state preparation and dental application of an organically modified calcium phosphate. *J. Mater. Sci.* 2009, 44, 346-349.

[167] Karlinsey, R.L., Mackey, A.C., Walker, E.R., Frederick, K.E. Preparation, characterization and *in vitro* efficacy of an acid-modified β-TCP material for dental hard-tissue remineralization. *Acta Biomater.* 2010, 6, 969-978.

[168] Karlinsey, R.L., Mackey, A.C., Walker, E.R., Frederick, K.E. Surfactant-modified β-TCP: structure, properties, and *in vitro* remineralization of subsurface enamel lesions. *J. Mater. Sci. Mater. Med.* 2010, 21, 2009-2020.

[169] Yashima, M., Sakai, A., Kamiyama, T., Hoshikawa, A. Crystal structure analysis of β-tricalcium phosphate $Ca_3(PO_4)_2$ by neutron powder diffraction. *J. Solid State Chem.* 2003, 175, 272-277.
[170] Yin, X., Stott, M.J., Rubio, A. α- and β-tricalcium phosphate: a density functional study. *Phys. Rev.* B 2003, 68, 205205 (7 pages).
[171] Liang, L., Rulis, P., Ching, W.Y. Mechanical properties, electronic structure and bonding of α- and β-tricalcium phosphates with surface characterization. *Acta Biomater.* 2010, 6, 3763-3771.
[172] Kumar, A.R., Kalainathan, S. Microhardness studies on calcium hydrogen phosphate (brushite) crystals. *Mater. Res. Bull.* 2010, 45, 1664-1667.
[173] Pan, H.B., Darvell, B.W. Solubility of TTCP and β-TCP by solid titration. *Arch. Oral Biol.* 2009, 54, 671-677.
[174] Wang, W., Itoh, S., Yamamoto, N., Okawa, A., Nagai, A., Yamashita, K. Electrical polarization of β-tricalcium phosphate ceramics. *J. Am. Ceram. Soc.* 2010, 93, 2175-2177.
[175] Frondel, C. Whitlockite: a new calcium phosphate, $Ca_3(PO_4)_2$. *Am. Mineral.* 1941, 26, 145-152.
[176] Li, X., Ito, A., Sogo, Y., Wang, X., LeGeros, R.Z. Solubility of Mg-containing β-tricalcium phosphate at 25 °C. *Acta Biomater.* 2009, 5, 508-517.
[177] Kodaka, T., Debari, K., Higashi, S. Magnesium-containing crystals in human dental calculus. *J. Electron Microsc.* (Tokyo) 1988, 37, 73-80.
[178] Reid, J.D., Andersen, M.E. Medial calcification (whitlockite) in the aorta. *Atherosclerosis* 1993, 101, 213-224.
[179] Scotchford, C.A., Ali, S.Y. Magnesium whitlockite deposition in articular cartilage: a study of 80 specimens from 70 patients. *Ann. Rheum. Dis.* 1995, 54, 339-344.
[180] P'ng, C.H., Boadle, R., Horton, M., Bilous, M., Bonar, F. Magnesium whitlockite of the aorta. *Pathology 2008*, 40, 539-540.
[181] Horch, H.H., Sader, R., Pautke, C., Neff, A., Deppe, H., Kolk, A. Synthetic, pure-phase beta-tricalcium phosphate ceramic granules (Cerasorb®) for bone regeneration in the reconstructive surgery of the jaws. *Int. J. Oral Maxillofac. Surg.* 2006, 35, 708-713.
[182] Ogose, A., Kondo, N., Umezu, H., Hotta, T., Kawashima, H., Tokunaga, K., Ito, T., Kudo, N., Hoshino, M., Gu, W., Endo, N. Histological assessment in grafts of highly purified beta-tricalcium phosphate (OSferion®) in human bones. *Biomaterials* 2006, 27, 1542-1549.
[183] Kamitakahara, M., Ohtsuki, C., Miyazaki, T. Review paper: Behavior of ceramic biomaterials derived from tricalcium phosphate in physiological condition. *J. Biomater. Appl.* 2008, 23, 197-212.
[184] Liu, Y., Pei, G.X., Shan, J., Ren, G.H. New porous beta-tricalcium phosphate as a scaffold for bone tissue engineering. *J. Clin. Rehabil. Tiss. Eng. Res.* 2008, 12, 4563-4567.
[185] Epstein, N.E. Beta tricalcium phosphate: observation of use in 100 posterolateral lumbar instrumented fusions. *Spine J.* 2009, 9, 630-638.
[186] Güngörmüş, C., Kiliç, A., Akay, M.T., Kolankaya, D. The effects of maternal exposure to food additive E341 (tricalcium phosphate) on foetal development of rats. *Environmental Toxicology and Pharmacology* 2010, 29, 111-116.

[187] Welch, J.H., Gutt, W. High-temperature studies of the system calcium oxide–phosphorus pentoxide. *J. Chem. Soc.* 1961, 4442-4444.

[188] Jokic, B., Jankovic-Castvan, I., Veljovic, D.J., Bucevac, D., Obradovic-Djuricic, K., Petrovic, R., Janackovic, D.J. Synthesis and settings behavior of α-TCP from calcium deficient hydroxyapatite obtained by hydrothermal method. *J. Optoelectron. Adv. Mater.* 2007, 9, 1904-1910.

[189] Nurse, R.W., Welch, J.B., Gun, W. High-temperature phase equilibria in the system dicalcium silicate – tricalcium phosphate. *J. Chem. Soc. 1959*, 1077-1083.

[190] Sayer, M., Stratilatov, A.D., Reid, J.W., Calderin, L., Stott, M.J., Yin, X., MacKenzie, M., Smith, T.J.N., Hendry, J.A., Langstaff, S.D. Structure and composition of silicon-stabilized tricalcium phosphate. *Biomaterials* 2003, 24, 369-382.

[191] Reid, J.W., Pietak, A.M., Sayer, M., Dunfield, D., Smith, T.J.N. Phase formation and evolution in the silicon substituted tricalcium phosphate / apatite system. *Biomaterials* 2005, 26, 2887-2897.

[192] Reid, J.W., Tuck, L., Sayer, M., Fargo, K., Hendry, J.A. Synthesis and characterization of single-phase silicon substituted α-tricalcium phosphate. *Biomaterials* 2006, 27, 2916-2925.

[193] Kanazawa, T., Umegaki, T., Uchiyama, N. Thermal crystallisation of amorphous calcium phosphate to α-tricalcium phosphate. *J. Chem. Tech. Biotechnol.* 1982, 32, 399-406.

[194] TenHuisen, K.S., Brown, P.W. Formation of calcium-deficient hydroxyapatite from α-tricalcium phosphate. *Biomaterials* 1998, 19, 2209-2217.

[195] Durucan, C., Brown, P.W. α-tricalcium phosphate hydrolysis to hydroxyapatite at and near physiological temperature. *J. Mater. Sci. Mater. Med.* 2000, 11, 365-371.

[196] Durucan, C., Brown, P.W. Kinetic model for α-tricalcium phosphate hydrolysis. *J. Am. Ceram. Soc.* 2002, 85, 2013-2018.

[197] Camiré, C.L., Gbureck, U., Hirsiger, W., Bohner, M. Correlating crystallinity and reactivity in an α-tricalcium phosphate. *Biomaterials* 2005, 26, 2787-2794.

[198] Yin, X., Stott, M.J. Theoretical insights into bone grafting Si-stabilized α-tricalcium phosphate. *J. Chem. Phys.* 2005, 122, 024709 (9 pages).

[199] Yin, X., Stott, M.J. Surface and adsorption properties of α-tricalcium phosphate. *J. Chem. Phys.* 2006, 124, 124701 (9 pages).

[200] Jones, H.B. Contributions to the chemistry of the urine. On the variations in the alkaline and earthy phosphates in the healthy state, and on the alkalescence of the urine from fixed alkalies. *Phil. Trans. R. Soc. Lond.* 1845, 135, 335-349.

[201] Dorozhkin S.V. Amorphous calcium orthophosphates: nature, chemistry and biomedical applications. *Int. J. Mater. Chem.* 2012, 2, 19-46.

[202] Blumenthal, N.C., Posner, A.S., Holmes, J.M. Effect of preparation conditions on the properties and transformation of amorphous calcium phosphate. *Mater. Res. Bull.* 1972, 7, 1181-1190.

[203] Termine, J.D., Peckauskas, R.A., Posner, A.S. Calcium phosphate formation *in vitro*. II. Effects of environment on amorphous-crystalline transformation. *Arch. Biochem. Biophys.* 1970, 140, 318-325.

[204] Posner, A.S., Betts, F. Synthetic amorphous calcium phosphate and its relation to bone mineral structure. *Acc. Chem. Res.* 1975, 8, 273-281.

[205] Elliott, J.C. Recent studies of apatites and other calcium orthophosphates. In: Les matériaux en phosphate de calcium. Aspects fondamentaux. / *Calcium phosphate materials. Fundamentals.* Brès, E., Hardouin, P., Eds. Sauramps Medical: Montpellier, France, 1998: pp. 25-66.

[206] Li, Y., Weng, W. *In vitro* synthesis and characterization of amorphous calcium phosphates with various Ca/P atomic ratios. J. *Mater. Sci. Mater. Med.* 2007, 18, 2303-2308.

[207] Tadic, D., Peters, F., Epple, M. Continuous synthesis of amorphous carbonated apatites. *Biomaterials* 2002, 23, 2553-2559.

[208] Keller, L., Dollase, W.A. X-ray determination of crystalline hydroxyapatite to amorphous calcium-phosphate ratio in plasma sprayed coatings. *J. Biomed. Mater. Res.* 2000, 49, 244-249.

[209] Kumar, R., Cheang, P., Khor, K.A. Phase composition and heat of crystallization of amorphous calcium phosphate in ultra-fine radio frequency suspension plasma sprayed hydroxyapatite powders. *Acta Mater.* 2004, 52, 1171-1181.

[210] Harries, J.E., Hukins, D.W.L., Hasnain, S.S. Analysis of the EXAFS spectrum of hydroxyapatite. *J. Phys. C: Solid State Phys.* 1986, 19, 6859-6872.

[211] Harries, J.E., Hukins, D.W.L., Holt, C., Hasnain, S.S. Conversion of amorphous calcium phosphate into hydroxyapatite investigated by EXAFS spectroscopy. *J. Cryst. Growth* 1987, 84, 563-570.

[212] Taylor, M.G., Simkiss, K., Simmons, J., Wu, L.N.Y., Wuthier, R.E. Structural studies of a phosphatidyl serine-amorphous calcium phosphate complex. *Cell. Mol. Life Sci.* 1998, 54, 192-202.

[213] Peters, F., Schwarz, K., Epple, M. The structure of bone studied with synchrotron X-ray diffraction, X-ray absorption spectroscopy and thermal analysis. *Thermochim. Acta* 2000, 361, 131-138.

[214] Posner, A.S., Betts, F., Blumenthal, N.C. Formation and structure of synthetic and bone hydroxyapatite. *Progr. Cryst. Growth Char.* 1980, 3, 49-64.

[215] Boskey, A.L. Amorphous calcium phosphate: the contention of bone. *J. Dent. Res.* 1997, 76, 1433-1436.

[216] Onuma, K., Ito, A. Cluster growth model for hydroxyapatite. *Chem. Mater.* 1998, 10, 3346-3351.

[217] Tadic, D., Epple, M. Amorphous calcium phosphates as bone substitution materials. *Eur. J. Trauma* 2002, 28, 136-137.

[218] Combes, C, Rey, C. Amorphous calcium phosphates: synthesis, properties and uses in biomaterials. *Acta Biomater* 2010, 6, 3362-3378.

[219] Bache, F. *A system of chemistry for the use of students of medicine. Printed and published for the author.* William Fry, Printer. Philadelphia, 1819, 624 pp.

[220] Wilson, R.M., Elliott, J.C., Dowker, S.E.P., Rodriguez-Lorenzo, L.M. Rietveld refinements and spectroscopic studies of the structure of Ca-deficient apatite. *Biomaterials* 2005, 26, 1317-1327.

[221] Zahn, D., Hochrein, O. On the composition and atomic arrangement of calcium-deficient hydroxyapatite: an *ab-initio* analysis. *J. Solid State Chem.* 2008, 181, 1712-1716.

[222] Ivanova, T. I., Frank-Kamenetskaya, O.V., Kol'tsov, A.B., Ugolkov, V.L. Crystal structure of calcium-deficient carbonated hydroxyapatite thermal decomposition. *J. Solid State Chem.* 2001, 160, 340-349.

[223] Sinha, A., Nayar, S., Agrawal, A., Bhattacharyya, D., Ramachandrarao, P. Synthesis of nanosized and microporous precipitated hydroxyapatite in synthetic polymers and biopolymers. *J. Am. Ceram. Soc.* 2003, 86, 357-359.

[224] Mayer, I., Jacobsohn, O., Niazov, T., Werckmann, J., Iliescu, M., Richard-Plouet, M., Burghaus, O., Reinen, D. Manganese in precipitated hydroxyapatites. *Eur. J. Inorg. Chem.* 2003, 1445-1451.

[225] Vallet-Regí, M., Rodríguez-Lorenzo, L.M., Salinas, A.J. Synthesis and characterisation of calcium deficient apatite. *Solid State Ionics* 1997, 101-103, 1279-1285.

[226] Siddharthan, A., Seshadri, S.K., Kumar, T.S.S. Microwave accelerated synthesis of nanosized calcium deficient hydroxyapatite. *J. Mater. Sci. Mater. Med.* 2004, 15, 1279-1284.

[227] Hutchens, S.A., Benson, R.S., Evans, B.R., O'Neill, H.M., Rawn, C.J. Biomimetic synthesis of calcium-deficient hydroxyapatite in a natural hydrogel. *Biomaterials* 2006, 27, 4661-4670.

[228] Mochales, C., Wilson, R.M., Dowker, S.E.P., Ginebra, M.P. Dry mechanosynthesis of nanocrystalline calcium deficient hydroxyapatite: structural characterization. *J. Alloy Compd.* 2011, 509, 7389-7394.

[229] Brès, E.F., Duhoo, T., Leroy, N., Lemaitre, J. Evidence of a transient phase during the hydrolysis of calcium-deficient hydroxyapatite. Zeitschrift fuer Metallkunde / *Mater. Res. Adv. Tech.* 2005, 96, 503-506

[230] Dorozhkin, S.V. Biphasic, triphasic and multiphasic calcium orthophosphates. *Acta Biomater.* 2012, 8, 963-977.

[231] Dorozhkina, E.I., Dorozhkin, S.V. Mechanism of the solid-state transformation of a calcium-deficient hydroxyapatite (CDHA) into biphasic calcium phosphate (BCP) at elevated temperatures. *Chem. Mater.* 2002, 14, 4267-4272.

[232] Dorozhkin, S.V. Mechanism of solid-state conversion of non-stoichiometric hydroxyapatite to diphase calcium phosphate. *Russ. Chem. Bull.* (Int. Ed.) 2003, 52, 2369-2375.

[233] Rodríguez-Lorenzo, L. Studies on calcium deficient apatites structure by means of MAS-NMR spectroscopy. *J. Mater. Sci. Mater. Med.* 2005, 16, 393-398.

[234] Brown, P.W., Martin, R.I. An analysis of hydroxyapatite surface layer formation. *J. Phys. Chem.* B 1999, 103, 1671-1675.

[235] Honghui, Z., Hui, L., Linghong, G. Molecular and crystal structure characterization of calcium-deficient apatite. *Key Eng. Mater.* 2007, 330-332, 119-122.

[236] Viswanath, B., Shastry, V.V., Ramamurty, U., Ravishankar, N. Effect of calcium deficiency on the mechanical properties of hydroxyapatite crystals. *Acta Mater.* 2010, 58, 4841-4848.

[237] Matsunaga, K. Theoretical investigation of the defect formation mechanism relevant to nonstoichiometry in hydroxyapatite. *Phys. Rev.* B 2008, 77, 104106 (14 pages).

[238] Bourgeois, B., Laboux, O., Obadia, L., Gauthier, O., Betti, E., Aguado, E., Daculsi, G., Bouler, J.M. Calcium-deficient apatite: a first *in vivo* study concerning bone ingrowth. *J. Biomed. Mater. Res.* A 2003, 65A, 402-408.

[239] Liu, T.Y., Chen, S.Y., Liu, D.M., Liou, S.C. On the study of BSA-loaded calcium-deficient hydroxyapatite nano-carriers for controlled drug delivery. *J. Control. Release* 2005, 107, 112-121.

[240] Tsuchida, T., Yoshioka, T., Sakuma, S., Takeguchi, T., Ueda, W. Synthesis of biogasoline from ethanol over hydroxyapatite catalyst. *Ind. Eng. Chem. Res.* 2008, 47, 1443-1452.

[241] Elliott, J.C., Mackie, P.E., Young, R.A. Monoclinic hydroxyapatite. *Science* 1973, 180, 1055-1057.

[242] Rangavittal, N., Landa-Cánovas, A.R., González-Calbet, J.M., Vallet-Regí, M. Structural study and stability of hydroxyapatite and β-tricalcium phosphate: two important bioceramics. *J. Biomed. Mater. Res.* 2000, 51, 660-668.

[243] Kim, J.Y., Fenton, R.R., Hunter, B.A., Kennedy, B.J. Powder diffraction studies of synthetic calcium and lead apatites. Austr. *J. Chem.* 2000, 53, 679-686.

[244] Kay, M.I., Young, R.A., Posner, A.S. Crystal structure of hydroxyapatite. *Nature* 1964, 204, 1050-1052.

[245] Calderin, L., Stott, M.J., Rubio, A. Electronic and crystallographic structure of apatites. *Phys. Rev.* B, 2003, 67, 134106 (6 pages).

[246] Rulis, P., Ouyang, L., Ching, W.Y. Electronic structure and bonding in calcium apatite crystals: hydroxyapatite, fluorapatite, chlorapatite and bromapatite. *Phys. Rev.* B 2004, 70, 155104 (8 pages).

[247] Snyders, R., Music, D., Sigumonrong, D., Schelnberger, B., Jensen, J., Schneider, J.M. Experimental and ab initio study of the mechanical properties of hydroxyapatite. *Appl. Phys. Lett.* 2007, 90, 193902 (3 pages).

[248] Ching, W.Y., Rulis, P., Misra, A. *Ab initio* elastic properties and tensile strength of crystalline hydroxyapatite. *Acta Biomater.* 2009, 5, 3067-3075.

[249] Treboux, G., Layrolle, P., Kanzaki, N., Onuma, K., Ito, A. Symmetry of Posner's cluster. *J. Am. Chem. Soc.* 2000, 122, 8323-8324.

[250] Yin, X., Stott, M.J. Biological calcium phosphates and Posner's cluster. *J. Chem. Phys.* 2003, 118, 3717-3723.

[251] Kanzaki, N., Treboux, G., Onuma, K., Tsutsumi, S., Ito, A. Calcium phosphate clusters. *Biomaterials* 2001, 22, 2921-2929.

[252] Calderin, L., Dunfield, D., Stott, M.J. Shell-model study of the lattice dynamics of hydroxyapatite. *Phys. Rev.* B 2005, 72, 224304 (12 pages).

[253] Tanaka, Y., Iwasaki, T., Nakamura, M., Nagai, A., Katayama, K., Yamashita, K. Polarization and microstructural effects of ceramic hydroxyapatite electrets. *J. Appl. Phys.* 2010, 107, 014107 (10 pages).

[254] Tanaka, Y., Iwasaki, T., Katayama, K., Hojo, J., Yamashita, K. Effect of ionic polarization on crystal structure of hydroxyapatite ceramic with hydroxide nonstoichiometry. *J. Jpn. Soc. Powder and Powder Metall.* 2010, 57, 520-528.

[255] Tofail, S.A.M., Baldisserri, C., Haverty, D., McMonagle, J.B., Erhart, J. Pyroelectric surface charge in hydroxyapatite ceramics. *J. Appl. Phys.* 2009, 106, 106104 (3 pages).

[256] Kawabata, K., Yamamoto, T. First-principles calculations of the elastic properties of hydroxyapatite doped with divalent ions. *J. Ceram. Soc. Jpn.* 2010, 118, 548-549.

[257] Matsunaga, K., Kuwabara, A. First-principles study of vacancy formation in hydroxyapatite. *Phys. Rev.* B 2007, 75, 014102 (9 pages).

[258] de Leeuw, N.H. Computer simulations of structures and properties of the biomaterial hydroxyapatite. *J. Mater. Chem.* 2010, 20, 5376-5389.

[259] Corno, M., Rimola, A., Bolis, V., Ugliengo, P. Hydroxyapatite as a key biomaterial: quantum-mechanical simulation of its surfaces in interaction with biomolecules. *Phys. Chem. Chem. Phys.* 2010, 12, 6309-6329.

[260] Briak-Ben, E., Abdeslam, H., Ginebra, M.P., Vert, M., Boudeville, P. Wet or dry mechanochemical synthesis of calcium phosphates? Influence of the water content on DCPD – CaO reaction kinetics. *Acta Biomater.* 2008, 4, 378-386.

[261] Markovic, M., Fowler, B.O., Tung, M.S. Preparation and comprehensive characterization of a calcium hydroxyapatite reference material. *J. Res. Natl. Inst. Stand. Technol.* 2004, 109, 553-568.

[262] Bonel, G., Heughebaert, J.C., Heughebaert, M., Lacout, J.L., Lebugle, A. Apatitic calcium orthophosphates and related compounds for biomaterials preparation. *Ann. NY Acad. Sci.* 1988, 523, 115-130.

[263] Narasaraju, T.S.B., Phebe, D.E. Some physico-chemical aspects of hydroxylapatite. *J. Mater. Sci.* 1996, 31, 1-21.

[264] Riman, R.E., Suchanek, W.L., Byrappa, K., Chen, C.W., Shuk, P., Oakes, C.S. Solution synthesis of hydroxyapatite designer particulates. *Solid State Ionics* 2002, 151, 393-402.

[265] Koutsopoulos, S. Synthesis and characterization of hydroxyapatite crystals: a review study on the analytical methods. *J. Biomed. Mater. Res.* 2002, 62, 600-612.

[266] Norton, J., Malik, K.R., Darr, J.A., Rehman, I.U. Recent developments in processing and surface modification of hydroxyapatite. *Adv. Appl. Ceram.* 2006, 105, 113-139.

[267] Sadat-Shojai, M., Khorasani, M.T., Dinpanah-Khoshdargi, E., Jamshidi, A. Synthesis methods for nanosized hydroxyapatite with diverse structures. *Acta Biomater.* 2013, 9, 7591-7621.

[268] Rakovan, J. Growth and surface properties of apatite. In: Phosphates: geochemical, geobiological and materials importance. Series: *Reviews in Mineralogy and Geochemistry.* Vol. 48. Hughes, J.M., Kohn, M., Rakovan, J., Eds. Mineralogical Society of America: Washington, D.C., USA, 2002; pp. 51-86.

[269] Dorozhkin, S.V. Dissolution mechanism of calcium apatites in acids: a review of literature. *World J. Methodol.* 2012, 2, 1-17.

[270] Suchanek, W., Yoshimura, M. Processing and properties of hydroxyapatite-based biomaterials for use as hard tissue replacement implants. *J. Mater. Res.* 1998, 13, 94-117.

[271] Sun, L., Berndt, C.C., Gross, K.A., Kucuk, A. Review: material fundamentals and clinical performance of plasma sprayed hydroxyapatite coatings. *J. Biomed. Mater. Res. (Appl. Biomater.)* 2001, 58, 570-592.

[272] Ong, J.L., Chan, D.C.N. Hydroxyapatites and their use as coatings in dental implants: a review. *Crit. Rev. Biomed. Eng.* 1999, 28, 667-707.

[273] Dey, A., Mukhopadhyay, A.K., Gangadharan, S., Sinha, M.K., Basu, D. Characterization of microplasma sprayed hydroxyapatite coating. *J. Therm. Spray Technol.* 2009, 18, 578-592.

[274] Dorozhkin, S.V. Calcium orthophosphate coatings, films and layers. *Prog. Biomater. 2012,* 1, 1-40.

[275] Yuan, Y., Huang, P., Peng, Q., Zhang, C., Weng, J. Osteogenesis of porous bioceramics scaffolds consisted of hydroxyapatite spherules after implanted in different non-osseous sites. *Mater. Sci. Forum* 2009, 610-613, 1335-1338.

[276] Engin, N.O., Tas, A.C. Manufacture of macroporous calcium hydroxyapatite bioceramics. *J. Eur. Ceram. Soc.* 1999, 19, 2569-2572.

[277] Mangano, C., Piattelli, A., Perrotti, V., Iezzi, G. Dense hydroxyapatite inserted into postextraction sockets: a histologic and histomorphometric 20-year case report. *J. Periodontol.* 2008, 79, 929-933.

[278] Bernardi, G. Chromatography of nucleic acids on hydroxyapatite. *Nature* 1965, 206, 779-783.

[279] Doonan, S. Chromatography on hydroxyapatite. *Methods Mol. Biol.* 2004, 244, 191-194.

[280] Brand, M., Rampalli, S., Chaturvedi, C.P., Dilworth, F.J. Analysis of epigenetic modifications of chromatin at specific gene loci by native chromatin immunoprecipitation of nucleosomes isolated using hydroxyapatite chromatography. *Nat. Protoc.* 2008, 3, 398-409.

[281] Hou, Y., Morrison, C.J., Cramer, S.M. Classification of protein binding in hydroxyapatite chromatography: synergistic interactions on the molecular scale. *Anal. Chem.* 2011, 83, 3709-3716.

[282] Hilbrig, F., Freitag, R. Isolation and purification of recombinant proteins, antibodies and plasmid DNA with hydroxyapatite chromatography. *Biotechnol. J.* 2012, 7, 90-102.

[283] Uskoković, V., Uskoković, D.P. Nanosized hydroxyapatite and other calcium phosphates: chemistry of formation and application as drug and gene delivery agents. *J. Biomed. Mater. Res. B (Appl. Biomater.)* 2011, 96B, 152-191.

[284] Chen, F., Huang, P., Zhu, Y.J., Wu, J., Zhang, C.L., Cui, D.X. The photoluminescence, drug delivery and imaging properties of multifunctional Eu^{3+}/Gd^{3+} dual-doped hydroxyapatite nanorods. *Biomaterials* 2011, 32, 9031-9039.

[285] Li, D., Liang, Z., Chen, J., Yu, J., Xu, R. AIE luminogen bridged hollow hydroxyapatite nanocapsules for drug delivery. *Dalton Transact.* 2013, 42, 9877-9883.

[286] Long, T., Guo, Y.P., Liu, Y.Z., Zhu, Z.A. Hierarchically nanostructured mesoporous carbonated hydroxyapatite microspheres for drug delivery systems with high drug-loading capacity. *RSC Adv.* 2013, 3, 24169-24176.

[287] Feng, D., Shi, J., Wang, X., Zhang, L., Cao, S. Hollow hybrid hydroxyapatite microparticles with sustained and pH-responsive drug delivery properties. *RSC Adv.* 2013, 3, 24975-24982.

[288] Niwa, M., Sato, T., Li, W., Aoki, H., Aoki, H., Daisaku, T. Polishing and whitening properties of toothpaste containing hydroxyapatite. *J. Mater. Sci. Mater. Med.* 2001, 12, 277-281.

[289] Kim, B.I., Jeong, S.H., Jang, S.O., Kim, K.N., Kwon, H.K., Park, Y.D. Tooth whitening effect of toothpastes containing nano-hydroxyapatite. *Key Eng. Mater.* 2006, 309-311, 541-544.

[290] Pietrasik, J., Szustakiewicz, K., Zaborski, M., Haberko, K. Hydroxyapatite: an environmentally friendly filler for elastomers. *Mol. Cryst. Liq. Cryst.* 2008, 483, 172-178.

[291] Bailliez, S., Nzihou, A., Bèche, E., Flamant, G. Removal of lead (Pb) by hydroxyapatite sorbent. *Process Saf. Environ.* 2004, 82, 175-180.
[292] Corami, A., Mignardi, S., Ferrini, V. Cadmium removal from single- and multi-metal (Cd plus Pb plus Zn plus Cu) solutions by sorption on hydroxyapatite. *J. Colloid Interf. Sci.* 2008, 317, 402-408.
[293] Phonthammachai, N., Ziyi, Z., Jun, G., Fan, H.Y., White, T.J. Synthesis of high performance hydroxyapatite-gold catalysts for CO oxidation. *Gold Bull.* 2008, 41, 42-50.
[294] Chen, W., Huang, Z.L., Liu, Y., He, Q.J. Preparation and characterization of a novel solid base catalyst hydroxyapatite loaded with strontium. *Catalysis Comm.* 2008, 9, 516-521.
[295] Domínguez, M.I., Romero-Sarria, F., Centeno, M.A., Odriozola, J.A. Gold/hydroxyapatite catalysts. Synthesis, characterization and catalytic activity to CO oxidation. *Appl. Catalysis* B 2009, 87, 245-251.
[296] Xu, J., White, T., Li, P., He, C., Han, Y.F. Hydroxyapatite foam as a catalyst for formaldehyde combustion at room temperature. *J. Am. Chem. Soc.* 2010, 132, 13172-13173.
[297] Encyclopaedia; or, a dictionary of arts, sciences, and miscellaneous literature; Conſtructed on a Plan, by which the different sciences and arts are digeſted into the Form of diſtinct treatises or systems, comprihending the history, theory, and practice, of each, According to the Lateſt Diſcoveries and Improvements; and full explanations given of the various detached parts of knowledge, where relating to Natural and Artificial Objects, or to Matters Ecclesiastical, Civil, Military, Commercial, &c. Including Elucidations of the moſt important Topics relative to Religion, Morals, Manners, and the Oeconomy of Life. Together with A Description of all the Countries, Cities, principle Mountains, Seas, Rivers, &c. throughout the World; A General History, *Ancient* and *Modern*, of the different Empires, Kingdoms, and States; and An Account of the Lives of the moſt Eminent Perſons in every Nation, from the earliest ages down to the preſent times. The first American edition, in eighteen volumes, greatly improved. Vol. XIV. PAS – PLA Philadelphia: printed by Thomas Dobson, at the Stone-house, N° 41, South second-street. 1798, 797 pp.
[298] Busch, S., Dolhaine, H., Duchesne, A., Heinz, S., Hochrein, O., Laeri, F., Podebrad, O., Vietze, U., Weiland, T., Kniep, R. Biomimetic morphogenesis of fluorapatite-gelatin composites: fractal growth, the question of intrinsic electric fields, core / shell assemblies, hollow spheres and reorganization of denatured collagen. *Eur. J. Inorg. Chem.* 1999, 1643-1653.
[299] Wu, Y.J., Tseng, Y.H., Chan, J.C.C. Morphology control of fluorapatite crystallites by citrate ions. *Cryst. Growth Des.* 2010, 10, 4240-4242.
[300] Mazelsky, R., Hopkins, R.H., Kramer, W.E. Czochralski-growth of calcium fluorophosphates. *J. Cryst. Growth* 1968, 3-4, 260-264.
[301] Loutts, G.B., Chai, B.H.T. Growth of high-quality single crystals of FAP ($Ca_5(PO_4)_3F$) and its isomorphs. *Proc. SPIE – Int. Soc. Optical Eng.* 1993, 1863, 31-34.
[302] Warren, R.W. Defect centers in calcium fluorophosphate. *Phys. Rev.* B 1972, B6, 4679-4689.
[303] Mann, A.W., Turner, A.G. Excess calcium fluoride in fluorapatite. *Aust. J. Chem.* 1972, 25, 2701-2703.

[304] Dorozhkin, S.V. A hierarchical structure for apatite crystals. *J. Mater. Sci. Mater.* Med. 2007, 18, 363-366.
[305] Mehmel, M. *On the structure of apatite.* I. Z. Kristallogr. 1930, 75, 323-331.
[306] Naray-Szabo, S. *The structure of apatite* $(CaF)Ca_4(PO_4)_3$. Z. Kristallogr. 1930, 75, 387-398.
[307] Nikcevic, I., Jokanovic, V., Mitric, M., Nedic, Z., Makovec, D., Uskokovic, D. Mechanochemical synthesis of nanostructured fluorapatite / fluorhydroxyapatite and carbonated fluorapatite / fluorhydroxyapatite. *J. Solid State Chem.* 2004, 177, 2565-2574.
[308] Montazeri, L., Javadpour, J., Shokrgozar, M.A., Bonakdar, S., Javadian, S. Hydrothermal synthesis and characterization of hydroxyapatite and fluorhydroxyapatite nano-size powders. *Biomed. Mater.* 2010, 5, 045004.
[309] Zhu, Q., Jiang, W., Wang, H., Shao, C. Preparing fluorhydroxyapatite by aqueous precipitation method. *Adv. Mater. Res.* 2012, 412, 167-170.
[310] Rodríguez-Lorenzo, L.M., Hart, J.N., Gross, K.A. Structural and chemical analysis of well-crystallized hydroxyfluorapatites. *J. Phys. Chem.* B 2003, 107, 8316-8320.
[311] Azami, M., Jalilifiroozinezhad, S., Mozafari, M. Calcium fluoride/hydroxyfluorapatite nanocrystals as novel biphasic solid solution for tooth tissue engineering and regenerative dentistry. *Key Eng. Mater.* 2012, 493-494, 626-631.
[312] Daculsi, G., Kerebel, L.M. Ultrastructural study and comparative analysis of fluoride content of enameloid in sea-water and fresh-water sharks. *Arch. Oral Biol.* 1980, 25, 145-151.
[313] Daculsi, G., Kerebel, L.M., Kerebel, B. Effects of fluoride on human enamel and selachian enameloid *in vitro*: a high-resolution TEM and electron diffraction study. *Calcif. Tissue Int.* 1981, 33, 9-13.
[314] Prostak, K.S., Seifert, P., Skobe, Z. Enameloid formation in two tetraodontiform fish species with high and low fluoride contents in enameloid. *Arch. Oral Biol.* 1993, 38, 1031-1044.
[315] Dahm, S., Risnes, S. A comparative infrared spectroscopic study of hydroxide and carbonate absorption bands in spectra of shark enameloid, shark dentine, and a geological apatite. *Calcif. Tissue Int.* 1999, 65, 459-465.
[316] Carr, A., Kemp, A., Tibbetts, I., Truss, R., Drennan, J. Microstructure of pharyngeal tooth enameloid in the parrotfish Scarus rivulatus (Pisces: Scaridae*). J. Microscopy* 2006, 221, 8-16.
[317] Leveque, I., Cusack, M., Davis, S.A., Mann, S. Promotion of fluorapatite crystallization by soluble-matrix proteins from *Lingula Anatina* shells. *Angew. Chem. Int. Ed. Engl.* 2004, 43, 885-888.
[318] Schemehorn, B.R., Wood, G.D., McHale, W., Winston, A.E. Comparison of fluoride uptake into tooth enamel from two fluoride varnishes containing different calcium phosphate sources. *J. Clin. Dent.* 2011, 22, 51-54.
[319] Hattab, F.N. Remineralisation of carious lesions and fluoride uptake by enamel exposed to various fluoride dentifrices *in vitro*. *Oral Health Prev. Dent.* 2013, 11, 3, 281-290.
[320] Heling, L., Heindel, R., Merin, B. Calcium-fluorapatite. A new material for bone implants. *J. Oral Implantol.* 1981, 9, 548-555.

[321] Gineste, L., Gineste, M., Ranz, X., Ellefterion, A., Guilhem, A., Rouquet, N., Frayssinet, P. Degradation of hydroxylapatite, fluorapatite and fluorhydroxyapatite coatings of dental implants in dogs. *J. Biomed. Mater. Res.* 1999, 48, 224-234.

[322] Agathopoulos, S., Tulyaganov, D.U., Marques, P.A.A.P., Ferro, M.C., Fernandes, M.H., Correia, R.N. The fluorapatite – anorthite system in biomedicine. *Biomaterials* 2003, 24, 1317-1331.

[323] Yoon, B.H., Kim, H.W., Lee, S.H., Bae, C.J., Koh, Y.H., Kong, Y.M., Kim, H.E. Stability and cellular responses to fluorapatite-collagen composites. *Biomaterials* 2005, 26, 2957-2963.

[324] Bogdanov, B.I., Pashev, P.S., Hristov, J.H., Markovska, I.G. Bioactive fluorapatite-containing glass ceramics. *Ceram. Int.* 2009, 35, 1651-1655.

[325] Nordquist, W.D., Okudera, H., Kitamura, Y., Kimoto, K., Okudera, T., Krutchkoff, D.J. Part II: Crystalline fluorapatite-coated hydroxyapatite implant material: a dog study with histologic comparison of osteogenesis seen with FA-coated HA grafting material versus HA controls: potential bacteriostatic effect of fluoridated HA. *J. Oral Implant.* 2011, 37, Pages 35-42.

[326] Kheradmandfard, M., Fathi, M.H., Ahangarian, M., Zahrani, E.M. *In vitro* bioactivity evaluation of magnesium-substituted fluorapatite nanopowders. *Ceram. Int.* 2012, 38, 169-175.

[327] Sharifnabi, A., Fathi, M.H., Yekta, B.E., Hossainalipour, M. The structural and bio-corrosion barrier performance of Mg-substituted fluorapatite coating on 316L stainless steel human body implant. *Appl. Surf. Sci.* 2014, 288, 331-340.

[328] Savarino, L., Fini, M., Ciapetti, G., Cenni, E., Granchi, D., Baldini, N., Greco, M., Rizzi, G., Giardino, R., Giunti, A. Biologic effects of surface roughness and fluorhydroxyapatite coating on osteointegration in external fixation systems: an *in vivo* experimental study. J. *Biomed. Mater. Res.* A 2003, 66 A, 652-661.

[329] Vitkovič, M., Noaman, M.S.M., Palou, M.T., Jantová, S. Potential applications of fluorhydroxyapatite as biomaterials in medicine. *Central Eur. J. Chem.* 2009, 7, 246-251.

[330] Chaari, K., Ayed, F.B., Bouaziz, J., Bouzouita, K. Elaboration and characterization of fluorapatite ceramic with controlled porosity. *Mater. Chem. Phys.* 2009, 113, 219-226.

[331] Qu, H., Wei, M. The effect of fluoride contents in fluoridated hydroxyapatite on osteoblast behavior. *Acta Biomater.* 2006, 2, 113-119.

[332] Bhadang, K.A., Holding, C.A., Thissen, H., McLean, K.M., Forsythe, J.S., Haynes, D.R. Biological responses of human osteoblasts and osteoclasts to flame-sprayed coatings of hydroxyapatite and fluorapatite blends. *Acta Biomater.* 2010, 6, 1575-1583.

[333] Theiszova, M., Jantova, S., Letasiova, S., Palou, M., Cipak, L. Cytotoxicity of hydroxyapatite, fluorapatite and fluor-hydroxyapatite: a comparative *in vitro* study. *Neoplasma* 2008, 55, 312-316.

[334] Davis, T.S., Kreidler, E.R., Parodi, J.A., Soules, T.F. The luminescent properties of antimony in calcium halophosphates. *J. Luminescence* 1971, 4, 48-62.

[335] Mazelsky, R., Ohlmann, R.C., Steinbruegge, K. Crystal growth of a new laser material, fluorapatite. J. Electrochem. *Soc. Solid State Sci,* 1968, 115, 68-70.

[336] Ohlmann, R.C., Steinbruegge, K.B., Mazelsky, R. Spectroscopic and laser characteristics of neodymium-doped calcium fluorophosphate. *Appl. Optics* 1968, 7, 905-914.

[337] An, L., Zhang, L., Zou, J. Fluorapatite: an effective solid base catalyst for Michael addition of indole/pyrrole to nitroalkenes under solventless condition. *Chin. J. Chem.* 2009, 27, 2223-2228.

[338] Rogers, A.F. Dahllite (podolite) from Tonopah, Nevada: voelckerite, a new basic calcium phosphate; remarks on the chemical composition of apatite and phosphate rock. *Am. J. Sci., Ser.* 4, 1912, 33, 475-482.

[339] Rogers, A.F. A new locality for voelckerite and the validity of voelckerite as a mineral species. *Miner. Mag.* 1914, 17, 155-162.

[340] Voelcker, J.A. Die chemische Zusammensetzung des Apatits nach eigenen vollständigen Analysen. *Ber. Dtsch. Chem. Ges.* 1883, 16, 2460-2464.

[341] Gross, K.A., Berndt, C.C., Dinnebier, R., Stephens, P. Oxyapatite in hydroxyapatite coatings, *J. Mater. Sci. Mater. Med.* 1998, 33, 3985-3991.

[342] Hartmann, P., Jäger, C., Barth, S., Vogel, J., Meyer, K. Solid state NMR, X-ray diffraction, and infrared characterization of local structure in heat-treated oxyhydroxyapatite microcrystals: an analog of the thermal decomposition of hydroxyapatite during plasma-spray procedure. *J. Solid State Chem.* 2001, 160, 460-468.

[343] Alberius-Henning, P., Adolfsson, E., Grins, J., Fitch, A. Triclinic oxy-hydroxyapatite. *J. Mater. Sci.* 2001, 36, 663-668.

[344] Wang, T., Dorner-Reisel, A. Thermo-analytical investigations of the decomposition of oxyhydroxyapatite. *Mater. Lett.* 2004, 58, 3025-3028.

[345] Liu, Y., Shen, Z. Dehydroxylation of hydroxyapatite in dense bulk ceramics sintered by spark plasma sintering. *J. Eur. Ceram. Soc.* 2012, 32, 2691-2696.

[346] van't Hoen, C., Rheinberger, V., Höland, W., Apel, E. Crystallization of oxyapatite in glass-ceramics. *J. Eur. Ceram. Soc.* 2007, 27, 1579-1584.

[347] Duff, E.J. Orthophosphates – VII. Thermodynamical considerations concerning the stability of oxyapatite, $Ca_{10}O(PO_4)_6$, in aqueous media. *J. Inorg. Nuclear Chem.* 1972, 34, 853-857.

[348] de Leeuw, N.H., Bowe, J.R., Rabone, J.A.L. A computational investigation of stoichiometric and calcium-deficient oxy- and hydroxy-apatites. *Faraday Disc.* 2007, 134, 195-214.

[349] 349 Hilgenstock, G. Eine neue Verbindung von P_2O_5 und CaO. Stahl und Eisen 1883, 3, 498.

[350] 350. Hilgenstock, G. Das vierbasische Kalkphosphat und die Basicitätsstufe des Silicats in der Thomas-Schlacxke. *Stahl und Eisen* 1887, 7, 557-560.

[351] Kai, D., Fan, H., Li, D., Zhu, X., Zhang, X. Preparation of tetracalcium phosphate and the effect on the properties of calcium phosphate cement. *Mater. Sci. Forum* 2009, 610-613, 1356-1359.

[352] Romeo, H.E., Fanovich, M.A. Synthesis of tetracalcium phosphate from mechanochemically activated reactants and assessment as a component of bone cements. *J. Mater. Sci. Mater. Med.* 2008, 19, 2751-2760.

[353] Jalota, S., Tas, A.C., Bhaduri, S.B. Synthesis of HA-seeded TTCP ($Ca_4(PO_4)_2O$) powders at 1230 °C from $Ca(CH_3COO)_2 \cdot H_2O$ and $NH_4H_2PO_4$. *J. Am. Ceram. Soc.* 2005, 88, 3353-3360.

[354] Moseke, C., Gbureck, U. Tetracalcium phosphate: synthesis, properties and biomedical applications. *Acta Biomater.* 2010, 6, 3815-3823.

[355] Martin, I., Brown, P.W. Hydration of tetracalcium phosphate. *Adv. Chem. Res.* 1993, 5, 115-125.

[356] Jillavenatesa, A., Condrate, Sr. R.A. The infrared and Raman spectra of tetracalcium phosphate ($Ca_4(PO_4)_2O$). *Spectrosc. Lett.* 1997, 30, 1561-1570.

[357] Nakano, T., Kaibara, K., Umakoshi, Y., Imazato, S., Ogata, K., Ehara, A., Ebisu, S., Okazaki, M. Change in microstructure and solubility improvement of HAp ceramics by heat-treatment in a vacuum. *Mater. Transact.* 2002, 43, 3105-3111.

[358] Ellinger, R.F., Nery, E.B., Lynch, K.L. Histological assessment of periodontal osseous defects following implantation of hydroxyapatite and biphasic calcium phosphate ceramics: a case report. *Int. J. Periodont. Restor. Dent.* 1986, 3, 22-33.

[359] Nery, E.B., Lynch, K.L., Hirthe, W.M., Mueller, K.H. Bioceramic implants in surgically produced infrabony defects. *J. Periodontol.* 1975, 46, 328-347.

[360] Pan, L., Li, Y., Weng, W., Cheng, K., Song, C., Du, P., Zhao, G., Shen, G., Wang, J., Han, G. Preparation of submicron biphasic α-TCP/HA powders. *Key Eng. Mater.* 2006, 309-311, 219-222.

[361] Kui, C. Slip casting derived α-TCP/HA biphasic ceramics. *Key Eng. Mater.* 2007, 330-332, 51-54.

[362] Li, Y., Kong, F., Weng, W. Preparation and characterization of novel biphasic calcium phosphate powders (α-TCP/HA) derived from carbonated amorphous calcium phosphates. *J. Biomed. Mater. Res.* B (Appl. Biomater.) 2009, 89B, 508-517.

[363] Wang, R., Weng, W., Deng, X., Cheng, K., Liu, X., Du, P., Shen, G., Han, G. Dissolution behavior of submicron biphasic tricalcium phosphate powders. *Key Eng. Mater.* 2006, 309-311, 223-226.

[364] Li, Y., Weng, W., Tam, K.C. Novel highly biodegradable biphasic tricalcium phosphates composed of α-tricalcium phosphate and β-tricalcium phosphate. *Acta Biomater.* 2007, 3, 251-254.

[365] Li, Y., Li, D., Weng, W. *In vitro* dissolution behavior of biphasic tricalcium phosphate composite powders composed of α-tricalcium phosphate and β-tricalcium phosphate. *Key Eng. Mater.* 2008, 368-372, 1206-1208.

[366] Vani, R., Girija, E.K., Elayaraja, K., Parthiban, S.P., Kesavamoorthy, R., Kalkura, S.N. Hydrothermal synthesis of porous triphasic hydroxyapatite/(α and β) tricalcium phosphate. *J. Mater. Sci. Mater. Med.* 2009, 20, S43-S48.

[367] Lafon, J.P., Champion, E., Bernache-Assollant, D. Processing of AB-type carbonated hydroxyapatite $Ca_{10-x}(PO_4)_{(6-x)}(CO_3)_{(x)}(OH)_{(2-x-2y)}(CO_3)_{(y)}$ ceramics with controlled composition. *J. Eur. Ceram. Soc.* 2008, 28, 139-147.

[368] Tonegawa, T., Ikoma, T., Yoshioka, T., Hanagata, N., Tanaka, J. Crystal structure refinement of A-type carbonate apatite by X-ray powder diffraction. *J. Mater. Sci.* 2010, 45, 2419-2426.

[369] Yahia, F.B.H., Jemal, M. Synthesis, structural analysis and thermochemistry of B-type carbonate apatites. *Thermochim. Acta* 2010, 50, 22-32.

[370] Kannan, S., Rebelo, A., Lemos, A.F., Barba, A., Ferreira, J.M.F. Synthesis and mechanical behaviour of chlorapatite and chlorapatite / β-TCP composites. *J. Eur. Ceram. Soc.* 2007, 27, 2287-2294.

[371] García-Tuñón, E., Couceiro, R., Franco, J., Saiz, E., Guitián, F. Synthesis and characterisation of large chlorapatite single-crystals with controlled morphology and surface roughness. *J. Mater. Sci. Mater. Med.* 2012, 23, 2471-2482.

[372] Zhao, P., Zhao, T., Ren, X., Zhao, S., Yang, K. Molten salt synthesis of chlorapatite whiskers. *Adv. Mater. Res.* 2013, 668, 302-304.

[373] Kannan, S., Goetz-Neunhoeffer, F., Neubauer, J., Ferreira, J.M.F. Ionic substitutions in biphasic hydroxyapatite and β-tricalcium phosphate mixtures: structural analysis by Rietveld refinement. *J. Am. Ceram. Soc.* 2008, 91, 1-12.

[374] Boanini, E., Gazzano, M., Bigi, A. Ionic substitutions in calcium phosphates synthesized at low temperature. *Acta Biomater.* 2010, 6, 1882-1894.

[375] Pushpakanth, S., Srinivasan, B., Sastry, T.P., Mandal, A.B. Biocompatible and antibacterial properties of silver-doped hydroxyapatite. *J. Biomed. Nanotechnol.* 2008, 4, 62-66.

[376] Zhang, Y., Yin, Q.S., Zhang, Y., Xia, H., Ai, F.Z., Jiao, Y.P., Chen, X.Q. Determination of antibacterial properties and cytocompatibility of silver-loaded coral hydroxyapatite. *J. Mater. Sci. Mater. Med.* 2010, 21, 2453-2462.

[377] Kim, T.N., Feng, Q.L., Kim, J.O., Wu, J., Wang, H., Chen, G.C., Cui, F.Z. Antimicrobial effects of metal ions (Ag^{+}, Cu^{2+}, Zn^{2+}) in hydroxyapatite. *J. Mater. Sci. Mater. Med.* 1998, 9, 129-134.

[378] Stanić, V., Dimitrijević, S., Antić-Stanković, J., Mitrić, M., Jokić, B., Plećaš, I.B., Raičević, S. Synthesis, characterization and antimicrobial activity of copper and zinc-doped hydroxyapatite nanopowders. *Appl. Surf. Sci.* 2010, 256, 6083-6089.

[379] Li, Y., Ho, J., Ooi, C.P. Antibacterial efficacy and cytotoxicity studies of copper (II) and titanium (IV) substituted hydroxyapatite nanoparticles. *Mater. Sci. Eng. C* 2010, 30, 1137-1144.

[380] Thomas, S., Assi, P., Marycel, B., Correa, M., Liberato, W., Brito, V. Yttrium 90-hydroxyapatite, a new radioisotope for chronic synovitis in hemophilia. *Haemophilia* 2008, 14, 77.

[381] Chinol, M., Vallabhajosula, S., Goldsmith, S.J., Klein, M.J., Deutsch, K.F., Chinen, L.K., Brodack, J.W., Deutsch, E.A., Watson, B.A., Tofe, A.J. Chemistry and biological behavior of samarium-153 and rhenium-186-labeled hydroxyapatite particles: potential radiopharmaceuticals for radiation synovectomy. *J. Nucl. Med.* 1993, 34, 1536-1542.

[382] Argüelles, M.G., Berlanga, I.S.L., Torres, E.A. Preparation of ^{153}Sm-particles for radiosynovectomy. *J. Radioanal. Nucl. Chem.* 1999, 240, 509-511.

[383] O'Duffy, E., Clunie, G., Lui, D., Edwards, J., Ell, P. Double blind glucocorticoid controlled trial of samarium-153 particulate hydroxyapatite radiation synovectomy for chronic knee synovitis. *Ann. Rheum. Dis.* 1999, 58, 554-558.

[384] Ishikawa, K. Bone substitute fabrication based on dissolution-precipitation reactions. *Materials* 2010, 3, 1138-1155.

In: Apatite
Editors: M. Iafisco and J. M. Delgado-López
ISBN: 978-1-63321-500-9

Chapter 2

NANOCRYSTALLINE APATITES: SYNTHESIS, PHYSICAL-CHEMICAL AND THERMODYNAMIC CHARACTERIZATION

***Michele Iafisco*[1]*, José Manuel Delgado-López*[2] *and Christophe Drouet*[3*]**

[1]National Research Council (CNR). Institute of Science and Technology for Ceramics (ISTEC), Faenza (RA), Italy
[2]Laboratorio de Estudios Cristalográficos, Instituto Andaluz de Ciencias de la Tierra (IACT, CSIC-UGR), Armilla (Granada), Spain
[3]CIRIMAT Carnot Institute, UMR CNRS/INPT/UPS 5085, University of Toulouse, Ensiacet, Toulouse cedex 4, France

ABSTRACT

Nanocrystalline nonstoichiometric apatites constitute the mineral part of bones, and synthetic approaches have been developed to prepare "biomimetic analogs" via "soft" chemical routes. These compounds exhibit many differences to stoichiometric hydroxyapatite, in that they are nanocrystalline, hydrated, nonstoichiometric, plate-like shaped, and thermodynamically metastable. All these features can in fact explain the many peculiarities of their physical-chemical (and thus biological) properties as well as their ability to evolve spontaneously in solution. After reviewing the occurrence of apatite nanocrystals *in vivo*, we present in this contribution the main synthesis routes to prepare biomimetic analogs, and we then address some of the latest developments in physical-chemical and thermodynamic characterization of nanocrystalline apatites, focusing on advanced total scattering, vibrational spectroscopies, and calorimetry techniques. As in bone, apatite nanocrystals exhibit a very peculiar surface state that includes highly-labile ions confined in an ionic and hydrated metastable layer on the surface of the nanocrystals. These labile ions then take part to surface ion exchanges and adsorption mechanisms relevant for their biological activity.

[*] Corresponding Author address. Email: christophe.drouet@ensiacet.fr.

Keywords: Apatites, nanocrystals, advanced total scattering, vibrational spectroscopy, calorimetry, biomineralization

1. INTRODUCTION

Nanocrystalline calcium phosphates with apatitic structure (or "apatites") constitute the mineral part of most hard tissues in vertebrates. The structure and properties of apatites open several possibilities in medical sciences and also play an important role in living organisms for biomineralization processes. Besides their excellent properties as biomaterial, one of the most interesting characteristics of nanocrystalline apatites is their surface reactivity and capability to bind a wide variety of (bio)molecules due to their high surface-to-volume ratio and to a peculiar surface state involving a non-apatitic hydrated ionic layer with labile ions. Surface properties as well as other structural and chemical properties such as crystal dimensions and morphology, can be exploited and tailored by materials scientists to obtain nanostructured apatite-based bioactive biomaterials, in particular, in view of drug delivery and tissue engineering applications. The thermodynamic properties of nanocrystalline apatites are also bound to differ from stoichiometric counterparts. The present chapter outlines the recent progresses on the preparation of potentially-multifunctional nanocrystalline apatites with tailored properties as well as on their physical-chemical and thermodynamic properties. The chapter is structured in 3 main parts. The first part (section 2) is devoted to the description of biologically-related calcium phosphates in bone, teeth and pathological calcifications. The next section (section 3) deals with the latest published works regarding the preparation and crystallization mechanisms of biomimetic apatites. In the last part (sections 4-5-6), we report recent advances in the study of structural and thermodynamic characteristics of nanocrystalline apatites by conventional and less-conventional techniques such as advanced total scattering, vibrational spectroscopies (IR and Raman) and calorimetry, respectively.

2. APATITE NANOCRYSTALS IN BIOLOGICAL SYSTEMS

In biological systems, calcium phosphates are the main constituent of most normal (bone, dentine, fish scales, antlers…) and pathological (e.g. dental and urinary calculi, tendon mineralization, calcification of blood vessels) calcifications [1]. Except for small portions of the inner ear, all hard tissues of the human body are made of calcium phosphates [2-4]. Structurally, with the exception of enamel, they occur in the form of poorly crystallized non-stoichiometric and carbonate-substituted apatite nanocrystals [2-4]. In this section, details about the principal characteristics of bone, tooth and pathological calcifications will be given, highlighting the fact that the living organisms can produce amazing high-performance materials and Nature can be a powerful tool as source of inspiration for materials scientists [5].

Bones are the main mineralized tissues of the human body and their function is to move, support and protect organs of the body, produce red and white blood cells and store mineral elements [6]. Bones appear in a variety of shapes and have a complex internal and external structure; they are lightweight, yet strong and hard, in addition to satisfying their many other functions. There are 206 bones in the adult body and about 300 bones in the infant body. The diversity of structures within the bone family reflects the fine-tuning or adaptation of the structure to its function [7]. However, all have in common the basic building block which is comprised primarily of collagen type I fibrous matrix mineralized with carbonated apatite nanocrystals [8]. Bone tissue is a hierarchically-structured composite material that has continually fascinated scientists because of its particular structure and unique mechanical properties and remodeling capability. In fact, bone is a living tissue, far from being inert, which continuously undergoes remodeling and repair processes.

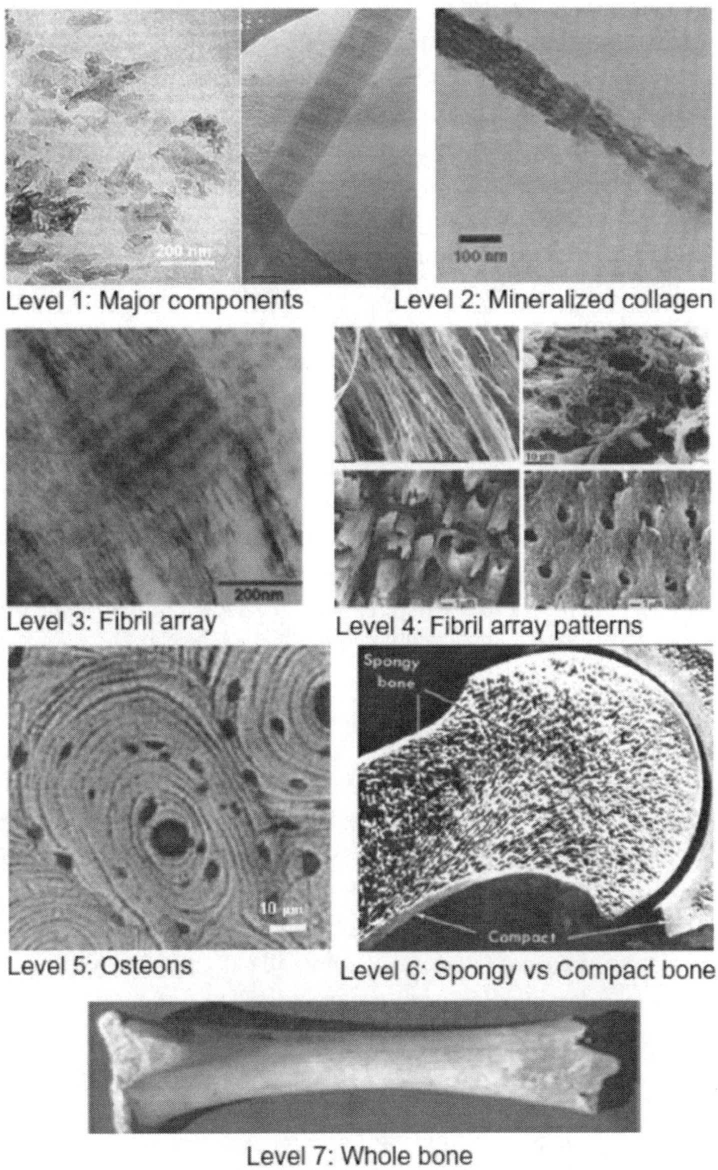

Figure 1. Seven hierarchical levels of organization of the bone family of materials. From reference [5] with permission.

To understand well the intricate bone architecture, several hierarchical models have been proposed in the past. The most known is the model of Weiner and Wagner [7] who described seven hierarchical levels that range from the nano- to the macro-scale (Figure 1).

The first level consists of the single molecular-level components such as type I collagen fibrils, carbonated apatite nanocrystals, water and some non-collagenous bone proteins. There are two hundred or more so-called non-collagenous proteins (NCPs), such as osteonectin, osteocalcin, osteopontin, bone morphogenetic proteins, bone proteoglycan, and bone sialoprotein, but together they generally comprise less than 10% of the total protein content [9]. Type I collagen, as the major organic component of bone, is made up of three polypeptide strands called alpha peptides forming a triple helical assembly approximately 300 nm long and 1.5 nm in diameter [10]. The trimeric polypeptidic molecule consists of two α1 and one α2 peptide chains that comprise a repeating glycine (Gly)–X–Y triplet, in which X and Y can be a residue but are usually proline (Pro) and hydroxyproline (Hyp), respectively [10]. The triple helical collagen molecules self-assemble with their long axes parallel to each other into a staggered arrangement in which each molecule is shifted with respect to its neighboring molecules forming a characteristic D-periodic cross-striated pattern (where D = 67 nm, the characteristic axial periodicity of collagen) [11]. The successive molecules in same axial dimension are 40 nm apart, named "hole" or "gap" zone [12]. The typical 67 nm period of cross-striation pattern of collagen assembly is widely observable by transmission electron microscope (TEM), atomic force microscope (AFM), and X-ray diffraction (XRD) [13-15]. The most widely accepted model for packing of collagen molecules involves five triple helices aligned hexagonally in cross section and longitudinally with approximately a quarter of the molecular length of staggered arrangement to form the five-stranded microfibrils. The diameter of the collagen fibrils in the five-stranded packing model should be about 3.6 nm according to 1.5 nm of diameter of single collagen molecule, which has been verified by TEM examination of native collagen fibrils [16]. The collagen microfibrils are then assembled into collagen fibrils in diameter from 35–500 nm that are further combined, oriented and laid up to form ordered structures with particular morphologies for tissues [17]. The main function of collagen concerns integrity and mechanical reinforcement of both soft and hard connective tissue.

The second level is formed by the mineralized collagen fibrils. In such fibrils, tiny plate-shaped crystallites of carbonated apatite are organized into parallel arrays with their *c*-axes co-aligned with the long axes of the collagen fibril [17, 18]. The first-formed minerals nucleate initially in the "hole" zones of assembled collagen fibrils with continuous growth, and then penetrate into the overlap regions of the fibrils [19]. There are more mineral crystals in the gap than in the overlap regions of the mineralized fibrils leading to a banded periodic appearance, with the repetition motif corresponding to the D-period of the fibrils [19].

The third level is composed of arrays of mineralized collagen fibrils. These fibrils are almost always associated as bundles or other arrangements, often aligned along their long axis. However, the organization of these fibrils in different bone types varies significantly, reflecting the differences in their functional properties [7].

The fourth level is composed of the different patterns of fibril arrays. At this level the conspicuous diversity in structure occurs, with fibril arrays organized in a variety of patterns. These patterns can be parallel arrays, woven arrangements, plywood like structures, and radial arrays [7].

Cylindrical structures called osteons make up the fifth level. Bones themselves, as opposed to dentin, cementum, scales, and most mineralized tendons, often undergo internal remodeling [20]. This process involves the excavating out of large tunnels by teams of specialized cells called osteoclasts. These tunnels are then refilled by osteoblasts, starting with the deposition of a thin layer of cement on the existing excavated surface, followed by layers of lamellar bone. The process stops when the tunnel is almost completely filled, leaving a narrow channel at the center that functions as a blood vessel [21]. In fact, other even smaller capillary-like features (canaliculi) are also built into the structure. These canaliculi are numerous and house the cells (osteocytes) that remain within the bone material itself. The canaliculi tend to radiate out from the central blood vessel. Thus the structure of an osteon is basically onion-like in cross-section with layers of lamellae surrounding a central hole; in longitudinal section they are cylindrical [22].

The sixth level is formed by spongy (trabecular or cancellous) or compact (cortical) bone tissues. Cancellous bone is extremely porous (75–95 wt.% porosity) and provides space for marrow and blood vessels, but has consequently low compressive strength [23]. Cortical bone is otherwise the dense and tough outer layer (5–10 wt.% porosity) that allows many functions of bone [24].

The seventh level is simply the whole bone.

Let us now inspect further the constitutive apatite nanocrystals. From a chemical-physical point of view, bone is composed of about 65 wt.% of mineral phase (nanosized crystals of apatite), 25 wt.% of organic phase (basically type-I collagen, fewer non-collagenous proteins (NCPs) and minor organic molecules such as citrate) and about 10 wt.% of water [1, 3]. Bone apatite nanocrystals are calcium- (and hydroxide-) deficient, with a molar ratio of $Ca/(P+C) < 1.67$ (where P and C respectively stand for phosphate and carbonate ions in phosphate crystallographic sites), which is the theoretical value for the stoichiometric hydroxyapatite mineral ($Ca_{10}(PO_4)_6(OH)_2$). Other interesting features of bone apatite are the low crystallinity degree, the presence of ionic substituents in its crystal structure (4–6 wt.% carbonate, 0.9 wt.% Na, 0.5 wt.% Mg and others) and a typical plate-like crystal morphology [25]. The reported dimensions of nanocrystals can change according to the different technique (and sample preparation) employed, and on the fact that by definition bone is recurrently partly remodeled thus being systematically composed of crystals with various "aging" states. Using transmission electron TEM, XRD and small-angle X-ray scattering (SAXS), the characterized dimensions generally vary in the ranges: 20–50 nm in length, 15–30 nm in width and thickness of 1.5–4 nm [26-28]. It should however be mentioned that many characterizations are often carried out after drastic drying of the bone specimen, thus allowing structural alterations such as dehydration. Some AFM studies revealed that widths and lengths of bone crystals range from 30 to 200 nm [29].

It is well known that the organic component of bone acts as an important regulator of lattice orientation, particle size, and size distribution in biomineralization processes, but the molecular recognition details at inorganic crystal interfaces are poorly understood [30]. Recently, most investigations have been focused on the effects of macromolecules such as carboxylate-rich proteins on hydroxyapatite nucleation and growth because complex interactions and changes of conformation are involved [31]. However, the role of small organic molecules such as citrate was deemphasized and was assumed to be simply that of adsorption on the crystal growth steps, resulting in inhibition of crystal growth [32]. New insight about the role of citrate is emerging since it has been recently identified as playing critical roles in interfering with crystal thickening and stabilizing apatite nanocrystal sizes in bone [33]. As it will be shown in section 4 this citrate-binding model may help understanding the mechanism of apatite morphology control *in vivo* (plate-like morphology in bone rather than rod-like in tooth enamel).

Although apatites are sometimes – wrongly – presented as inert chemical compounds with a passive role in the mechanical properties of bone, it must be highlighted that the nanocrystals fulfill very important biological roles, including the "ion reservoir" function. In fact, they are thought to participate actively in bone homeostasis through an exchange of the mineral ions or charged organic molecules with the biological environment [34]. This role has been attributed to apatite nanocrystals due to their very high surface area and to the presence of a hydrated (and ionic) surface layer with strong abilities for ions exchange and adsorption of organic molecules [35]. Due to their variable and complex composition, biological apatites were sometimes (in former literature works) considered as a mixture of calcium phosphates and carbonates phases, which is not the case as pointed out by techniques such as X-ray diffraction. However, this technique cannot provide information on the structural details of their surface. Recent progress on wet surface characterization of synthetic nanocrystalline apatites has occurred thanks to the use of spectroscopic methods. The data obtained by Fourier transformed infrared spectroscopy (FTIR) and solid state nuclear magnetic resonance (NMR) revealed that freshly precipitated nanocrystals show specific signals that disappear after drying and are assigned to the presence of a hydrated structured surface layer that are non-apatitic in nature [36]. Some of these spectroscopic features have some similarities (e.g. some IR and XRD features) with those of triclinic octacalcium phosphate (OCP), but a detailed analysis can clearly identify differences [60], and other techniques such as solid state NMR can even more clearly distinguish nanocrystalline apatite from OCP. The surface layer on apatite nanocrystals progressively disappears upon aging (maturation) in solution, in favor of a development of more stable apatitic domains of the crystals. The loosely-bound ions of this layer can be easily and reversibly substituted. Another feature is that the adsorption properties and reactivity of nanocrystalline apatites strongly depend on the extent and composition of this layer.

As previously described the vertebrates body contains numerous mineralized tissues, but enamel is the tissue with the most robust mechanical properties. Enamel is the hardest material formed by vertebrates and is the most mineralized tissue present in the body [37]. Mature enamel is composed of highly crystalline carbonated-hydroxyapatite crystals up to 95-97% by weight and less than 1% of organic material. The high degree of mineralization makes enamel a fascinating model for understanding fundamental mineralization processes that occur within an extracellular matrix. It is different from bone and dentin in terms of architecture, pathology and the biological mechanisms mediating its formation. Additionally mature enamel is acellular and it is not able to be resorbed or remodeled. As a result, enamel regeneration cannot occur *in vivo* and is therefore an attractive biological target for tissue regeneration approaches.

The mammalian tooth is composed of four distinct structures: enamel, dentin, pulp and cementum [38]. Enamel makes up the uppermost 1-2 mm of the tooth crown and contains a high mineral amount, giving it a high modulus but also making it susceptible to cracking. Dentin lies below the enamel and is tougher, forming the bulk of the tooth and absorbing stresses from enamel, preventing its fracture [39]. The inorganic composition of dentin is similar to that of bone. The cementum is the mineralized layer that surrounds the root of the tooth covering the dentin layer and some of the enamel layer. The cementum allows for the anchoring of the tooth to the alveolar bone (jawbone) through the periodontal ligament. The enamel and dentin tissues give rise to a tough, crack-tolerant and abrasion-resistant tissue through their unique architectures and mineral compositions. Enamel is highly patterned and consists of organized interwoven bundles of crystallites (called rods or prisms). It has a reported toughness higher than that of well-crystallized hydroxyapatite, indicating that the organization of the crystallites is essential for enamel function [40]. Due to the high mineral content and minimal presence of organic matter, enamel is brittle. Interestingly, the architecture of enamel crystallites can deflect a propagating crack preventing it from reaching the dentin-enamel junction (DEJ), which also has been shown to resist delamination of the tissues despite their differences in composition [41]. The mechanical properties of enamel, dentin and the DEJ are not completely understood and they comprise at present a significant area of research. The comprehension of the properties of these tissues could serve to motivate further studies to develop more robust materials for dental applications as well as to inspire fabrication of non-biologically-oriented materials.

Similarly to bone, enamel possesses a complex architecture. It can be divided into several hierarchical levels from the nano- to the macro-scale [42]. On the nanoscale, the protein-protein and protein-mineral interactions in the presence of supersaturated ions create a highly organized array of crystallites of hydroxyapatite that grow preferentially along their *c*-axis [43]. The sizes of these crystallites depend on the stage of the mineralization. The crystallites grow primarily in length during the secretory stage and continue to grow in width and thickness during the maturation stage. The assembly of amelogenin proteins has been shown to be crucial for the proper development of enamel crystallites. Disruption of the assembly alters formation at the nanoscale, subsequently affecting larger length scales and giving rise to a diseased or malformed enamel phenotype. On the mesoscale level, there are three main structural components: the rod, the interrod and the aprismatic enamel. The main component of enamel on the mesoscale includes rods, which are bundles of aligned crystallites that are "woven" into intricate architectures that are approximately 3-5 μm in diameter (Figure 2) [44, 45].

The second structural component of the enamel matrix is the interrod (or interprismatic) enamel, which surrounds and packs between the rods. The difference between the rod and the interrod is the orientation of hydroxyapatite crystals; the rod contains aligned crystallites, whereas the mineral in the interrod is less ordered. These structures coalesce to form the tissue of enamel, which can resist high forces and damage by crack deflection. The third structure, aprismatic enamel, refers to the structures containing hydroxyapatite crystals that show no mesoscale or macroscale alignment. The macroscale architecture includes specific zones of enamel that have unique characteristics, which contribute to the whole tissue. The enamel adjacent to the DEJ exhibits a gradual transition from dentin to enamel. Aprismatic regions of enamel have been proposed to be primitive areas of the tooth serving as a toughening mechanism due to their flexible nature. Several authors have identified these aprismatic areas to be located adjacent to the DEJ and at the incisal surface of both deciduous and permanent human enamel [46]. The Tomes' process, a unique structure present at the secretory pole of an enamel-forming cell, is responsible for aligned mineral formation in the prismatic enamel. The absence of this process may give rise to the aprismatic zone in the tooth.

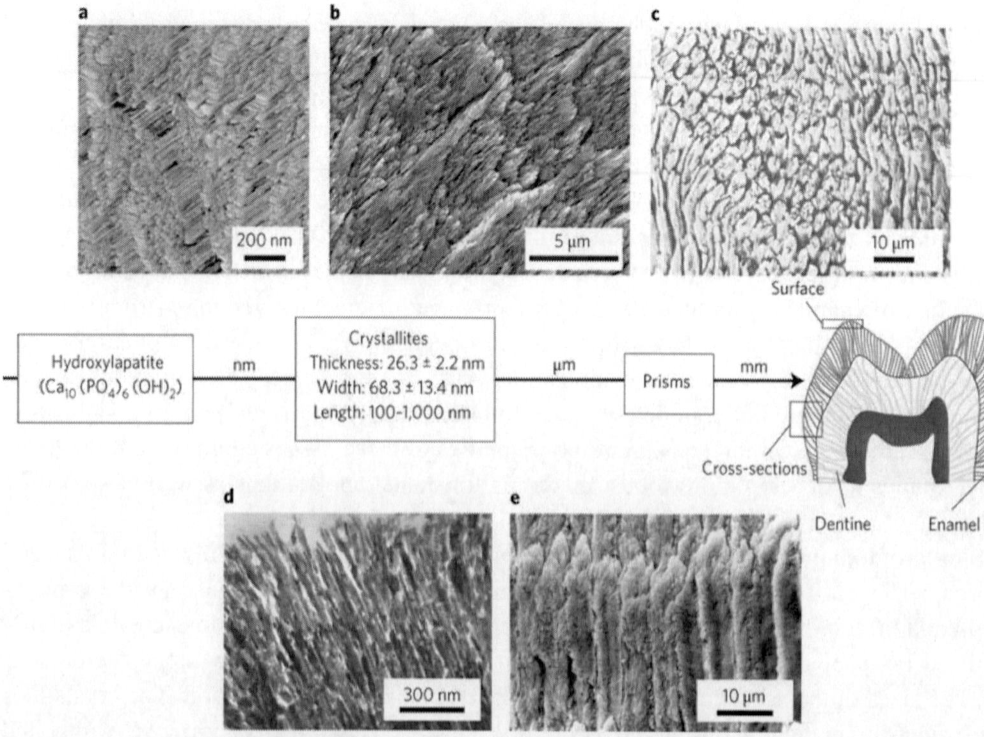

Figure 2. The enamel is composed of three-dimensionally organized nanosized hydroxyapatite crystallites (a,b,d) that are arranged into micrometer-sized prisms (c,e). a, Atomic force microscope and b,c, scanning electron microscope images of the enamel surface. d, Transmission electron microscope and e, scanning electron microscope images of a cross-section of the enamel. From reference [45] with permission.

In the body of mammals, osteoblasts and odontoblasts cells fix calcium and phosphate ions and then precipitate biological apatite onto organic matrices. This is the process of physiological biomineralization that is restricted (in the absence of specific pathologies) to the specific sites in the skeletal tissues, including growth plate cartilage, bones and teeth. Unfortunately, owing to ageing, to various diseases and under certain pathological conditions, blood vessels and some internal organs may get calcified as well [47], among other biological locations. This process is called pathological calcification or ectopic mineralization and leads to increased morbidity/mortality. In general, any type of abnormal accumulation of calcium phosphates in wrong places is accounted for by a disruption of systemic defense mechanism against calcification [48].

Unwanted depositions are always connected to various diseases (whether they are at their origin or their consequence), for instance: soft tissue calcification (in damaged joints, blood vessels, dysfunctional areas in the brain, diseased organs, scleroderma, prostate stones) [49], kidney and urinary stones [50], dental pulp stones and dental calculus [51], salivary stones, gall stones, pineal gland calcification, atherosclerotic arteries and veins [52], coronary calcification, cardiac skeleton, damaged cardiac valves, calcification on artificial heart valves [53], carpal tunnel, cataracts, malacoplakia, calcified menisci, dermatomyositis and other. All these cases are examples of a calcinosis, which might be described as a formation of mineral (often calcium phosphate) deposits in any soft tissue [54]. Contrary to the mineral phases of the normal calcifications, which consist of only one type of calcium phosphate, the mineral phases of abnormal and/or pathological calcifications are found to occur as single or mixed phases of other types of calcium phosphates and/or other phosphatic and non-phosphatic compounds, in addition to or in place of biological apatite [55]. This occurs, at least in part, because the pH of the solution is often relatively low in the places of pathological calcifications. However, in some cases, the chemical composition of an unwanted inorganic phase might depend on the age of the pathological calcification and its location. It is interesting to note that the mineral phases of animal calculus (e.g. from dog) was found to consist of calcium carbonate and biological apatite, while human calculi do not contain calcium carbonate. Some findings suggested that the mechanisms and factors regulating the physiological biomineralization might be similar to those influencing the ectopic mineralization: both were initiated by various organics (i.e., membrane-enclosed particles released from the plasma membrane of mineralization competent cells), that were present [56].

3. Preparation and Crystallization Methods of Apatite Nanocrystals

Due to their structural similarity with the mineral phase of bone and dentin, synthetic (nanocrystalline) apatites exhibit excellent biological properties such as biocompatibility, bioactivity, lack of toxicity, absence of inflammatory and immune responses and relatively high bioresorbability [57, 58]. These properties can be significantly enhanced by improving their biomimetism, that is, the preparation of synthetic analogs, so-called "biomimetic", able to precisely mimic the morphological and physical-chemical features of biological apatite compounds [25, 59]. Nanocrystalline apatites should not be confused with other calcium phosphate phases, including octacalcium phosphate; a recent review on the obtainment of apatite and its conclusive characterization has been recently been published [60].

In contrast to hydroxyapatite [HA, $Ca_{10}(PO_4)_6(OH)_2$], which is the stoichiometric calcium-phosphate apatitic phase that is the most stable and least soluble under physiological conditions, biomimetic nanocrystalline apatites are non-stoichiometric (Ca/P ratio lower than 1.67) and calcium- and OH-deficient, and may incorporate substituted ions in the crystal lattice (e.g. Na^+, Mg^{2+}, K^+, Sr^{2+}, Zn^{2+}, carbonate, etc.) [61]. The low degree of crystallinity, the small crystal sizes and the Ca and OH deficiency together with the foreign ions substitution are responsible for the relatively high solubility and specific biological activity of biomimetic apatite with respect to stoichiometric hydroxyapatite (see also the section on apatite thermodynamics in this chapter). They also mature when submitted to humid environments; similarly "mature" bone crystals in vertebrates are less soluble and reactive than embryonic (young) bone mineral crystals [62]. From a chemical and structural point of view, the composition of nanocrystalline apatites differs significantly from that of hydroxyapatite. The global chemical composition of biological apatites (or their synthetic analogues) has been a somewhat controversial topic but their chemical formula can generally be described as:

$$Ca_{10-x}\,(PO_4)_{6-x}\,(HPO_4 \text{ or } CO_3)_x\,(OH \text{ or } \tfrac{1}{2}CO_3)_{2-x} \text{ with } 0 \leq x \leq 2$$

In the last years, many different strategies have been employed and developed for the preparation of synthetic nanosized apatite crystals. A common preparation method is the stoichiometric titration of an aqueous calcium solution with a solution of phosphoric acid up to neutrality [63-65]. Several additional methods have been successfully employed in the synthesis of nanocrystalline apatites, including wet-chemical precipitation in various conditions [66-68, 70, 95], sol–gel synthesis [69], hydrothermal synthesis [71-73], rapid or continuous precipitation from solution [74], mechanochemical synthesis [75], microwave processing [76, 77], vapor diffusion [78-80], emulsion-based syntheses [81] and other methodologies producing (nano)crystals of various shapes and sizes [82, 83]. However, the preparation of biomimetic nanocrystalline apatites is still considered a scientific and technological challenge. As explained previously, biological apatites are known for their high content of defects affecting the lattice parameters, crystal morphology, crystallinity, solubility and thermal stability [84]. Moreover, the surface of biological apatite crystals is rarely smooth, a characteristic that is strictly connected to its biological significance. It is hypothesized that the exceptional roughness of these crystals, comprising surface irregularities in the order of size of single unit cells, may correspond to the tendency to increase protein binding in the biomineralization process [85, 86]. Recent studies have also demonstrated that a rough surface improves the biocompatibility of the material and has a positive effect on inflammatory reactions [87].

Using as main criteria the control of apatite structure and morphology, apatite crystallization methods can be rationally divided into high and low temperature approaches. The synthesis at high temperature usually involves the homogenization of precursor compounds, including for example $Ca_3(PO_4)_2$ and $Ca(OH)_2$, and their annealing at about 1000 °C. The advantage is the possibility to set the final stoichiometry of the mixture, whereas the main downsides are the long processing times and high annealing temperatures. In general, when using this method the Ca/P ratio is a fundamental parameter; in fact, if the initial molar ratio of Ca/P is not well configured to 1.67, other calcium phosphate phases are bound to appear such as α- or β-tricalcium phosphate (TCP) when the Ca/P value is lower than the stoichiometric ratio and CaO when the Ca/P value is higher. The α-TCP phase is normally formed at temperatures around 1200 °C whereas the β-TCP phase is formed at lower temperatures, up to 900 °C. In addition to the high levels of energy consumption, another major downside of high temperature methods is the difficulty to prepare uniform nanosized crystals [88], especially due to grains coarsening at hight temperatures.

On the other hand, the synthetic methods at low temperature offer the advantage to produce nanosized (and hydrated) crystals, but one disadvantage may be in some cases the presence of transient and metastable phases in the final product. To investigate the effect of temperature on the chemical-physical features of nanosized apatite, Sakhno et al. have characterized two apatites synthesized at 40 °C and 95 °C [89]. The apatite synthesized at low temperature displayed platelet-like morphology and was constituted by a crystalline core coated by an amorphous-like surface layer of 1–2 nm. Increasing the temperature of synthesis, the platelet morphology was retained but the apatite nanoparticles exhibited a higher degree of crystallinity (evaluated by XRD). HR-TEM observations revealed that, in this case, the crystalline order was extended up to the particles surfaces, exhibiting clearly the planes (010), (100), and (001). Infrared spectroscopy was used to investigate the surface hydration of both materials, in terms of adsorbed H_2O molecules and surface hydroxyl groups, as well as the Lewis acidity of surface cations, by removing the adsorbed water and adsorbing the molecular probe CO [89]. For both features, strong similarities between amorphous and crystalline surfaces were found. However, interestingly the apatite synthesized at 95°C having a more crystalline surface appeared able to physisorb multilayers of water in a larger extent than the less crystallized samples.

The disadvantage of synthetic methods involving low temperature is the possible crystallization of transient phases. To explain why these extraneous phases can be crystallized the calcium phosphates solubility diagram (at 37°C and ionic strength 0.1 M) published by Johnsson and Nancollas can be considered (Figure 3A) [90, 91]. According to this diagram, the calcium phosphate phases decrease their solubility with the increase of pH. At pH above 4.0, hydroxyapatite is the most stable phase followed by TCP and OCP, whereas at pH values lower than 4.0 the dicalcium phosphate dihydrate or brushite (DCPD) phase is more stable than hydroxyapatite. On the other hand, the solubility diagram calculated by solid titration published in 2009 by Pan and Darvell considers that DCPD is not the most stable phase below pH 4.2, since calcium-deficient hydroxyapatite is less soluble (Figure 3B) [92]. This fact is due to the metastability of DCPD, which nucleates more easily than hydroxyapatite at low pH. This discrepancy highlights the importance of kinetic factors when defining the pH domains in which the calcium phosphate phase can be crystallized.

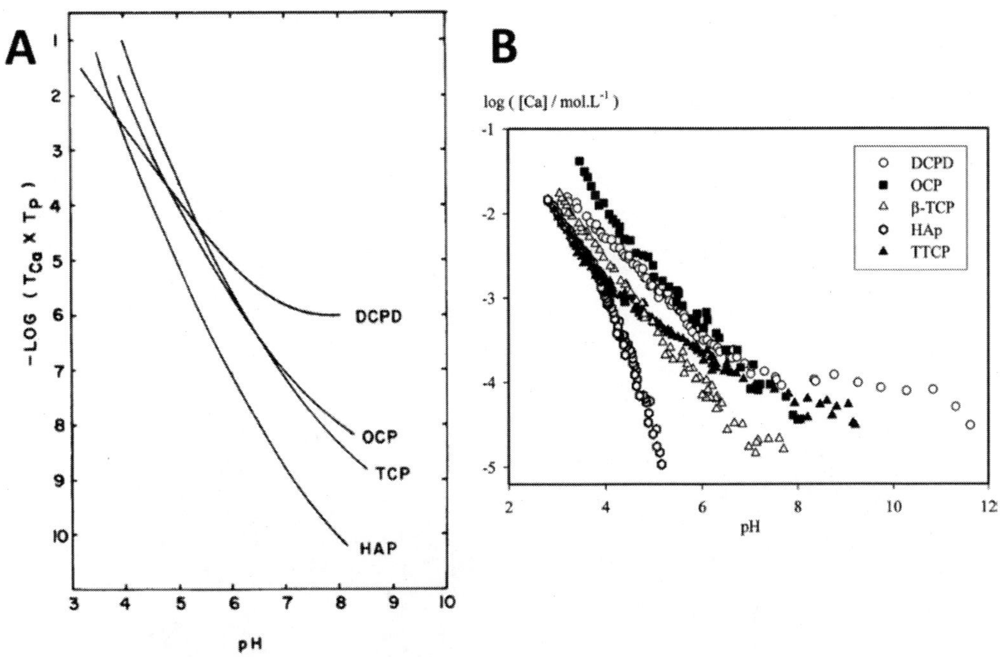

Figure 3. Solubility isotherms of calcium phosphate phases (A) according to reference [91] at and (B) according to reference [92].

The use of organic compounds as templates for the generation of inorganic structures and materials with nanosized dimensions is another interesting strategy which has received increasing attention over the last decade. For instance, the role of citrate molecular ions in controlling and stabilizing the size of bone apatite nanocrystals can be a biologically inspired lesson from nature, which can be developed into an advanced strategy to control nanomaterial fabrication. To this regard, Delgado-Lopez and co-wokers recently proposed a novel protocol to obtain bioinspired citrate-covered nanocrystalline apatite [93]. This method basically consists in the thermal decomplexing of metastable calcium/citrate/phosphate solutions. Brieftly, two solutions (1:1 v/v) of (i) 0.1 M $CaCl_2$ + 0.4 M $Na_3(Cit)$ and (ii) 0.12 M $Na_2HPO_4 + x$ mM Na_2CO_3 ($x = 0$ or 100) were mixed at 4 °C and then the temperature is rised up to 80 °C. This temperature increase causes the precipitation of apatite nanocrystals, which are then matured (up to 96 hours) in the same solution at 80 °C. Citrate anions were used as a calcium complexing agent to prepare homogeneous metastable solutions in order to avoid the instantaneous calcium carbonate or calcium phosphate precipitation. Indeed, as shown in Figure 4a, the activity of the Ca-citrate complex ($[Ca(Cit)]^-$) decreases at increasing temperature. The destabilization of this complex provides a gradual release of Ca^{2+} ions into the solution, which in the presence of phosphate groups causes the formation of $[CaPO_4]^-$ and $[CaHPO_4]^0$, repeatedly proposed as growth units for (hydroxy)apatite [94]. The size, crystallinity and composition of the nanoparticles can be tailored by the maturation time as well as the concentration of sodium carbonate in the mother solution. Figure 4b and c show TEM images of nanocrystalline apatites corresponding to different maturation times, i. e., 5 min and 96 hours, respectively. At low maturation times (5 min), poorly crystalline 100 nm-mean length nanocrystalline apatites with low carbonation degree (1.5% w/w, mainly as B-type substitutions, *i.e.* carbonate replacing phosphate groups) and covered by citrate (5.9% w/w) were obtained. Interestingly, this citrate content is close to that recently measured on bone apatite [33]. On increasing the aging time up to 96 hours, the mean length and the citrate content progressively decrease while the nanoparticles become better crystallized (Fig. 4b and c). In addition, the presence of carbonate ions in the solution favors their incorporation in the crystal lattice up to 3.1% (both, in A and B positions) giving rise also to shorter and more isometric nanoparticles [93].

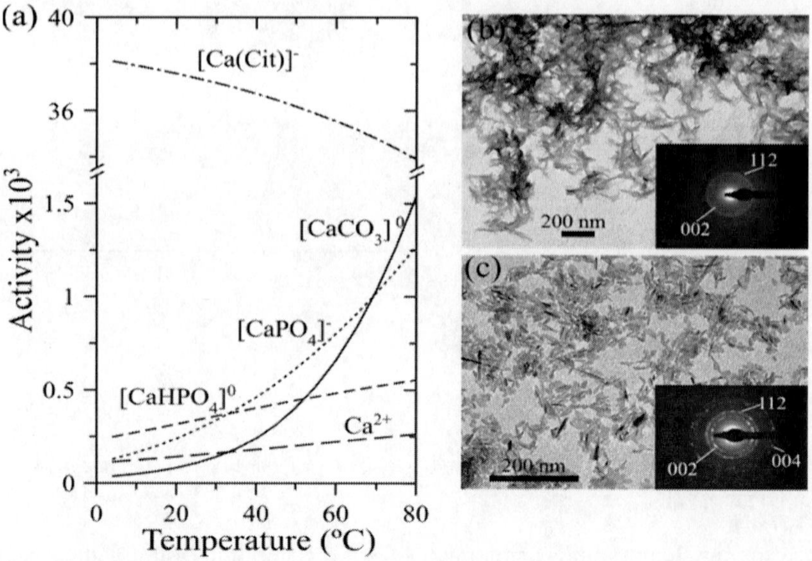

Figure 4. (a) Calculated activities, as a function of the temperature, of the main ionic and neutral species in the mother solution in the presence of 50 mM Na_2CO_3 (Note: Cit = $C_6H_5O_7$). The activities were calculated with Visual MINTEQ 3.0 speciation software [93]. Reprinted from reference [93], Copyright (2012), with permission from Elsevier. TEM micrographs of citrate-functionalized apatite nanoparticles precipitated after 5 minutes (b) and 96 hours (c) of ageing time.

To sum up, parameters such as pH, ionic strength, temperature, aging (maturation) time, Ca/P molar ratios and supersaturation are very important and they have to be considered for the control of the formation of apatites. The effects of synthesis and post-synthesis conditions on the physical-chemical characteristics of precipitated biomimetic nanocrystalline apatites have been recently overviewed [95]. The pH significantly affects the precipitation of apatite as it influences the amount of free OH^- groups and the balance of phosphate species. A shift to low pH decreases the concentration of free OH^- groups and shifts the balance of phosphate species from PO_4^{3-} to HPO_4^{2-} to $H_2PO_4^-$ to H_3PO_4 [91] (the species found in physiological conditions being only HPO_4^{2-} and $H_2PO_4^-$, forming an acido-basic couple of $pK_a \sim 7.2$). At low pH the phosphate groups are highly protonated, and the precipitation is less favored. Moreover, the pH can shift the surface charge of the interacting particles by changing the distribution of proton and hydroxyl groups at the interface. Although H_3O^+ and OH^- are usually considered as charge-determining ions in the case of apatite particles, other ions can modify the surface charge, and in particular, in calcium-containing solutions, Ca^{2+} ions may bind to the negatively-charged apatite surface at pH > IEP (isoelectric point), leaving the surface neutral rather than negative. The opposite effect can take place in phosphate-rich solutions when binding of HPO_4^{2-} species at pH < IEP may result in a negatively-charged apatite particle surface rather than the positive [91].

One of the most accepted models for the crystallization of apatite is the Ostwald–Lussac model which predicts the highest nucleation rate for the least stable phase for which the supersaturation limit is exceeded under given conditions [96]. This model implies that amorphous calcium phosphate (ACP) will be the first phase to precipitate, followed by the solid state transformation to OCP and then to apatite. $Ca(OH)_2$ or TCP may be secondary phases depending on the exact stoichiometric ratio between the precursor calcium and phosphate ions in the solution. However, the exact chemical pathways, the transformation mechanism (dissolution/recrystallization or bulk rearrangement of ions within the prime crystal lattice), the transient compounds, and time frames for the nucleation and growth of the single phases are still not well clear. In any case, the crystallization of apatite is a process that is expected to require some time (at moderate temperature) due to ion diffusion limitations in the solution and to the fact that the hydroxyapatite unit cell involves a total of 18 ions that have to organize. In addition, the phosphate ions theoretically needed to crystallize apatite are dehydrated while only protonated species exist in physiological or biomimetic-like conditions. Therefore, the progression of crystallization is anticipated to be rather slow, requiring the deprotonation of phosphate ions and their 3D organization with calcium ions, thus leading to a situation where a rather crystalline crystal core remains covered by a surface hydrated layer containing ionic species still composed mostly divalent ions like protonated phosphate HPO_4^{2-} and Ca^{2+}. The role of pH, maturation time in solution (aging) and maturation temperature is thus prominent on the characteristics of apatite nanocrystals obtained, as explained previously [95].

The most important parameter used to describe the precipitation of apatite (at a given temperature) is the supersaturation ratio $S = Q/K_{sp}$, where Q is the product of ionic activities of precursor ions in the solution for the given stoichiometry (for half a unit cell of hydroxyapatite, $Q=[Ca^{2+}]^5[PO_4^{3-}]^3[OH^-]$), and where K_{sp} is the product of ionic activities for the given compound at the saturation level (solubility product). In view of the continuous transfer of the matter across the solid/liquid interface in both directions, K_{sp} could be also defined as the product of activities of dissolved ions of a solid substance in equilibrium between the dissolved ions precipitating and the precipitated ions dissolving. Ostwald–Lussac rule postulates that the most soluble phase (i.e., the least stable) for which $S > 1$ is the first to precipitate, which will be successively followed by precipitation of less soluble phases. The reason for this is that the thermodynamic barrier between the state occupied by dissolved ionic species and the solid phase will be the lowest for the thermodynamically most unstable phase [97, 98].

An important contributor to this effect is the surface/interfacial energy, that is, the work required to increase the surface area of a substance by one area unit, which is the obstacle that must be overcome when forming a solid phase. An amorphous phase is less ordered than a crystalline phase and it has a lower interfacial energy (particularly if it is hydrated and thus if it is rather similar in chemistry to the surrounding aqueous environment) which implies that it tends to be the first to precipitate prior to subsequently transforming into a more stable, crystalline modification. In the case of the precipitation of calcium phosphates, this means that apatite could be the last phase to form. The initial precipitation of ACP is normally followed by nucleation of OCP at a certain stage. However, only the phases with $S > 1$ are involved in this successive precipitation. Those for which $S < 1$ are assumed not to be precipitated at any stage of this process, and their appearance may only be transitory during phase transitions that involve rearrangements of ions in the solid state. This kinetic rule was empirically observed, although it can be nowadays supported by theoretical arguments. It was shown that in simulated body fluid (SBF), whose S is normalized per growth unit ($n = 9$ for apatite) and it is equal to 19.5 with respect to apatite, the nucleation rate of OCP is higher than the one for apatite, implying that OCP would be the previous crystalline phase to form [99]. Finally it can be concluded that the intermediate phases are expected to play a crucial role in defining the morphology, the interfacial properties and the growth mechanism of the apatite. For instance, a topotactic transformation of OCP platelets into apatite platelets could be one possible explanation, among others, of the plate-like morphology of bone apatite as opposed to the expected hexagonal hydroxyapatite symmetry. The presence of OCP as transient phase does not however seem to always occur, as even after short aging periods (e.g. a few minutes), the characteristic FTIR bands of OCP do not appear detectable in apatites precipitated at pH ~ 7.2 and 20°C from calcium nitrate and ammonium hydrogenphosphate in excess (internal communication).

Another model based on aggregation of roughly spherical units of composition $Ca_9(PO_4)_6$ called "Posner's clusters" (PC) was proposed in the 60's to explain the formation of apatite via an amorphous precursor in neutral and basic conditions. The chemical analysis of the precursor amorphous phase indicated it to be a hydrated calcium phosphate ($Ca_3(PO_4)_2 \cdot xH_2O$) with a Ca/P ratio of 1.50, consisting of close-packed Posner's clusters to form larger spherical particles with water in the interstices [100-102].

After these works several researchers found evidences that strongly support an aggregation mechanism to explain apatite formation. The mechanism proposed by Melikhov [103] can be divided into the following steps: (i) homogeneous nucleation of ACP; (ii) aggregation of primary ACP particles into typically spherical units; (iii) aggregation of spheres into chain-like structures; (iv) growth of these structures; (v) secondary precipitation and phase transformation. The initially precipitated particles of the amorphous phase were observed to be round-shaped with 20–30 nm in size (although they can reach 120 nm in size) [103], but composed of smaller particles of 4 nm in size on average. It was also observed that an increase in the ripening time implied aggregation of spherical singlets and formation of needle-shaped calcium phosphates particles of about 20 nm in length [94].

4. Advanced Total Scattering for the Characterization of Nanocrystalline Apatites

The large potential of Total Scattering (TS) methods to provide quantitative information, at the atomic and nanometer length scales, on both the structure and the shape of apatite nanocrystals is still essentially under-estimated. As an example of their capabilities, we will point herein to quantitative characterization in terms of structure, composition, size and morphology by TS of bio-inspired citrate-functionalized apatite nanocrystals precipitated during different maturation times by the thermal decomplexing method described above either in the presence (cAp samples) or in the absence (Ap samples) of sodium carbonate (CO_3^{2-}, being the major phosphate substituent in biological apatites [104]). These bio-inspired apatites were used to explore the origin of their platy crystal morphology (typical of bone apatite), breaking the hexagonal symmetry of the apatite crystal structure (space group $P6_3/m$) [105].

TS methods offer the advantage, compared to conventional diffraction methods, of treating equally Bragg and diffuse scatterings (originating from long-range order and short-range effects, respectively). They can be applied both in direct (radial or pair distribution function) [106] and reciprocal space by using the Debye equation [107, 108]. While the first approach provides a 1D plot of density peaks at atomic separations, the second one makes use of interatomic distances derived from model nanoparticles to reproduce the experimental pattern [105]. The Debye Function Analysis (DFA) used to analyze the above described bioinspired citrate-apatite nanocrystlas can be summarized in the following steps: *i)* a population of atomistic nanocrystal models of increasing size (bottom-up approach) was generated by using the crystal unit cell of a Ca-deficient hydroxyapatite [109] as the building block. To this aim, hexagonal prisms (Figure 5c) were built layer-by-layer under the assumption of two independent "growth" directions, one along the six-fold symmetry ***c*** axis and the second one in the orthogonal *ab* plane, up to 15 nm in *ab* and 50 nm along *c*. *ii)* In order to deal with the platy hydroxyapatite shapes, *bivariate* populations of crystals grown along *c*, and, in *ab*, by considering anisotropic, parallelogram-shaped bases with fixed 1:2 *a:b* ratio (Figure 5d), were built. These populations were used *together* with the hexagonal base prisms defined in *i)* and constrained to the same nanocrystal size distribution and stoichiometry values, in a *biphasic* pattern model. When both the hexagonal and platy morphologies were used, the goodness of fit obtained by matching the experimental and the model patterns was systematically better than the model relying on the hexagonally shaped apatite only. *iii)* The sampled interatomic distances of each nanocrystal were computed and stored in suitable databases to be used in the subsequent DFA analysis. *iv)* To best match the experimental diffraction traces, a number of adjustable parameters were included in the pattern model [105]. This approximation allowed the quantification of the shape anisotropy of the *ab*-base (platy shape) in addition to the anisotropy due to the crystal elongation in the *c*-axis direction without introducing too many unnecessary parameters [105].

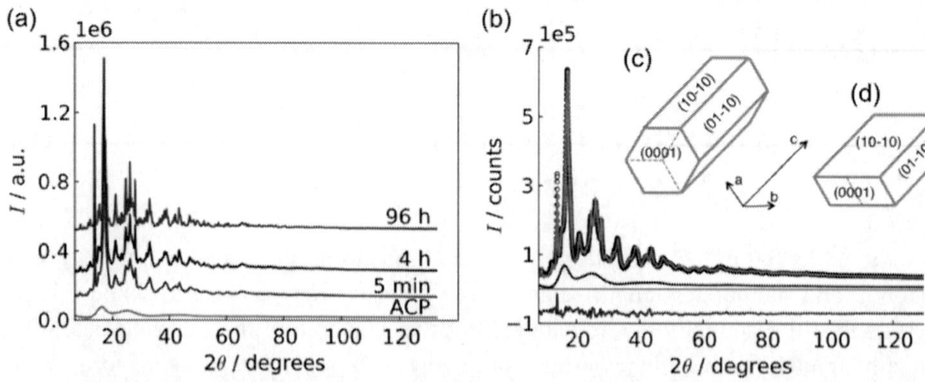

Figure 5. Synchrotron diffraction data of cAp (a) samples at different maturation times (5 min, 4 h, 96 h) collected with a wavelenght of 0.82699 Å. The ACP sample was nstantaneously precipitated (cAp conditions) and shows the typical pattern features of an amorphous material. (b) Best fit of the cAp 5 min sample provided by the DFA method. The ACP trace is also represented (agreement indices: GoF = 5.34, R_{wp} = 3.80). The inset in (b) show the hexagonal (c) and platy morphologies (d) of Ca-deficient hydroxyapatite crystals used in the DFA modeling. *(h, k, -h-k, l)* indices of the most relevant crystal facets are provided. Adapted from reference [105], Copyright (2014), with permission from John Wiley & Sons.

Figure 5a shows the synchrotron X-ray scattering data of the citrate-functionalized samples obtained in the presence of carbonate (cAp). They were processed according to the DFA model desceibed above by considering two bivariate populations of Ca-deficient hydroxyapatite nanocrystals, having hexagonal [110] and platy shapes (Fig. 5c and d), respectively, and an additional amorphous phase modeled by suitably scaling the experimental ACP diffraction curve (Fig. 5b). The best pattern model of the cAp maturated during 5 min is shown in Figure 5b. For each dataset, this analysis provided relevant structural and microstructural information such as the relative abundance of ACP and hexagonally/platy shaped nanocrystals and the bivariate size distribution maps of apatite nanocrystals and the average crystal size and shape [105].

Figure 6a reports the variation of the weight percentage (wt%) of each of the three phases (ACP, hexagonal and platy apatites) in Ap and cAp samples at increasing maturation time. ACP is a minor component slowly decreasing upon maturation; nanocrystals of platy shape, varying from 82.2 to 38.1 wt% (Ap) and from 76.6 to 28.1 wt% (cAp), represent the total or nearly the total of the crystalline components at 5 min of maturation; they tend to decrease with time, the hexagonally shaped ones becoming the major crystalline component after 96 h in both Ap (55.8 wt%) and cAp (62.3 wt%) samples [105].

Average crystal sizes (and size distributions, σ) have been also derived. For the hexagonal shape, values are given as $<D_{ab}>$, σ_{ab} and $<L_c>$, σ_c (Figure 6b) while for the platy shape, the average crystal width (<W>) and thickness (<T>) in the *ab*-plane are shown in Figure 6c. Upon maturation, crystals of progressively larger sizes along the two (or three, for platelets) growth directions were found; in cAp, crystals are systematically smaller and less anisotropic than in Ap, in agreement with previously published results [93, 104].

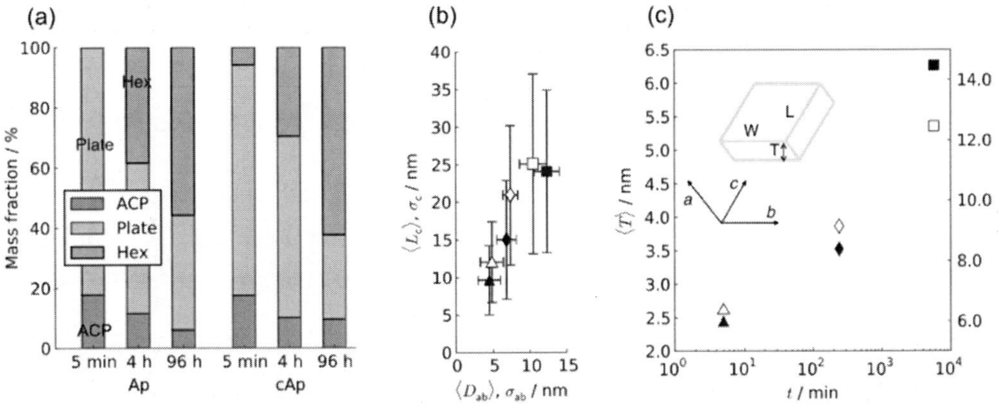

Figure 6. a) Relative abundance of amorphous (ACP) and crystalline (apatite) phases showing hexagonal or platy shape, as estimated by DFA for each Ap and cAp sample at growing maturation time. b) Average sizes of Ca-deficient hydroxyapatite crystals of hexagonal shape at growing maturation time: ▲ 5 min; ♦ 4 h; ■ 96 h (open symbols refer to Ap samples, solid symbols to cAp samples). Values are given as crystal length ($<L_c>$, σ_c as error bars) and diameter of the circle of equivalent area ($<D_{ab}>$, σ_{ab} as error bars). b) Average thickness (T) and width (W) of platy crystals (the crystal length L is the same as in b). The inset shows the assignment of T, W, L to both crystal domains and nanoparticles. Adapted from reference [105], Copyright (2014), with permission from John Wiley & Sons.

The size and morphology evolution of Ap and cAp nanoparticles was also assessed by *ex situ* Atomic Force Microscopy (AFM) observations. The sub-nanometer vertical resolution (better than 0.1 nm) of the AFM images was exploited, in particular, to obtain reliable values of the nanoparticles thickness. Amorphous calcium phosphate nanoparticles showed platy morphology which is a fairly unusual finding [105]. In all samples, the nanoparticles are platelets increasing their thickness upon maturation time. Figure 7 shows the comparisson of the average thickness (T_{DFA}), width (W_{DFA}) and length (L_{DFA}) of *crystal domains* provided by DFA, and those determined by AFM on whole *single nanoparticles* (T_{AFM}, W_{AFM}, L_{AFM}), disregarding their amorphous or crystalline nature. Importantly, crystal thickness systematically fits that of nanoparticles ($T_{DFA} \approx T_{AFM}$) until 4 h maturation (Fig. 7a), while width and length are systematically smaller in crystals than in nanoparticles ($W_{DFA} < W_{AFM}$ and $L_{DFA} < L_{AFM}$, Figure 7b,c). Therefore, at low and medium maturation time, the nanoparticles *must be* single crystal domains along the thickness and multiple crystal domains along the width and length. The coincidence of T_{DFA} and T_{AFM}, moreover, indicates that crystals are oriented with their (01-10) plane parallel to the nanoparticle surface imaged by AFM [105]. This enabled to infer the anisotropic growth along the crystallographic *a*- and *b*-axes (breaking the hexagonal symmetry and giving rise to platy crystals). Indeed, the growth is expected to stop in one direction because of the finite thickness of the parent amorphous platelets (T_{AFM}) and the further binding of citrate on the surface, but to freely continue in the other directions within each nanoparticle, therefore increasing W_{DFA} and L_{DFA}, as later substantiated.

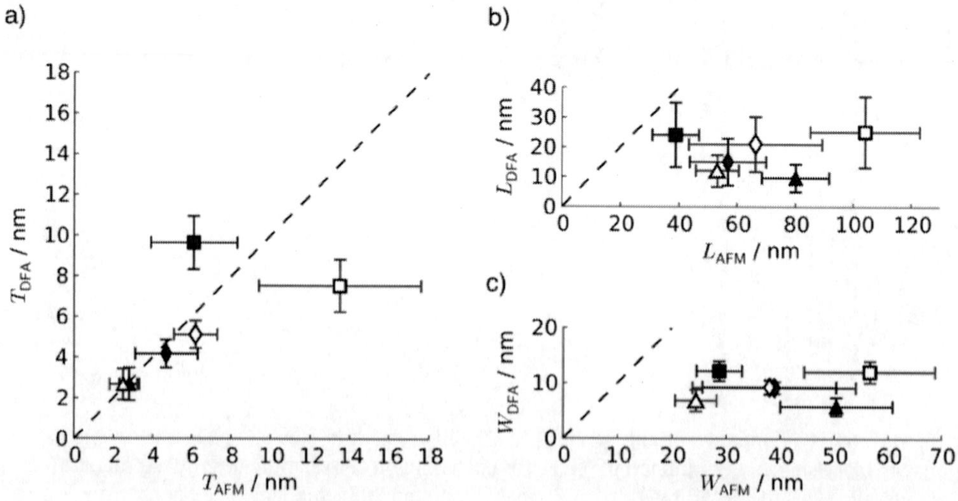

Figure 7. a) Average thickness of crystal domains provided by method DFA method (T_{DFA}) *vs* the average thickness of nanoparticles provided by AFM (T_{AFM}); b) Average length of crystal domains (L_{DFA}) *vs* the average length of nanoparticles (L_{AFM}) and c) average width of crystal domains (W_{DFA}) *vs* the average width of NPs (W_{AFM}). The vertical and horizontal bars depict the DFA and AFM-derived distributions about each average. The broken line is a guide to the eye to easily identify samples having very similar DFA- and AFM-derived dimensions. Adapted from reference [105], Copyright (2014), with permission from John Wiley & Sons.

On the basis of the above described results, a plausible mechanism was proposed for the formation of apatite platelets (Figure 8) [105]. In this model, citrate ions seem to play multiple roles. The sodium citrate crystals detected at the earliest stages [111] are thought to act as a template surface which, through adsorption of ionic species (Ca^{2+}, HPO_4^{2-} and OH^-, neglecting the minor ones), might trigger the heterogeneous nucleation process through surface formation of "clusters" that, laterally aggregating, form ACP particles with unusual platy morphology [105]. The drop of pH and the temperature changes in the initial stages [105] produce partial dissolution of the template leading to the formation of $Hcit^{2-}/cit^{3-}$, which partially bind to the ACP platelets. The presence of adsorbed citrate on the surface slows down the plate thickening by inhibiting further ion adsorption and nanoparticle aggregation. Apatite crystals start to grow from multiple nuclei *within* each ACP platelet. The hexagonal crystal symmetry enables independent growth of each nucleus along the *c*-axis and in the orthogonal *ab*-plane (isotropically along the *a*- and *b*- axes) (stage 1) until the platelet surface is reached. At this point (stage 2) citrate ions, strongly bound to surface Ca^{2+} ions on the free (10-10) crystal facet, stop any further growth in this direction, which anyway continues in the other two (in the presence of as yet untransformed amorphous material) (stage 3) until separately nucleated crystal domains touch each other. Platy crystals of the same thickness as ACP platelets are thus formed at low maturation times. After 4 h (stage 4), this mechanism is still active in controlling the average crystal thickness. However, the hexagonal crystal morphology begins to appear (or, in other words, the average crystal anisotropy in the *ab*-plane starts to decrease), concomitantly with dissolution, aggregation, recrystallization phenomena and becomes dominant after 96 h, forming larger (more stoichiometric) apatite crystals.

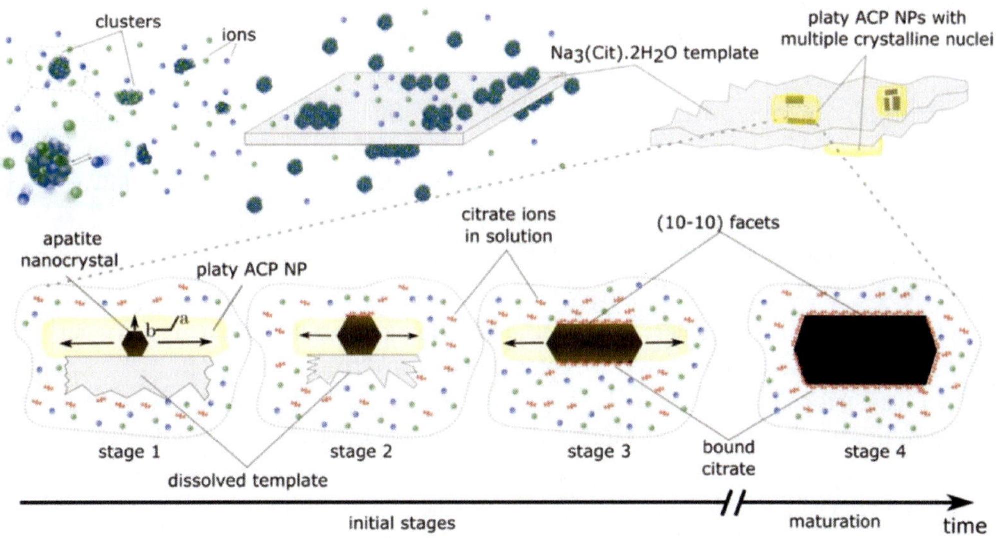

Figure 8. Schematics of the mechanism forming apatite platelets during the ACP-to-crystal transformation. A time scale of minutes applies to the initial stages and hours to the maturation process. Reprinted from reference [105], Copyright (2014), with permission from John Wiley & Sons.

This amorphous-to-crystalline transformation process and the appealing mechanism inducing the platy morphology in bio-inspired apatite nanocrystals strongly support the potential role of citrate in driving the formation of platy apatite nanocrystals similar to those formed in bone under the control of acidic non-collagenous proteins, collagen fibrils and, likely, citrate ions. Moreover, the mechanism described, can be of great relevance in tuning the properties of this important class of biomaterials for advanced biomedical applications. Additional possible origins to the platy morphology of apatite nanocrystals formed *in vivo* or in biomimetic-like syntheses may also be linked to the transient formation of an octacalcium phosphate (OCP) precursor [60], which also exhibits a plate-like morphology and which could hydrolyze to apatite in a topotactic process, thus without major morphological changes.

5. Spectroscopic Investigations of Apatite Nanocrystals

Vibrational spectroscopies (typically IR and Raman) have the advantage over diffraction techniques to give detailed information on local chemical environments, therefore proving to be complementary. For instance, in apatite compounds, various ionic entities (the most important being PO_4^{3-}, HPO_4^{2-}, CO_3^{2-}, or OH^-) lead to specific vibrational signatures that relate to the position of these species relatively to the apatitic lattice (or to the non-apatitic surface layer on the nanocrystals), and also give information on their relative abundance in the solid by comparing bands intensities. When associated to microscopy observations, it also becomes possible to draw even more "local" indications and also to inspect the distribution of chemical species by way of mapping analyses.

The vibrational features of phosphate and carbonate ions within apatitic compounds have been the object of a wealth of literature reports. A review is for example available [112]. The vibrational signature of these ions when included in an apatitic solid phase remains rather close (although not identical) to that of the free ions. Therefore, the nomenclature of vibration modes is generally used in reference to these isolated ionic states. For HPO_4^{2-} ions, the additional proton as compared to PO_4^{3-} alters, as expected, the vibration of the phosphate entity due to a weakening of the P-(OH) bond and a strengthening of the other three P-O bonds, therefore leading to wavenumber shifts. Taking into account the incorporation of such ions in a solid lattice rather than in a free ionic state is at the origin of a lowering of symmetry, which in turn modifies degenerated vibration modes to non-degenerated ones. This phenomenon is for example at the origin of the splitting of the ν_2, ν_3 and ν_4 phosphate bands or of the (low) activity in infrared of the ν_1 and ν_2 phosphate modes. Libration bands can also be visible in the solid state, as is the case for the apatitic OH band at 632 cm^{-1}. Typical band assignments for hydroxyapatite have for example been listed (e.g. [113, 114]).

For nonstoichiometric as well as multi-substituted apatite compounds, as is generally the case in hard tissues or for synthetic biomimetic analogs, additional spectral modifications are to be expected due to the presence of vacancies and of varying substituents leading to a possible distortion of local site symmetry. This may then be traduced, in the spectra, by shifts in band positions (e.g. [115]) or by additional contributions (e.g. [116]). In contrast to the case of stoichiometric hydroxyapatite, deficient apatites generally involve the presence of protonated phosphate groups in the chemical composition, under the form HPO_4^{2-}. These ions are thus expected to generate additional vibrations, which is indeed verified experimentally, in particular by a shoulder to the low wavenumbers on the $\nu_4(PO_4)$ band, in the region 530-550 cm^{-1}, or a shoulder to the $\nu_3(PO_4)$ band around 1145 cm^{-1}. The peculiar structuration of immature nanocrystalline apatites, with nanocrystals composed of an apatitic core (often nonstoichiometric) and a non-apatitic surface layer (containing ions and water molecules), also leads to specificities in vibrational spectral signature. It is possible to apply curve-fitting procedures, especially in the $\nu_2\nu_4(PO_4)$ domain, so as to better identify the main contributions making up the global spectrum. Stoichiometric hydroxyapatite exhibits in this area 5 contributions (the $\nu_2(PO_4)$ band, the $\nu_{lib}(OH)$ libration band and 3 $\nu_4(PO_4)$ bands). In contrast, nonstoichiometric immature apatites (e.g. bone mineral or biomimetic analogs) show bands relating to apatitic HPO_4^{2-} ions around 550 cm^{-1}, as well as non-apatitic HPO_4^{2-} ions around 535 cm^{-1} and presumably also non-apatitic PO_4^{3-} ions around 617 cm^{-1}, in addition to the 5 previously cited bands which are found respectively around 470, 632, 601, 575 and 560 cm^{-1}. This phenomenon had been discussed on many occasions in modern literature; an overview of these spectral data has been recently published [95] with also comparisons with other calcium phosphates [60]. In the case of nanocrystals (large surface-to-volume ratio), the surface species represent indeed a non-negligible proportion of chemical species and may thus be also analyzed by vibrational spectroscopies.

For phosphate chemical environments, the analysis of the $\nu_2\nu_4(PO_4)$ domain is thus very informative. Other vibrational domains can however also be explored. The $\nu_3(PO_4)$ region is one other domain of potential interest. However, in this case, it appears rather complex to decompose into a set of well-determined contributions. Therefore, Fourier self-deconvolution methods are often preferred in this case to full curve-fittings. This has been applied to biological apatites and some correlations between local maxima in the $\nu_3(PO_4)$ domain and apatitic or non-apatitic contributions found in the $\nu_4(PO_4)$ region have been discussed. A contribution around 1020 cm^{-1} seems for example associated to immature phosphates whereas the contribution at 1030 cm^{-1} increases upon maturation. A similar tendency seems to be found respectively for maxima at 1110cm^{-1}/1100 cm^{-1}, although the former seems also to be linked to the presence of B-type carbonates. Another intermediate maximum around 1125 cm^{-1} appears associated to non-apatitic phosphate environments, and a shoulder around 1145 cm^{-1} is indicative of HPO_4^{2-} ions. The analysis of the $\nu_3(PO_4)$ spectral domain has recently been applied to the analysis of skeletal remains with various datations and preservation conditions in archeological and forensics contexts [117] showing that the exploration of this specteral domain, in addition to the $\nu_4(PO_4)$ band, can also be relevant in the investigation of ancient skeletal specimens.

In Raman spectroscopy, the spectra obtained for apatites are globally less complex than for IR. The main band is related to the $\nu_1(PO_4)$ phosphate vibration while other bands are significantly less intense. A shoulder to this main band, in the lower wavenumbers side (around 955 cm^{-1}), however becomes detectable for immature compounds, probably due to non-apatitic PO_4^{3-} ions.

Alterations of vibrational features have been shown to occur depending on synthesis or post-synthesis conditions [95]. For instance, a fine structuration of IR bands can be observed on freshly precipitated apatites, with features resembling those of octacalcium phosphate (while not being identical, see [60]), while these spectral features progressively smooth out upon drying. This phenomenon is thought to be related to a partial amorphization of the surface of the nanocrystals (this is not a separate amorphous phase!) due to the removal of some water molecules, thus leaving the top-surface non-apatitic ions in a more disturbed environment [95]. These modifications are however (at least partly) reversible, in much the same way as in the case for ionic exchanges [34].

By IR or Raman techniques, it is also possible to inspect the adsorption of molecular species on the surface of apatite crystals (e.g. [118]). In this view, the use of Raman (micro)spectroscopy is particularly interesting as it also allows to work directly in aqueous systems, not being disturbed by water vibrations. Therefore, it allows *in situ* observations, which could in particular be interesting to analyze nanocrystals in their wet state and thus closer to the "in use" state *in vivo*. IR and/or Raman techniques may help confirming the presence of specific (bio)molecules on the crystal surface, and in some cases may even allow quantitative assessments.

Beside a rather "conventional" IR and Raman use, other declinations have also been addressed at times. This is for example the case of near-infrared spectroscopy (NIR); water absorption in the NIR region being characterized by weaker extinction coefficients than in the mid-IR region. Surface hydration of nanocrystalline apatites has been explored by this method. The $\delta+\nu_{asym}$ H_2O band for water adsorbed on nanocrystalline apatite [119] was found to be involved several contributions: a main absorption around 5170 cm^{-1} attributable to H_2O molecules acting as donors of two equivalent H-bonds, and two shoulders around 5300 (narrow) and 5000 (broad) cm^{-1} attributed to the OH moieties (non H-bonded and H-bonded, respectively) of water molecules acting as donors of a single H bond ([119, 120]). By comparing apatite samples enriched or not with Mg^{2+}, the authors unveiled an increased tendency to fix water when Mg was present on the surface of the nanocrystals in favor or a greater interaction of Mg^{2+} ions with H_2O molecules as compared to Ca^{2+}. NIR findings also pointed out the co-existence of several types of adsorbed water on nanocrystalline apatites, possibly forming overlayers.

Vibrational spectroscopies thus should be seen, probably along with other spectroscopy techniques such as solid state NMR, as powerful tools for inspecting, to a local scale, the chemical environments constituting apatite compounds, with very specific features detected for nanocrystalline ones.

6. Thermodynamic Aspects Relating to Apatite Nanocrystals

As mentioned previously, apatites are ubiquitous compounds within natural biominerals, especially teeth and bone tissues, as well as in synthetic bone repair materials. However, thanks to a rich physicochemistry, apatite systems can display a variety of properties depending on parameters such as their stoichiometry and crystal dimensions among other factors. This adaptability allows them to adapt to various functions in the organism. In this view, the evolution of animal species from primitive organisms with exoskeletons based on calcium carbonate or silica up to more advanced calcium phosphate (apatite) based endoskeletons may be seen as a major progression in evolution, permitting to various areas of the skeleton to play different functions. For instance, tooth enamel is required to remain inert and to tolerate the recurrent bacterial acidic attacks arising from the mouth environment. Enamel should thus exhibit a high thermodynamic stability, which is ensured (in Humans) by a chemical composition for mature enamel very close to stoichiometric hydroxyapatite as was mentioned previously (solubility product $pK_{sp} \sim 118$ [121]). In contrast, bone mineral has to continually remain highly reactive, in particular for allowing microfracture auto-healing and for playing an active role in homeostasis (i.e. helping body fluids to keep optimal ionic concentrations). The apatite phase constituting the mineralized part of bones is thus composed of nanocrystals exhibiting a low degree of crystallinity [122] and departing from stoichiometry by way of cation and anion vacancies, and exposing on their surface a non-apatitic surface layer with exchangeable (labile) reactive ions, as already mentioned.

The preparation of synthetic analogs to biological apatites (i.e. mimicking compositional, crystallographic and microstructural features) is possible under “mild” synthetic conditions, for example by precipitation at ambient temperature and physiological pH, and variations in synthesis conditions (temperature, pH, maturation time, etc.) [95] also make it possible to prepare apatite systems that can mimic either immature or mature bone mineral.

A decrease in surface reactivity of biological or synthetic apatites has been evidenced, for example through ion exchange or adsorption studies, upon crystal ageing in solution (e.g. [123-125]). During this aging process, some ions of the surface layer are progressively incorporated into the apatitic core of the nanocrystals, thus decreasing their surface reactivity.

Obviously, the reactivity and thermodynamic stability of apatite nanocrystals are bound to be closely related. However, despite the large number of physical-chemical investigations of apatitic systems, thermodynamic studies remain scarce. The vast majority of literature data relating to apatites concerned in fact stoichiometric end-members (e.g. [126-128]) mostly of $M_{10}(PO_4)_6X_2$ apatites, where M represents a divalent cation and X a monovalent anion (hydroxide or halide). In some instances, the energetics of synthetic (well crystallized) carbonated apatites has also been punctually addressed [129]. But these systems generally differ significantly from biological apatites.

As a general trend, it may be remarked in such stoichiometric apatitic systems that the Gibbs free energy of formation $\Delta G_f°$ is largely guided by enthalpy contribution $\Delta H_f°$, whereas the entropy contribution $T\cdot\Delta S_f°$ remains of much lower impact. Therefore, it is generally possible to determine the relative stability ranking of apatite phases (presenting different chemical compositions) on the basis of the ranking in $\Delta H_f°$ (which may be accessed by calorimetry techniques) rather than in terms of $\Delta G_f°$.

Based on a review of published thermodynamic properties of $M_{10}(PO_4)_6X_2$ apatites, Drouet (Drouet 2014, to be published) recently pointed out that for a given X^- anion, the thermodynamic stability was first linked to the electropositive character of the M^{2+} cation, and only secondly to its ionic radius; while for a given M^{2+} cation, the variation of X was both strongly impacted by the electronegativity of X and its ionic size. However, the apatitic systems found in biological apatites generally depart from "ideal" end-member stoichiometric compositions: not only nonstoichiometric phases are encountered in immature enamel or in all bone apatites, but also solid-solutions accommodating in the same phase several cations and/or anions are observed. Also, due to the immature character of bone apatite (or their synthetic analogs), a hydrated feature is expected to exist also in these systems within a non-apatitic surface layer on the nanocrystals. The case of stoichiometric solid-solutions has only sporadically been addressed in terms of thermodynamic, such as calcium-cadmium hydroxyl-, fluor- or chlor-apatites or other substitutions [130-133], but not for biomimetic apatites.

The experimental exploration of the thermodynamics of nanocrystalline apatites mimicking physical-chemical features of biological apatites (nonstoichiometry, hydrated character, nanocrystalline feature) has only been addressed so far, to the best of our knowledge, in one experimental study [134]. In that work, the enthalpy of formation of nanocrystalline apatites exhibiting various maturation stages was determined by drop solution calorimetry. Adding entropy considerations, the Gibbs free energies of formation of such apatites could then be evaluated. Very interestingly, a clear trend was unveiled and quantified by the authors, with the stability of the apatite phase increasing systematically for increasing maturation states. The variation in free energy of maturation $\Delta G_{maturation}(i \rightarrow f)$, relating to the maturation process in solution, from an initial maturation stage "i" to a more advanced stage "f", has also been estimated from these data [134]. The clearly negative numbers obtained for $\Delta G_{maturation}$ confirmed (qualitatively but also quantitatively) the existence of a *thermodynamic driving force* along which immature apatites inexorably evolve, while in solution, toward more advanced maturation states closer from stoichiometry (Figure 9). If this evolutionary tendency is very advantageous in the case of enamel maturation (prior to tooth eruption) which should be as chemically stable as possible, it is on the contrary not desirable in the case of bone mineral since the latter is expected to play a permanent role in homeostasis. The authors then correlated these thermodynamic findings to the *necessity of bone remodeling* in vertebrates: because bone mineral is expected to act as an ion reservoir for homeostasis and is also involved in self-repair processes, then bone remodeling becomes a necessity so as to get rid of mature less-reactive apatite crystals in favor of the neo-formation of highly-reactive immature apatite nanocrystals. This represents thus a *physical necessity* for bone remodeling in vertebrates, and this idea changes the paradigm along which this process would have only biological origins.

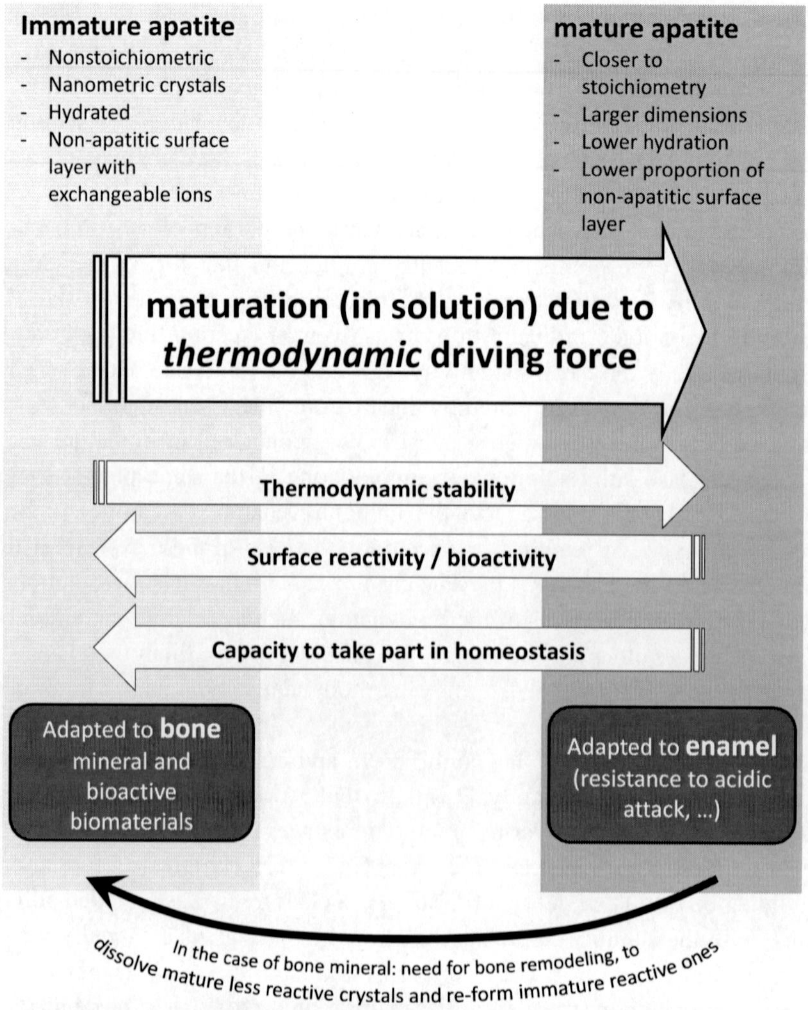

Figure 9. Illustration of evolutionary trends of nanocrystalline apatite.

Finally, since the thermodynamic properties of apatite compounds are bound to depend closely on their exact composition, there is a need to make accessible more data on the formation energetics of complex apatitic compositions that may include several types of cations and anions and in stoichiometric or nonstoichiometric settings (as is the case *in vivo* or for some existing or upcoming apatite-based biomaterials). In this context, Drouet (2014, to be published) has recently developed a predictive approach of the thermodynamics of apatite compounds based on an additive contributional model. In such a system, each ion from the unit formula contributes, in the structure, in an additive manner; the addition of all fitted contributions ultimately allowing one to estimate the energetics of the whole system. This method (involving so far entities such as Ca^{2+}, Mg^{2+}, Sr^{2+}, Ba^{2+}, Cu^{2+}, Zn^{2+}, Cd^{2+}, Pb^{2+}, OH^-, F^-, Cl^-, Br^-, PO_4^{3-}, HPO_4^{2-} or else H_2O) was found to be applicable to end-members but also to solid-solutions and to nonstoichiometric compounds, hydrated or not, and these features (solid-solutions, nonstoichiometry, eventual hydration) are important characteristics of apatites encountered in practice. In this view, a calculation program (C. Drouet, P. Alphonse, Therm'AP calculation program, www.christophedrouet.com/thermAP.html) has been made available to estimate, in first approximation, the $\Delta H_f°$, S°, $\Delta G_f°$, and derived solubility product of apatite compounds based on their chemical composition.

A good knowledge of the thermodynamic properties of apatite systems, especially in their nanocrystalline metastable form, appears as essential in view of better understanding evolution features in dry or wet conditions (e.g. during synthesis or post-synthesis treatments, or after implantation of a biomaterial *in vivo*).

REFERENCES

[1] Gómez-Morales J, Iafisco M, Delgado-López JM, Sarda S, Drouet C. Progress on the preparation of nanocrystalline apatites and surface characterization: Overview of fundamental and applied aspects. *Prog Cryst Growth Charact Mater* 2013;59(1):1-46.

[2] Dorozhkin SV. Bioceramics of calcium orthophosphates. *Biomaterials.* 2010;31(7):1465-85.

[3] Dorozhkin SV. Nanosized and nanocrystalline calcium orthophosphates. *Acta Biomater.* 2010;6(3):715-34.

[4] Dorozhkin SV, Epple M. Biological and Medical Significance of Calcium Phosphates. *Angewandte Chemie International Edition.* 2002;41(17):3130-46.

[5] Roveri N, Iafisco M. Evolving application of biomimetic nanostructured hydroxyapatite. *Nanotechnol Sci Appl* 2010. p. 107-25.

[6] Loveridge N. Bone: More Than a Stick. *J Anim Sci.* 1999 January 1, 1999;77(E-Suppl 2):190-6.

[7] Weiner S, Wagner HD. The Material Bone: Structure-Mechanical Function Relations. *Annu Rev Mater Sci.* 1998;28(1):271-98.

[8] Beniash E. Biominerals—hierarchical nanocomposites: the example of bone. *Wiley Interdisciplinary Reviews: Nanomedicine and Nanobiotechnology.* 2011;3(1):47-69.

[9] Palmer LC, Newcomb CJ, Kaltz SR, Spoerke ED, Stupp SI. Biomimetic Systems for Hydroxyapatite Mineralization Inspired By Bone and Enamel. *Chem. Rev.* 2008;108(11):4754-83.

[10] Shoulders MD, Raines RT. Collagen Structure and Stability. *Annual Review of Biochemistry.* 2009;78(1):929-58.

[11] Bozec L, van der Heijden G, Horton M. Collagen Fibrils: Nanoscale Ropes. *Biophysical Journal.*92(1):70-5.

[12] Hulmes DJS. Building collagen molecules, fibrils, and suprafibrillar structures. *Journal of Structural Biology.* 2002;137(1-2):2-10.

[13] Landis WJ, Silver FH. The structure and function of normally mineralizing avian tendons. *Comparative Biochemistry and Physiology - A Molecular and Integrative Physiology.* 2002;133(4):1135-57.

[14] Smith JW. Molecular pattern in native collagen. *Nature.* 1968;219(5150):157-8.

[15] Wang XM, Cui FZ, Ge J, Wang Y. Hierarchical structural comparisons of bones from wild-type and liliputdtc232 gene-mutated Zebrafish. *Journal of Structural Biology.* 2004;145(3):236-45.

[16] Miller A. Collagen: the organic matrix of bone. *Philosophical transactions of the Royal Society of London Series B: Biological sciences.* 1984;304(1121):455-77.

[17] Cui FZ, Li Y, Ge J. Self-assembly of mineralized collagen composites. *Materials Science and Engineering R: Reports*. 2007;57(1-6):1-27.

[18] Traub W, Arad T, Weiner S. Three-dimensional ordered distribution of crystals in turkey tendon collagen fibers. *Proceedings of the National Academy of Sciences.* 1989 December 1, 1989;86(24):9822-6.

[19] Olszta MJ, Cheng X, Jee SS, Kumar R, Kim YY, Kaufman MJ, et al. Bone structure and formation: A new perspective. *Materials Science and Engineering R: Reports.* 2007;58(3-5):77-116.

[20] Hadjidakis DJ, Androulakis II. Bone Remodeling. *Annals of the New York Academy of Sciences. 2006*;1092(1):385-96.

[21] Manolagas SC. Birth and Death of Bone Cells: Basic Regulatory Mechanisms and Implications for the Pathogenesis and Treatment of Osteoporosis. *Endocrine Reviews.* 2000;21(2):115-37.

[22] McNamara LM, Majeska RJ, Weinbaum S, Friedrich V, Schaffler MB. Attachment of Osteocyte Cell Processes to the Bone Matrix. *The Anatomical Record: Advances in Integrative Anatomy and Evolutionary Biology.* 2009;292(3):355-63.

[23] Rho JY, Kuhn-Spearing L, Zioupos P. Mechanical properties and the hierarchical structure of bone. *Medical Engineering and Physics.* 1998;20(2):92-102.

[24] Fratzl P, Gupta HS, Paschalis EP, Roschger P. Structure and mechanical quality of the collagen-mineral nano-composite in bone. *J Mater Chem.* 2004;14(14):2115-23.

[25] Roveri N, Palazzo B, Iafisco M. The role of biomimetism in developing nanostructured inorganic matrices for drug delivery. *Expert Opin Drug Deliv.* 2008;5(8):861-77.

[26] Weiner S, Price PA. DISAGGREGATION OF BONE INTO CRYSTALS. Calcif Tissue Int. 1986;39(6):365-75..

[27] Jackson SA, Cartwright AG, Lewis D. The morphology of bone mineral crystals. *Calcified Tissue Research.* 1978;25(1):217-22.

[28] Guagliardi A, Giannini C, Cedola A, Mastrogiacomo M, Ladisa M, Cancedda R. Toward the X-ray microdiffraction imaging of bone and tissue-engineered bone. *Tissue Engineering - Part B: Reviews.* 2009;15(4):423-42.

[29] Hassenkam T, Fantner GE, Cutroni JA, Weaver JC, Morse DE, Hansma PK. High-resolution AFM imaging of intact and fractured trabecular bone. *Bone.* 2004;35(1):4-10.

[30] Tang R, Wang L, Nancollas GH. Size-effects in the dissolution of hydroxyapatite: an understanding of biological demineralization. *J Mater Chem.* 2004;14(14):2341-6.

[31] Nudelman F, Pieterse K, George A, Bomans PHH, Friedrich H, Brylka LJ, et al. The role of collagen in bone apatite formation in the presence of hydroxyapatite nucleation inhibitors. *Nat Mater.* 2010;9(12):1004-9..

[32] Hartles RL. Citrate in Mineralized Tissues. *Adv Oral Biol.* 1964;1:225-53.

[33] Hu YY, Rawal A, Schmidt-Rohr K. Strongly bound citrate stabilizes the apatite nanocrystals in bone. *Proc Natl Acad Sci USA*. 2010;107(52):22425-9..

[34] Rey C, Combes C, Drouet C, Lebugle A, Sfihi H, Barroug A. Nanocrystalline apatites in biological systems: characterisation, structure and properties. *Mater-wiss u Werkstofftech.* 2007;38(12):996-1002.

[35] Cazalbou S, Eichert D, Ranz X, Drouet C, Combes C, Harmand MF, et al. Ion exchanges in apatites for biomedical application. *J Mater Sci Mater Med.* 2005;16(5):405-9.

[36] Cazalbou S, Combes C, Eichert D, Rey C, Glimcher MJ. Poorly crystalline apatites: evolution and maturation in vitro and in vivo. *J Bone Miner Metab.* 2004;22(4):310-7.

[37] Whittaker DK. Structural variations in the surface zone of human tooth enamel observed by scanning electron microscopy. *Archives of Oral Biology.* 1982;27(5):383-92.

[38] Tamerler C, Sarikaya M. Molecular Biomimetics: Genetic Synthesis, Assembly, and Formation of Materials Using Peptides. *MRS Bull.* 2008;33(05):504-12.

[39] Arsenault AL, Robinson BW. The dentino-enamel junction: a structural and microanalytical study of early mineralization. *Calcif Tissue Int.* 1989;45(2):111-21.

[40] White SN, Luo W, Paine ML, Fong H, Sarikaya M, Snead ML. Biological organization of hydroxyapatite crystallites into a fibrous continuum toughens and controls anisotropy in human enamel. *Journal of Dental Research.* 2001;80(1):321-6.

[41] Imbeni V, Kruzic JJ, Marshall GW, Marshall SJ, Ritchie RO. The dentin-enamel junction and the fracture of human teeth. *Nat Mater.* 2005;4(3):229-32.

[42] Paine ML, White SN, Luo W, Fong H, Sarikaya M, Snead ML. Regulated gene expression dictates enamel structure and tooth function. *Matrix Biology.* 2001;20(5-6):273-92.

[43] Wen HB, Moradian-Oldak J, Fincham AG. Dose-dependent modulation of octacalcium phosphate crystal habit by amelogenins. *Journal of Dental Research.* 2000;79(11):1902-6.

[44] Cui FZ, Ge J. New observations of the hierarchical structure of human enamel, from nanoscale to microscale. *Journal of Tissue Engineering and Regenerative Medicine.* 2007;1(3):185-91.

[45] Hannig M, Hannig C. Nanomaterials in preventive dentistry. *Nature Nanotechnology.* 2010;5(8):565-9.

[46] Kodaka T, Nakajima F, Higashi S. Structure of the So-Called 'Prismless' Enamel in Human Deciduous Teeth. *Caries Res.* 1989;23(5):290-6.

[47] Daculsi G, Bouler JM, Legeros RZ. *Adaptive crystal formation in normal and pathological calcifications in synthetic calcium phosphate and related biomaterials.* 1997. p. 129-91.

[48] Kazama JJ, Amizuka N, Fukagawa M. Ectopic Calcification as Abnormal Biomineralization. *Therapeutic Apheresis and Dialysis.* 2006;10:S34-S8.

[49] Brancaccio D, Cozzolino M. *The mechanism of calcium deposition in soft tissues.* 2005. p. 279-86.

[50] Achilles W, Jockel U, Schaper A, Burk M, Riedmiller H, Khan SR, et al. In vitro formation of 'urinary stones': Generation of spherulites of calcium phosphate in gel and overgrowth with calcium oxalate using a new flow model of crystallization. *Scanning Microscopy.* 1995;9(2):577-86.

[51] Hayashizaki J, Ban S, Nakagaki H, Okumura A, Yoshii S, Robinson C. Site specific mineral composition and microstructure of human supra-gingival dental calculus. *Archives of Oral Biology.* 2008;53(2):168-74.

[52] Marra SP, Daghlian CP, Fillinger MF, Kennedy FE. Elemental composition, morphology and mechanical properties of calcified deposits obtained from abdominal aortic aneurysms. *Acta Biomater.* 2006;2(5):515-20.

[53] Gigi A, Compostella L, Fichera AM, Foresti E, Gazzano M, Ripamonti A, et al. Structural and chemical characterization of inorganic deposits in calcified human mitral valve. *Journal of Inorganic Biochemistry.* 1988;34(2):75-82.

[54] Laird DF, Mucalo MR, Yokogawa Y. Growth of calcium hydroxyapatite (Ca-HAp) on cholesterol and cholestanol crystals from a simulated body fluid: A possible insight into the pathological calcifications associated with atherosclerosis. *J Colloid Interface Sci.* 2006;295(2):348-63.

[55] Wesson JA, Ward MD. Pathological biomineralization of kidney stones. *Elements.* 2007;3(6):415-21.

[56] Kirsch T. Determinants of pathological mineralization. *Current Opinion in Rheumatology.* 2006;18(2):174-80.

[57] Navarro M, Michiardi A, Castaño O, Planell JA. Biomaterials in orthopaedics. *Journal of the Royal Society Interface.* 2008;5(27):1137-58.

[58] Iafisco M, Margiotta N. Silica xerogels and hydroxyapatite nanocrystals for the local delivery of platinum–bisphosphonate complexes in the treatment of bone tumors: A mini-review. *Journal of Inorganic Biochemistry.* 2012;117(0):237-47.

[59] Cai Y, Tang R. Calcium phosphate nanoparticles in biomineralization and biomaterials. *J Mater Chem.* 2008;18(32):3775-87.

[60] Drouet C. Apatite formation: why it may not work as planned, and how to conclusively identify apatite compounds. *Biomed Research International.* 2013; Volume 2013, Article ID 490946, 12 pages, doi: 10.1155/2013/490946

[61] Iafisco M, Ruffini A, Adamiano A, Sprio S, Tampieri A. Biomimetic magnesium-carbonate-apatite nanocrystals endowed with strontium ions as anti-osteoporotic trigger. *Materials Science and Engineering C.* 2014;35(1):212-9.

[62] Rey C, Combes C, Drouet C, Sfihi H, Barroug A. Physico-chemical properties of nanocrystalline apatites: Implications for biominerals and biomaterials. *Materials Science & Engineering C-Biomimetic and Supramolecular Systems.* 2007;27(2):198-205.

[63] Iafisco M, Varoni E, Di Foggia M, Pietronave S, Fini M, Roveri N, et al. Conjugation of hydroxyapatite nanocrystals with human immunoglobulin G for nanomedical applications. *Colloid Surface B.* 2012;90(0):1-7.

[64] Iafisco M, Palazzo B, Martra G, Margiotta N, Piccinonna S, Natile G, et al. Nanocrystalline carbonate-apatites: role of Ca/P ratio on the upload and release of anticancer platinum bisphosphonates. *Nanoscale.* 2012;4(1):206-17.

[65] Iafisco M, Bosco R, Leeuwenburgh SCG, van den Beucken JJJP, Jansen JA, Prat M, et al. Electrostatic Spray Deposition of Biomimetic Nanocrystalline Apatite Coatings onto Titanium. *Adv Eng Mater.* 2012;14(3):B13-B20.

[66] Wang JWS, L. L. Morphology-enhanced low-temperature sintering of nanocrystalline hydroxyapatite. *Advanced Materials.* 2007;19(17):6.

[67] Ganesan K, Epple M. Calcium phosphate nanoparticles as nuclei for the preparation of colloidal calcium phytate. *New Journal of Chemistry.* 2008;32(8):1326-30.

[68] Zhang Y, Lu J. A simple method to tailor spherical nanocrystal hydroxyapatite at low temperature. *Journal of Nanoparticle Research.* 2007;9(4):589-94.

[69] Ben-Nissan B, Choi AH. Sol-gel production of bioactive nanocoatings for medical applications. Part 1: an introduction. *Nanomedicine.* 2006;1(3):311-9..

[70] Tas AC. Synthesis of biomimetic Ca-hydroxyapatite powders at 37∞C in synthetic body fluids. *Biomaterials*. 2000;21(14):1429-38.

[71] Ashok M, Kalkura SN, Sundaram NM, Arivuoli D. Growth and characterization of hydroxyapatite crystals by hydrothermal method. *J Mater Sci-Mater M*. 2007;18(5):895-8..

[72] Guo XY, Gough JE, Xiao P, Liu J, Shen ZJ. Fabrication of nanostructured hydroxyapatite and analysis of human osteoblastic cellular response. *J Biomed Mater Res* A. 2007;82A(4):1022-32..

[73] Chaudhry AA, Haque S, Kellici S, Boldrin P, Rehman I, Fazal AK, et al. Instant nano-hydroxyapatite: a continuous and rapid hydrothermal synthesis. *Chem Commun*. 2006 (21):2286-8..

[74] Gómez-Morales J, Torrent-Burgues J, Boix T, Fraile J, Rodríguez-Clemente R. Precipitation of stoichiometric hydroxyapatite by a continuous method. *Crystal Research and Technology*. 2001;36(1):15-26.

[75] Yeong KCB, Wang J, Ng SC. Mechanochemical synthesis of nanocrystalline hydroxyapatite from CaO and CaHPO4. *Biomaterials*. 2001;22(20):2705-12..

[76] Krishna DSR, Siddharthan A, Seshadri SK, Kumar TSS. A novel route for synthesis of nanocrystalline hydroxyapatite from eggshell waste. *J Mater Sci-Mater* M. 2007;18:1735-43..

[77] Lak A, Mazloumi M, Mohajerani MS, Zanganeh S, Shayegh MR, Kajbafvala A, et al. Rapid Formation of Mono-Dispersed Hydroxyapatite Nanorods with Narrow-Size Distribution via Microwave Irradiation. *J Am Ceram Soc*. 2008;91(11):3580-4. PubMed PMID: WOS:000261063600015.

[78] Iafisco M, Delgado-Lopez JM, Gomez-Morales J, Hernandez-Hernandez MA, Rodriguez-Ruiz I, Roveri N. Formation of calcium phosphates by vapour diffusion in highly concentrated ionic micro-droplets. *Crystal Research and Technology*. 2011;46(8):6.

[79] Gómez-Morales J, Delgado-López JM, Iafisco M, Hernández-Hernández A, Prat M. Amino Acidic Control of Calcium Phosphate Precipitation by Using the Vapor Diffusion Method in Microdroplets. *Cryst Growth Des*. 2011;11(11):4802-9.

[80] Iafisco M; Gomez-Morales, J.; Hernandez-Hernandez, M. A.; Garcia-Ruiz, J. M.; Roveri, N. Biomimetic Carbonate-Hydroxyapatite Nanocrystals Prepared by Vapor Diffusion. *Advanced Engineering Materials*. 2010;12(7):6.

[81] Phillips MJ, Darr JA, Luklinska ZB, Rehman I. Synthesis and characterization of nano-biomaterials with potential osteological applications. *J Mater Sci-Mater* M. 2003;14(10):875-82.

[82] Layrolle P, Lebugle A. Characterization and reactivity of nanosized calcium phosphates prepared in anhydrous ethanol. *Chem Mater*. 1994;6(11):1996-2004.

[83] Ye F, Guo HF, Zhang HJ. Biomimetic synthesis of oriented hydroxyapatite mediated by nonionic surfactants. *Nanotechnology*. 2008;19(24)..

[84] Iafisco M, Varoni E, Battistella E, Pietronave S, Prat M, Roveri N, et al. The cooperative effect of size and crystallinity degree on the resorption of biomimetic hydroxyapatite for soft tissue augmentation. *Int J Artif Organs*. 2010;33(11):765-74..

[85] Iafisco M, Sabatino P, Lesci IG, Prat M, Rimondini L, Roveri N. Conformational modifications of serum albumins adsorbed on different kinds of biomimetic hydroxyapatite nanocrystals. *Colloids Surf B Biointerfaces*. 2010;81(1):274-84.

[86] Iafisco M, Foggia MD, Bonora S, Prat M, Roveri N. Adsorption and spectroscopic characterization of lactoferrin on hydroxyapatite nanocrystals. *Dalton Trans.* 2011;40(4):820-7.

[87] Müller B. Tailoring biocompatibility: Benefitting patients. *Materials Today.* 2010;13(4):58.

[88] Uskoković V, Uskoković DP. Nanosized hydroxyapatite and other calcium phosphates: Chemistry of formation and application as drug and gene delivery agents. *Journal of Biomedical Materials Research Part B: Applied Biomaterials.* 2011;96B(1):152-91.

[89] Sakhno Y, Bertinetti L, Iafisco M, Tampieri A, Roveri N, Martra G. Surface Hydration and Cationic Sites of Nanohydroxyapatites with Amorphous or Crystalline Surfaces: A Comparative Study. *J Phys Chem C.* 2010;114(39):16640-8.

[90] Johnsson MS-A, Nancollas GH. The Role of Brushite and Octacalcium Phosphate in Apatite Formation. *Crit Rev Oral Biol Med.* 1992, 1992;3(1):61-82.

[91] Wang L, Nancollas GH. Calcium Orthophosphates: Crystallization and Dissolution. *Chem Rev.* 2008;108(11):4628-69.

[92] Pan HB, Darvell BW. Calcium Phosphate Solubility: The Need for Re-Evaluation. *Crystal Growth & Design.* 2008;9(2):639-45.

[93] Delgado-López JM, Iafisco M, Rodríguez I, Tampieri A, Prat M, Gómez-Morales J. Crystallization of bioinspired citrate-functionalized nanoapatite with tailored carbonate content. *Acta Biomater.* 2012;8(9):3491-9.

[94] Lazic S. Microcrystalline hydroxyapatite formation from alkaline solutions. *J Cryst Growth.* 1995;147(1-2):147-54..

[95] Vandecandelaere N, Rey C, Drouet C. Biomimetic apatite-based biomaterials: on the critical impact of synthesis and post-synthesis parameters. *Journal of Materials Science-Materials in Medicine.* 2012;23(11):2593-606..

[96] Gower LB. Biomimetic model systems for investigating the amorphous precursor pathway and its role in biomineralization. *Chem Rev.* 2008;108(11):4551-627.

[97] Cölfen H, Mann S. Higher-Order Organization by Mesoscale Self-Assembly and Transformation of Hybrid Nanostructures. *Angewandte Chemie International Edition.* 2003;42(21):2350-65.

[98] Xu A-W, Ma Y, Cölfen H. Biomimetic mineralization. *J Mater Chem.* 2007;17(5):415-49.

[99] Lu X, Leng Y. Theoretical analysis of calcium phosphate precipitation in simulated body fluid. *Biomaterials.* 2005;26(10):1097-108.

[100] Eanes ED, Gillessen IH, Posner AS. Intermediate States in the Precipitation of Hydroxyapatite. *Nature.* 1965;208(5008):365-7.

[101] Bienenstock A, Posner AS. Calculation of the X-Ray intensities from arrays of small crystallites of hydroxyapatite. *Archives of Biochemistry and Biophysics.* 1968;124(C):604-7.

[102] Posner AS, Betts F. Synthetic amorphous calcium phosphate and its relation to bone mineral structure. *Acc Chem Res.* 1975 1975/08/01;8(8):273-81.

[103] Melikhov IV. Crystallization behavior of nanodisperse phases. *Inorg Mater.* 2000;36(3):278-86.

[104] Legeros RZ, Trautz OR, Legeros JP, Klein E, Shirra WP. Apatite Crystallites: Effects of Carbonate on Morphology. *Science.* 1967, 1967;155(3768):1409-11.

[105] Delgado-López JM, Frison R, Cervellino A, Gómez-Morales J, Guagliardi A, Masciocchi N. Crystal Size, Morphology, and Growth Mechanism in Bio-Inspired Apatite Nanocrystals. *Adv Funct Mater.* 2014;24(8):1090-9.
[106] Underneath the Bragg Peaks: *Structural Analysis of Complex Materials.* Takeshi E, Simon JLB, editors: Pergamon; 2012.
[107] Debye P. Zerstreuung von Röntgenstrahlen. *Annalen der Physik.* 1915;351(6):809-23.
[108] Cervellino A, Giannini C, Guagliardi A. DEBUSSY: a Debye user system for nanocrystalline materials. *J Appl Crystallogr.* 2010;43(6):1543-7.
[109] Leventouri T, Bunaciu CE, Perdikatsis V. Neutron powder diffraction studies of silicon-substituted hydroxyapatite. *Biomaterials.* 2003;24(23):4205-11.
[110] Guagliardi A, Cedola A, Giannini C, Ladisa M, Cervellino A, Sorrentino A, et al. Debye function analysis and 2D imaging of nanoscaled engineered bone. *Biomaterials.* 2010;31(32):8289-98.
[111] López-Macipe A, Gómez-Morales J, Rodríguez-Clemente R. Nanosized hydroxyapatite precipitation from homogeneous calcium/citrate/phosphate solutions using microwave and conventional heating. *Adv Mater.* 1998;10(1):49.
[112] Rey C, Combes C, Christophe D, Grossin d. Bioactive Ceramics: Physical Chemistry. In: Ducheyne P, Healy KE, Hutmacher DW, Grainger DW, Kirkpatrick CJ, editors. *Comprehensive Biomaterials.* 1: Elsevier; 2011. p. 187-221.
[113] Tsuda H, Arends J. Orientational micro-raman spectroscopy on hydroxyapatite single-crystals and human enamel crystallites. *Journal of Dental Research.* 1994;73(11):1703-10..
[114] Markovic M, Fowler BO, Tung MS. Preparation and comprehensive characterization of a calcium hydroxyapatite reference material. *Journal of Research of the National Institute of Standards and Technology.* 2004;109(6):553-68.
[115] He Q, Liu X, Hu XM, Li SC, Wang HJ. Solid solution between lead fluorapatite and lead fluorvanadate apatite: mixing behavior, Raman feature and thermal expansivity. *Physics and Chemistry of Minerals.* 2011;38(10):741-52.
[116] Fleet ME, Liu X. Coupled substitution of type A and B carbonate in sodium-bearing apatite. *Biomaterials.* 2007;28(6):916-26..
[117] Grunenwald A, Keyser C, Sautereau A-M, Crubezy E, Ludes B, Drouet C. Novel contribution on the diagenetic physico-chemical features of bone and teeth minerals, as substrates for ancient DNA typing. *Analytical and Bioanalytical Chemistry.* 2014;(in press).
[118] Xie HN, Stevenson R, Stone N, Hernandez-Santana A, Faulds K, Graham D. Tracking Bisphosphonates through a 20 mm Thick Porcine Tissue by Using Surface-Enhanced Spatially Offset Raman Spectroscopy. *Angewandte Chemie-International Edition.* 2012;51(34):8509-11.
[119] Bertinetti L, Drouet C, Combes C, Rey C, Tampieri A, Coluccia S, et al. Surface Characteristics of Nanocrystalline Apatites: Effect of Mg Surface Enrichment on Morphology, Surface Hydration Species, and Cationic Environments. *Langmuir.* 2009 2009;25(10):5647-54.
[120] Burneau A, Barres O, Gallas JP, Lavalley JC. Comparative-study of the surface hydroxyl-groups of fumed and precipitated silicas .2. characterization by infrared-spectroscopy of the interactions with water. *Langmuir.* 1990;6(8):1364-72.
[121] Mc Dowell H, Gregory TM, Brown WE. *J Res Natl Bur Stand.* 1977;81A:273-81.

[122] Grynpas M. Crystallinity of bone-mineral. *Journal of Materials Science*. 1976;11(9):1691-6.

[123] Kühl G, Nebergall WH. *Z Anorg Allg Chem.* 1963;324:313.

[124] Eichert D, Drouet C, Sfihi H, Rey C, Combes C. Book Chapter: Nanocrystalline apatite based biomaterials: synthesis, processing and characterization. Book: *Biomaterials Research Advances: Nova Publisher;* 2008. p. 93-143.

[125] Legros R, Balmain N, Bonel G. Age-related changes in mineral of rat and bovine cortical bone. *Calcif Tissue Int.* 1987;41(3):137-44.

[126] Jemal M, Bencherifa A, Khattech I, Ntahomvukiye I. Standard enthalpies of formation and mixing of hydroxyapatites and fluorapatites. *Thermochimica Acta.* 199515;259(1):13-21.

[127] Jemal M. *Phosph Res Bull.* 2004;15.

[128] Ben Cherifa A, Jemal M. *Phosph Res Bull.* 2004;15.

[129] Bel Hadj Yahia F, Jemal M. Synthesis, structural analysis and thermochemistry of B-type carbonate apatites. *Thermochimica Acta.* 2010 6/10/;505(1–2):22-32.

[130] Khattech I, Jemal M. Thermochemistry of phosphate products .2. Standard enthalpies of formation and mixing of calcium and strontium fluorapatites. *Thermochimica Acta.* 1997;298(1-2):23-30.

[131] Khattech I, Jemal M. Thermochemistry of phosphate products .1. Standard enthalpy of formation of tristrontium phosphate and strontium chlorapatite. *Thermochimica Acta.* 1997;298(1-2):17-21.

[132] Ben Abdelkader S, Khattech I, Rey C, Jemal M. Synthesis, characterization and thermochemistry of calcium-magnesium hydroxyapatite and fluorapatite. *Thermochimica Acta.* 2001;376(1):25-36.

[133] Ben Cherifa A, Jemal M. Enthalpy of formation and mixing of calcium-cadmium chlorapatites. *Journal of Thermal Analysis and Calorimetry.* 2002;68(3):1035-44.

[134] Rollin-Martinet S, Navrotsky A, Champion E, Grossin D, Drouet C. Thermodynamic basis for evolution of apatite in calcified tissues. *The American Mineralogist.* 2013;98:2037-45.

In: Apatite
Editors: M. Iafisco and J. M. Delgado-López

ISBN: 978-1-63321-500-9

Chapter 3

BEHIND APATITE FORMATION: AN ATTEMPT AT CONCEPTUALIZING NATURE

Marianthi Manda-Guiba**, *Ioanna Stasinopoulou and Dimitris Christofilos
Physics Division, School of Technology, Aristotle University of Thessaloniki,
University Campus, Thessaloniki, Greece

ABSTRACT

The process of biomineralization in vertebrates has long been the leading paradigm for the implementation of new biomimetic strategies in the biomedical engineering field of hard (carbonate apatite-mineralized) tissue regeneration and the reconstruction of the organic-mineral tissue interfaces. By pursuing the principles which nature follows to treat itself, the pivotal significance of the hierarchically self-assembling processing route of the biological apatite building units, can be realized. However, while a substantial body of evidence on apatite mineralization has been accumulating over the years, the underlying mechanism still remains largely obscure, holding back the designing and manufacturing of new functional bioactive self-healing biomedical products. A thorough understanding of the apatite mineralization is imperative to stimulate the advance of bio-inspired treatments for tissue formation and healing. In this context, this chapter approaches the natural selection of apatite as the building block of bone and teeth in vertebrates; furthermore, the theories behind the apatite crystallization process are discussed and the experimental models that attempt to support the mechanism are addressed. Finally, the last part of the chapter focuses on the in vitro precipitation of apatite conditioned by bioactive materials; also a summary of the factors that exert control over apatite formation is presented.

Keywords: Biomineralization, Bioactivity, Apatite, Precursors, Silicates

* Physics Division, School of Technology, Aristotle University of Thessaloniki, University Campus, GR 54124, Thessaloniki, Greece; Email:mgmanta21@gmail.com.

INTRODUCTION

The structural hierarchy and self-assembling in a multi-scale design are the fundamental principles that nature follows in order to support the optimum functional and metabolic biological activity of its prototypes. Being part of the development and growth, this process has been adaptively recorded in the genome sequencing of living organisms, through dynamic and slow pathways of subtle and continuous sequential changes [1, 2]. The purpose of this natural selection is the evolution in the form of a healing mechanism, which copes with the divergent situations that restrict the possibility to survive [3]. The unique repeatability and the power-efficiency of nature in the conduction of the "experimental" procedures, makes it both gold standard and challenge for the science, given that the latter shares with nature the same goals [4]. However, what nature repeats in a systematically unique way, yet it is an issue to be inquired into or a subject of a trial-and-error undertaking in the scientific field; in fact, science had never so far reached the point of replacing nature by using conventional human-designed solutions made in harsh conditions. Therefore, it has been realized that, as opposed to just being compared to nature, attempting at uncovering and patenting the natural laws, getting inspired by and/or mimicking the nature itself, holds the truth that could stimulate the advance of science. In this context, bio-inspiration and bio-imitation have set new scientific disciplines and opened the gate to the emerging world of biomimetics, wherein science and engineering promote the process of interpretation and application of biological principles to materials and technologies [5, 6, 7, 8]. In a feat of successful transfer of natural concepts to several scientific fields (such as the self-cleaning lotus effect), biomimetics have drawn great attention in the field of biomedics too. Specifically, the implementation of new biomimetic strategies in the biomedical engineering field of hard (carbonate apatite-mineralized) tissue regeneration and the reconstruction of the organic-mineral tissue interfaces is one of the major goals that are being pursued. In this regard, the process of biomineralization has been the leading paradigm for the induction of bioactivity; the natural mineralized prototypes are used as archetypes that provide material science with a source of knowledge so as the principles of apatite precipitation and organization, to be revealed and established [9]. In its strict form, the biological mineralization refers to the inimitable process which organisms (bacteria, unicellular, plants, invertebrates and vertebrates) adopt to form inorganic structures as their functional parts in aqueous environments [10]. Even though, this is running over than 550 million years ago, how this actually happens and what processes are involved are still largely obscure [11]. From the amorphous silica of sponges and diatoms to the formation of dental enamel, the polymorphic phenotype of biological mineralization is certified and attributed to the biogenic setting of steps towards the organization of inorganic structures; from amorphous to highly crystalline [12]. This biogenic control divides biological mineralization into: a) biologically induced mineralization of bacteria, ancient and protist organisms which form random aggregations of crystals and b) biologically controlled mineralization of advanced organisms such as humans which achieve the configuration of spatially controlled crystal structures of high stability and favorable geometry [13, 14, 15, 16]. Despite the enigmatic underlying mechanism of the crystal deposition during biologically controlled mineralization, intensive research in the field has revealed that it is conditioned by a gene shortcut [17]; the deposition of the crystalline phase is determined indirectly from the organic matrix, while the dimensions, the shape and the orientation of crystals are associated with the type of organic macromolecules [18, 19, 20, 21]. Specifically, the induction of mineralization is a self-assembled phenomenon, during which the organic macromolecules function as a scaffold or template, by attracting ions and acting as nucleation and growth areas [22, 23, 24, 25, 26, 27, 28].

Apatite in Vertebrates

Structure and Properties

Natural selection has led mineralization in different life forms in such a way that its crystal -chemical variation to reflect the biological adaptation to environmental requirements. For this reason, a wide spectrum of natural mineral phases has been identified serving different purposes in different environments. For example, iron oxides produced by magnetotactic bacteria, allow them to move along the geomagnetic field of earth [29], silicate minerals synthesized in sponges help them to construct part of their skeleton [30], and mineral polymorphs of calcium carbonate (aragonite or calcite) allow invertebrates to secrete different types of exoskeletons [31, 32]. Similarly, ionic calcium orthophosphate substitutes found in vertebrates constitute the principal inorganic building block for bone and teeth [33] [34, 35]; Anhydrous Calcium Hydrate Phosphate or Monetite (DCPA), Dicalcium Phosphate Dihydrate (DCPD), Tricalcium Phosphate (TCP), Octacalcium Phosphate (OCP), Hydroxyapatite (HAp), and Fluoroapatite (FAp), have been identified as the calcium orthophosphate salts of mineralized skeleton of vertebrates. The distinction between the different orthophosphate phases of calcium is based on calcium/phosphorus ratio [36]. Particularly, in mineralized tissue of human species (bone, enamel, dentin, cementum, calcified tendon and cartilage), the crystal structure of calcium orthophosphate found, is hydroxyapatite (Figure 1); usually termed as biological carbonate hydroxyapatite (cHAp), it abstains from the typical formula of the stoichiometric apatite {Ca10 (PO_4) $6OH^2$}, due to the variability in the impurities [37]. The question arises as what are the differences between the two forms of apatite and what is the rationale of its natural selection as a fundamental component in the construction of the skeleton and teeth of vertebrates? The answer lies in the diversity of physico-chemical properties of apatite, which come from the flexibility of the crystallographic structure and define the functionality of the mineralized hard tissues of vertebrates in terms of structural integrity and metabolic activity [38]. The structure of stoichiometric apatite displays the Ca/P ratio of 1.67 and can be represented as a three-dimensional network of PO_4- tetrahedra, which are interconnected by columns of (9-fold coordinated) Ca1 atoms. The channels passing through the network have axes that coincide with the six-fold screw axes and contain the triangles (7-fold coordinated) of Ca atoms [39]. The channels, which are formed by the triangles of Ca, are the crystallographic positions, which are occupied mostly by monovalent anions OH-, F- or Cl- [39]. The symmetry of the apatitic crystal depends on the type of occupancy of the channels. More specifically, the presence of small F- ion keeps the symmetry of the crystal in its hexagonal arrangement (P63 / m), in contrast to non- spherical - OH, the presence of which disturbs the symmetry of the crystals and interprets the crystallization of HA to its monoclinic phase P21 / b, which is less symmetrical [40, 41].

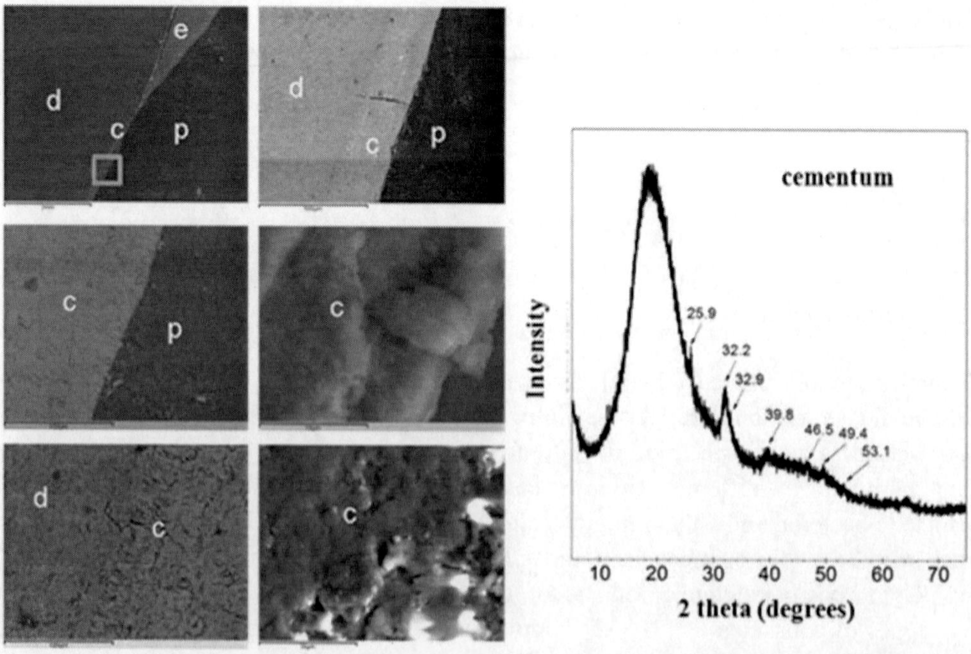

Figure 1. SEM images and X-Ray diffractogram of human dental apatite (d: dentin, e: enamel, c: cementum, p: periodontium).

The crystallographic flexibility of apatite is based on the ability of its lattice to function as part of a system that exchanges ions with the environment and refers to a process of maturation; a structured aqueous layer mediates between the surface of apatite crystals and its environment [42]. The aqueous layer is a pool of weakly interconnected metal, mostly divalent ions, which can easily be replaced by other ions of the solution, thus establishing equilibrium of exchange [43]. The corresponding equilibrium constants depend on the nature and the mobility of ions, the composition of apatite and maturation phase; the dynamics of the system creates a state of partial or total ionic substitution in several crystallographic positions of the apatitic lattice, not only resulting in the gradual spread of apatitic phase towards the environment, but also in the development of incomplete structures in terms of stoichiometry, with divergent Ca/P ratio (Figure 2) [44]. These structures represent the biological apatite and are best described by the following formulas: $Ca_{9.4\ 0.6}(PO_4)_{5.4}(HPO_4$ and $CO_3)_{0.6}(OH$ and ½ $CO_3)_{1.4\ 0.6}$ or $Ca_{8.3\ 1.7}$ $(PO_4)_{4.3}(HPO_4$ and $CO_3)_{1.7}(OH$ and/or ½ $CO_3)_{0.3\ 1.7}$ [45]. The ion substitution that drives the formation of biological apatite, usually termed as substitution B, is primarily related to the replacement of PO_4^{-3}- tetrahedral by the CO_3^{-2}- triangles. Notwithstanding, quite often, carbonate ions (CO_3^{-2}) substitute OH- ions in the lattice, in combination with other substitution ions (Mg^{2+}, Sr^{2+}, F^-, Cl^-, citrates and HPO_4^{-2}) [46, 47]. The aforementioned substitution ions are offset in the apatite lattice through a mechanism for the maintenance of the balance of electrical loads, by inducing gaps in the lattice and/or surrogate conjunctions. Although these compensatory phenomena barely alter the structure of the apatite, they generate a field of increased stresses, deformations and reactivity [48, 49], affecting the final properties of apatite; namely, the solubility, the hardness, the brittleness, the deformation, the thermal stability, and the refractivity [50, 51]. The variation of the apatite properties caused by carbonate ions may be due to: a) the variation of the strength of interatomic bonds; b) the diversification of the morphology and the dimensions during crystal growth; and c) the degree of hydroxylation. Namely, carbonate ions are strongly associated with the apatite crystal growth; since the type B substitution of phosphates implies a reduction of *a* and an increase of *c* parameter, the incorporation of carbonate ions into the apatitic lattice results in inhibition of crystal growth [43]. The content of apatite crystals in carbonate ions is related to the form of crystals too; the shape of the crystal fluctuates from the type of flat platelet to needle-shaped platelet, particle platelet, and small platelet and finally to the type of spherical particle according to occupied areas by the carbonate ions. This tendency is also associated with the progressively increasing controlled solubility of crystals [52]. In addition, the reduced crystal growth and the high level of atomic deformation, due to type B substitution, make the reincorporation of OH ions in the lattice energetically unfavorable; in this way, the appearance of gaps resulting from the removal of OH ions offsets the electrical imbalance and further increases solubility.

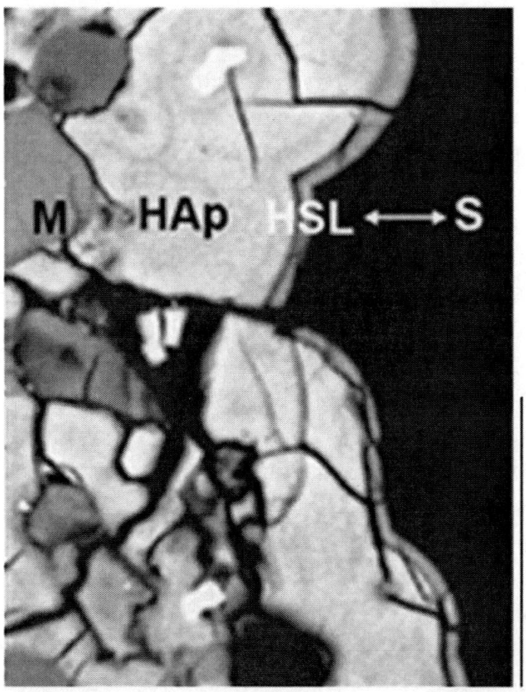

Figure 2. SEM image imprinting the Hap development towards the environment through the mediation of a less mature structure (M: material, HAp: apatite, HSL: hydrated structured layer, S: solution).

Furthermore, as opposed to the $Ca\text{-}PO_4$ bonds locating in the apatitic lattice, the generation of weakened $Ca\text{-}CO_3$ bonds after B substitution, also explains the higher solubility of biological apatite in acidic environments [50]. This interaction between the properties of apatite lattice with the structure and the chemical composition is reflected by the differences in the apatitic phases, which appear in the bone and the teeth of humans [53]. Specifically, as the organic component of mineralized tissues acts as an important regulator of crystal lattice, the nearly double carbon concentration in the environment of bone apatite compared to the enamel apatite, supports the dominance of B substitution in bone; this is outlined by the size of the crystals, that can be 10 to 100 times smaller than those of the enamel. In this way, due to the absence of carbonate ions in enamel, the reduction of non-apatite phases because of the growth of apatite structures is enhanced and accompanied by greater crystal growth and formation; the chemical formula of these crystals approximates the stoichiometric one. On the basis of the increased carbon ion concentration in bone environment, the plat-like morphology of bone apatite could be explained too, as a result of the effective interaction with the collagen fibrils. On the other hand, the rod-like crystals of enamel are associated with a small portion of the organic matrix. Hence, the presence of a high level of lattice vacancy, resulting from the removal of OH ions, in combination with the small dimensions of bone crystals and the increased solubility of the bone compared to the enamel apatite is related to the capacity of the bone to be absorbed and reformed during remodeling as opposed to the property of enamel to resist dissolution.

Theories of Apatite Formation

The formation of highly organized crystalline structures such as apatite is a solidification process (crystallization) of soluble metal elements in a supersaturated environment. Either viewed as the end-product of an equilibrium system or even as a long-lived metastable phase [54], apatite crystal formation passes through a chemical route, the characterization of which remains both quite controversial and provocative. Given that nature is inherently metastable, the interpretations of the underlying mechanisms and the likely phase-identification in a pre-crystalline stage of apatite formation become warrant. The theories and the experimental models which attempt to support and to standardize the process of apatite crystallization focus on: a) the investigation of reactions in terms of kinetic and thermodynamic aspects; and b) probing the correlation of apatitic nucleation and growth with several parameters such as the thermodynamic driving force, the pH and the supersaturation of the solution. The results of these experimental procedures can be summarized in Gibbs thermodynamic results and the development of the Classic Theory of Nucleation (CTN), and in the theory of Ostwald (Ostwald Step Rule Theory: OSRT) [55]. In conventional CNT model, the nucleation initiates from the formation of a critical sized cluster that grows through ion-by-ion deposition into the crystal phase [56]; no stable intermediates between the amorphous and the crystal phase are suggested and the discontinuous transition of one phase into another occurs upon overcoming a free-energy barrier [57]. Instead, Ostwald suggests the presence of metastable interim states, considering that when a system is in a metastable state, it is necessarily transformed into the most thermodynamically favorable phase. The OSRT is the basic model proposed for the interpretation of the early stages of apatite crystallization; described by two separate processes, OSRT involves: a) the nucleation coming from the formation of stable atomic or molecular aggregates; and b) the multiphase transformation into more mature crystalline structures. Thus, according to Ostwald, crystallization is a multiphase rather than a single phase process, during which, a sequence of amorphous precipitations and molecule precursors precede the stable, mature crystal, jumping from a state of high to a state of low free energy of the crystal lattice. The available energy space (energy landscape), which represents the energy control of interfacial energy and the chemical potential of crystallization, stimulates the transformation by reducing (activating) the thresholds of interfacial energy [58]. The molecule precursors utilize the available energy, while providing the necessary components to move to the next stage and to the final stage of maturation. However, OSRT cannot be considered to be a universal law; it is only a possible tendency that exists in nature, interpreted on the basis of reverse thermodynamics or combined statistical thermodynamics and structural variations with temperature. Thus, reasonable scientific questions and controversies arise, focusing on whether the transformation of phases during the crystallization and the OSRT constitute a generalized strategy, which comes from nature [59] [60]; in this regard, the intermediate apatite formation theory is disputed by the *de novo* apatite formation. The line of reasoning, both for and against, is based on: a) *in vivo* studies investigating the biological mineralization in invertebrates and vertebrates; and b) *in vitro* experiments searching the transformation of phases and the crystal growth; however, even the same data are several times interpreted in a different way. The theory of *de novo* apatite suggests that bone crystallization takes place through Ca- imperfect apatite crystals, the Ca-deficiency of which decreases during maturation. The arguments for this theory are build up on the data of several studies that failed to identify amorphous phases in the early stages of crystallization, either *in vitro* or *in vivo* [61, 62, 63]. However, it must be noted that these results cannot eliminate the possibility of precursor phases with certainty; this is because the results of studies are always associated with the lack of sensitivity of the cutting edge technologies to identify and characterize these phases. In particular, studies that aim to identify the intermediate phases suffer from deficiencies concerning the nature of these phases; but also the ability of microscopic and spectroscopic technologies to recognize intermediate stages is limited. Specifically, the thermodynamic instability of the intermediate phases and the rapid transformation within minutes or hours contrast with the time-consuming sample preparation procedure, which is required for the use of certain technologies (SEM, XRD, FTIR, NMR). Furthermore, the deformed amorphous nature of the intermediate phases makes them isotropic in polarized light and non- identifiable in XRD [64]. Another argument that could discredit the value of the *de novo* apatite formation is the variability of the spectral measurements; however, the supporters of this theory ascribe the variability to the large crystalline defects, the lack of stoichiometry, the presence of impurities, and the high proportion of crystalline surface-to-volume of the newly formed crystals, in the surface of which the population of non-apatite phosphate groups adheres, implying the misidentification. A substantial body of evidence on the route of apatite crystallization in vertebrates comes from studies that investigate crystal formation in invertebrates. Despite the differences between the various genders in different branches of the animal phylogenetic tree, these studies argue that biological mineralization by forming crystalline precursor phases is a generalized selection process among species [65]. In this regard, it is suggested that the identification of amorphous phases in the early mineralization-stages of invertebrates is indicative of a generalized mineralization process occurring in vertebrates too [66, 67, 68, 69, 70]. The amorphous ferrihydrite coming from the biogenic magnetite in the teeth of chitons (crustacean), the calcite and aragonite formed from the amorphous calcium carbonate in the shells of mollusks and the spine of echinoderms are only some examples in support of the precursor scenario during the biomineralization in invertebrates [71]. On the other hand, in vertebrates, scientific documentation referring to the pre-crystallic stage during the formation of bone and dental tissues is more limited. It was first according to Brown (1962), that OCP was proposed as the intermediate phase towards the crystallization to apatite; his theory was based on the crystallographic similarities between the two structures [72]. Later, Termine and Posner (1967) set up the theory of ACP as a precursor apatite phase during mineralization, supported by spectral differences between the bone of rabbit femurs and the crystalline apatite [73]. The lower reflected intensity of x-rays along with the low Ca/P ratio of bone were assigned by the Posner team at the contribution of an amorphous phase, which has the same characteristics as the synthetic ACP [74]. The *in vivo* studies, which support the multiphase theory in vertebrates, are based on observations on bone and dental tissue maturation, related to the age or the physiological condition. These studies contend that the formation of the apatite is the result of a progressive transformation of a deformed colloidal phase, which favors the direction and the formation of the mature phase. The study of Crane (2006) is a strong evidence for the presence of generalized natural selection in the multiphase biological mineralization in vertebrates; in this study, Raman spectroscopy was used to determine the intermediate phases in cranial sutures of muscles. The Raman spectra, which ensued from the tissues without previous specimen preparation, show features indicative of OCP during the early stages of the experiment [64]. The model precursor molecules-to-apatite was further confirmed by more recent studies of Mahamid on the dynamic formation of crystals during the development of the radii of the bone flaps of zebrafish and the enamel of the mandible incisors in mice [75, 76, 77]. Both studies suggest the presence of ACP in the early stages of crystallization (Figure 3); through these studies, Mahamid was provided with the proper dataset to support the intracellular formation of apatite-precursors inside granules and their infiltration into the collagen fibrils, which function as nucleation sites for the initiation of crystallization. Similarly, based on the results of high-TEM analysis in tilapia fish, the group of Okuda (2009) proposed the mechanism of preferential storage of ACP in gaps of the structure of collagen fibrils [78]. The credibility of theory of precursor molecules has recently been increased by the development of *in vitro* biological systems. In this context, information on apatite mineralization is being quarried by studying the putative transformation of the precursor phases to apatite through matrix-mineral interactions. Organic matrix in living organisms is an enriched-in-signals pool, associated with the regulation of mineralization in terms of nucleation and growth as well as crystal structure. Collagen fibrils and non-collagenous proteins (NCPs), that are known to get involved in bone and tooth formation, provide an intriguing signaling template, wherein mineralization takes place. In this regard, first, Olszta et al. (2007) proposed the formation of apatite through the development of high-water-content amorphous precursor phases in nano-structures of collagen fibrils systems [79]. The also known as polymer-induced liquid precursor (PILP) theory of Olszta highlights the role of polyanionic peptides in mineralization. Mimicking the aspartic acid proteins in natural matrix, polyaspartate is proposed to stabilize the ACP phases; apatite post-formation follows after ACP infiltration into collagen fibrils due to capillarity [79] or charge interactions [80, 81, 82]. Detection of transient mineral phases has also been reported in other *in vitro* studies investigating collagen mineralization directly mediated by specific NCPs found in organic matrix; dentin matrix protein 1, dentin phosphophoryn, amelogenin, fetuin have been related to the modulation of bone and tooth formation through the precursor pathway. Stabilization of precursors, crystal nucleation, crystal shape and size control, and mineralization inhibition by in fact promoting intrafibrillar mineralization are the pivotal roles that multifunctional NCPs are hypothesized to play; also depended on their concentration, degree of phosphorylation and the *in vitro* substrate [83]. In particular, the role of the ACP stabilization in apatite mineralization was further studied by Nudelman et al. (2012) reporting that the time-related formation and stability of ACP are associated to the apatite mineralization [84].

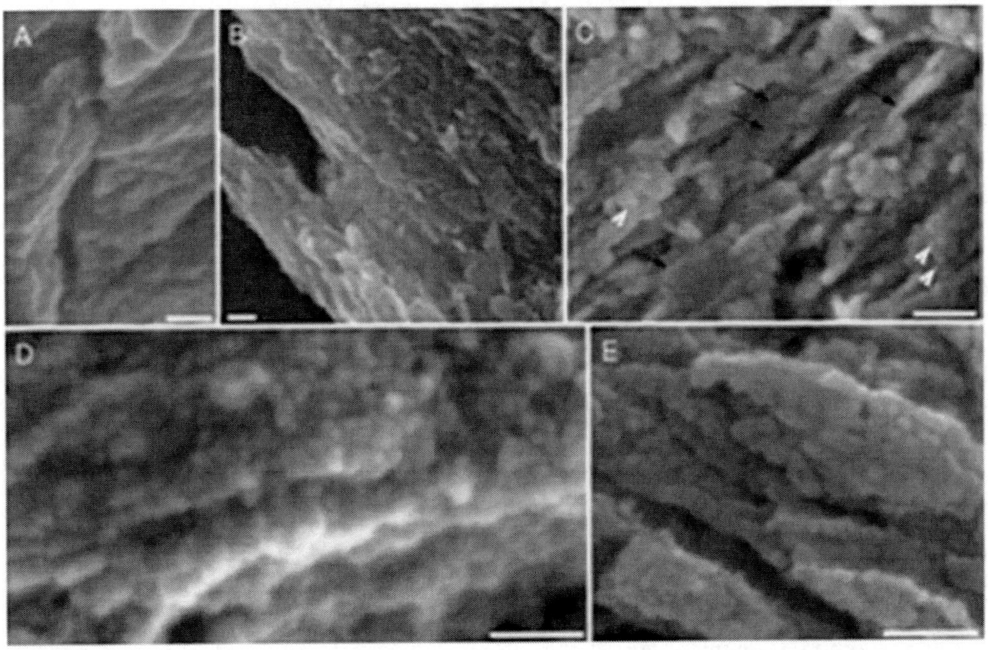

Figure 3. SEM images of mineral particle morphologies in different parts along the growing fin bony rays, representing different growth stages (Mahamid J et al. PNAS. 2008;105:12748-53.).

However, the orchestration of apatite mineralization is not a single event scenario or an individual's biomolecule operation; instead, in the native spatiotemporal environment of extracellular matrix, a cascade of signals and interactions may efficiently define the structure and the functions of organic matrix. In an attempt to overcome the inherent simplicity of the aforementioned *in vitro* models, demineralized tissue models have been employed as biomimetic analogs; sufficiently retaining the content, ratio and patterns of features found in the original tissues, these models provide a substrate that closer resembles the native organic matrix and increases the validity of the observations during the remineralization process [85]. Indeed, as the formation of ACP has been suggested to proceed apatite formation, Wang et al. (2013) further emphasized the key role of amorphous phases on the initiation of biomineralization by using a demineralized dentin biomimetic analog, wherein ACP was stabilized in a dose-depended manner by polyacrylic acid [86]; Lausch et al. (2013) associated time-related ACP formation with the apatite remineralization of demineralized mouse periodontium in metastable calcium phosphate solutions stabilized by pAsp [87]; as to mimicking bone formation, Manatee rib demineralized bone sections were used by Thula et al. (2011), enabling the remineralization of bone structure [88]. Overall, the formation of either *de novo* or amorphous apatite phases during the biomineralization in nature currently is under ongoing research, attempting to establish a major pool of data. Tracing the development of this research, it is clear that the elucidation of the biologically controlled mineralization of the organisms can be the natural guide to the preparation of clinically-effective bioactive materials.

In that scope, this chapter also attempts to clarify the *in vitro* apatite precipitation conditioned by bioactive materials, focusing on the example of silicates.

BIOACTIVITY: THE EXAMPLE OF SILICATES

The *in vitro* apatite formation was first identified by Clark (1976) [89] and later on it was further studied with the use of different analytical techniques [90, 91, 92, 93, 94]. Based on the results of the above experiments, the process of apatite precipitation takes place at the interface of material/solution via dissolution, proton formation and nucleation of the surface; hydrolysis and partial dissolution of the silica network occurs, followed by release of modifying cations and soluble silica in the form of $Si(OH)_4$ in the solution. This decay of Si-O-Si bonds results in the formation of silanols (Si-OH), which condense and repolymerize to a hydroxylated SiO_2-layer, enabling the precipitation of Ca/P. This initial ionic surface activity is critical for the crystallization of carbonate hydroxyapatite (Figure 4) [95]. However, despite the documented significance of the silicate network of the glassy phase (silanols, soluble silicon ions and soluble silica oligomers) for the manifestation of bioactivity, the suggested mechanism does not provide sufficient data for the assignation of the thermodynamic and the kinetic basis of the supposed stages of the overall mechanism [96]. The exact mechanisms by which the implant surfaces promote the precipitation of apatite is not fully known due to the difficulty of characterization of the early and unstable nano or micro phases and the inability of the experimental techniques to identify the structural and dynamic properties of materials at the atomic level [97]. The former hinders: a) the comprehension of the process by which simpler glasses react and dissolve in an aqueous medium; and b) the identification of the specific active regions in the silicate network, which are related to the manifestation of bioactivity. One more discerning interpretation of bioactivity is now achieved through classical and ab initio computer simulations, which recently have begun to reveal the individual characteristics of bioactive materials, verifying or rejecting the hypothetical conclusions of previous micro -macro *in vitro* experiments [98].

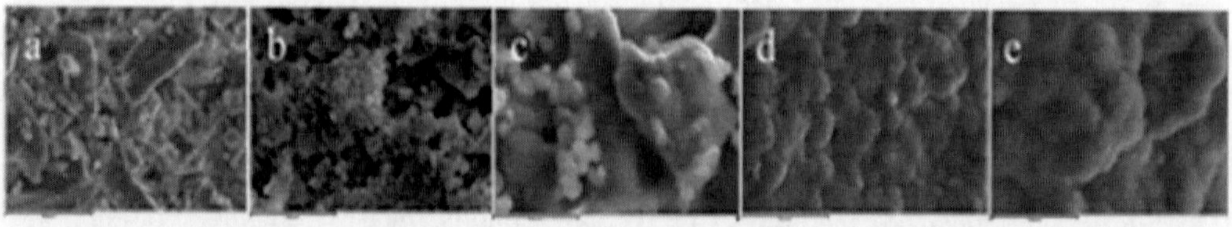

Figure 4. SEM images of *in vitro* apatite precipitation on the surface of bioglass 58S sintered at 930° C after immersion in DMEM solution for 6 hours to 18 days. (a: control, b: 6 hours, c: 1 days, d: 6 days, e: 18 days).

The formation of either *de novo* or amorphous apatite phases in the surface of bioactive silicate glasses or glass-ceramics is a thermodynamically driven phenomenon, the conduction of which requires the formation of initial nuclei in favorable regions [99]. The nucleation not only is a required condition for further crystalline increase, but it also determines the dimensions, the shape, the type, and the aggregation properties of crystals [100]. Taking place at the interface substrate/solution, this process initiates from the collision and random grouping of appropriate ions and free radicals to form a fetus [101, 102]. The unstable and soluble fetus is transformed into a stable and insoluble core, the structural units of which are adsorbed in the substrate [103]. Accordingly, the driving forces of the catalytic capacity of the substrate either to slow down or to accelerate the reaction and the supersaturation, determine the formation of nuclei and the relationship of the growing crystal with the substrate [104] [105, 106]. The natural role of the substrate in the heterogeneous nucleation is associated with the structure of the surface, the surface charge and the relevance of its network with this of the growing crystal in order the adsorption of the initial core to be achieved [107]. On the other hand, the thermodynamic role of the substrate is twofold; it induces the reduction of the free energy, accelerating the formation of the critical nucleus, compared to the homogeneous nucleation, while it reduces the active surface of collision of the structural units, decelerating the kinetics of reactions [108]. These inconsistent states overlap one another in different conditions of supersaturation, resulting in a different arrangement of the developing crystals in relation to the substrate [109]. Specifically, in high supersaturation conditions, nucleation occurs randomly in the body of the liquid phase, which greatly reduces the adsorption and structural synergy between the substrate and the crystals. Since, the activation energy of the nucleation is rather low, the active surface of collision becomes the dominant nucleating agent resulting in the formation of crystal with undefined orientation. In contrast, in low supersaturation conditions, wherein the activation energy is high, the development of well-oriented crystals in the surface of the substrate is promoted [110].

The formation of the active sites of *in vitro* nucleation for the apatite crystal formation and, therefore, the manifestation of bioactivity depend indirectly on the degree of solubility of the surface and directly on the silica-enriched layer, which is formed on the surface of the substrate. The solubility of the bioceramic material arises from the formation of silanols during: a) the detachment of cations from non-bridging oxygen atoms of Si-O- NBO (NBO: non-bridging oxygen atoms); b) the hydrolysis of Si-O-Si; and c) the detachment of silica parts as a result of the first mechanism [111]. The importance of the initial dissolution of the glass phase of bioactive materials for the bioactivity has been reported in a study of Arcos et al. (2003); the manifestation of bioactivity appears to be related to the activation energy required for the release of Si to the solution, which depends on the network connectivity (NC), the number and the mobility of silicate chains and the rings of bioceramic material [93]. The network connectivity is represented by the number NC, which corresponds to the average of oxygen atoms, which are linked to the Si atom, according to Hill hypothesis (Bridging oxygen: BO) [112]. According to this hypothesis, this number differentiates the bioactive from the biocompatible ceramics; if NC> 3, the ceramic is biocompatible [112] [113]. The incorporation of alkali and alkaline earth metals in silica matrix results in the reduction of bridging oxygen atoms, due to the network degradation, and in the increase of non-bridging oxygens, which are linked to these alkali elements [114]. Just because the alkali elements do not belong to the silica matrix, they actually facilitate their ionic exchange in the Si-O- NBO bonds with cations that are present in the solution. This results in an increase of the pH of solution, which favors the hydrolysis of Si-O to silanols. The ion exchange and the dissolution of the silica matrix requires a minimum number of non-bridging oxygens in order for a ceramic to be considered as bioactive; this number is expressed by the ratio of Si- O-NBO/Si-O-BO (BO: bridging oxygens), which must be greater than 1. Ceramics with a ratio lower than 1, appear an extremely decreased solubility [114]. The presence of Si-O-NBO is related to the mechanism responsible for the pH increase; moreover, due to the ionic exchange, it also participates in the increase of solubility through the formation of silanols. In detail, IR spectra document the presence of 2 types of Si-O- NBO (Si-O-1NBO and Si-O-2NBO); Si-O-2NBO seems to be related to the solubility through hydrolysis, by facilitating the absorption of water according to the theory of Helms & Poindexter (1994) [115].

However, the network connectivity hypothesis applies only to bioactive ceramic materials, produced by rapid fusion (such as 45S5). Instead, bioactive ceramics, derived from the technique of colloid gel, although they appear NC> 3, which corresponds to the NC of the 60S bioactive glass derived from fusion, exhibit bioactivity, which corresponds to 45S5. This fact is further confirmed by the $E\alpha_{Si}$ [93]. The increase of the bioactivity in these cases is due to factors related to the porosity and to the increase of the surface area, which further entails an increase of silanol groups. Another important parameter, which affects the solubility of ceramic and its bioactivity, is the type and number of the silicate structures that are present in the surface. Particularly, in 45S5, a critical number of silicate chains Q2 and Q3 has been identified (Qn: where Q represents the atoms of Si, which are linked to n C atoms), which does not appear in the biocompatible ceramics, the surface of which is dominated by silicate rings [116]. This justifies the fact that the Si atoms exhibit a smaller number of bridged oxygen atoms in silicate chains, compared to the rings, resulting in smaller energy requirements for the disruption of covalent Si-O bonds [117]. Moreover, for ceramics, which are derived from the sol-gel technique, when different temperatures of stabilization are selected during the process of their preparation, in conjunction with the glass chemical composition, a conversion of these silicate chains of surface into silica rings is observed; this leads to the stabilization of the glass and to the reduction of its solubility. This phenomenon is also reflected in the $E\alpha_{Si}$ value. Therefore, according to the Arcos hypothesis, the distinction between bioactive and bioactive ceramic materials can be more reliably determined by the $E\alpha_{Si}$ value; early studies showed that when EaSi <0.35 eV, the ceramic is bioactive [93]. As shown above, the structure of the silica matrix affects in a decisive way the solubility and the manifestation of the bioactivity of the ceramic. This means that any change in the proportion of the amorphous silica matrix and the growing crystals at the surface could inhibit or enhance the formation of apatite [118]. Specifically, while the solubility of the amorphous bioactive ceramics occurs at an homogeneous manner, the solubility of glass-crystalline bioactive ceramics varies in relation to their crystallinity, in such a way that crystals with different chemical composition and crystallographic orientation to exhibit different rates of solubility [119]. El Ghannam et al. (2001) studied the effect of heat-treating the transformation of the phases of the bioactive glass (45S5) and the association between this transformation, the solubility and the bioactivity [120]. According to the results of the study, when the bioactive glass is progressively heated, the glassy phase undergoes, at low temperatures, a biphasic separation (glass-in-glass phase separation), while at higher temperatures it crystallizes. The separation of the glass matrix is due to the tendency of P and Si ions to be concentrated in different phases, while the crystallization conduces to the development of non- phosphate phases characterized by strong bonds; the latter leaves one glassy matrix rich in phosphorus [121, 122]. Both situations induce the reduction of Ca and Si ions release and of the solubility, which is accompanied by reduced bioactivity. In addition, by comparing the bioactivity of bioactive ceramics with different crystalline phases, Peitl et al. (1996) reported that the bioactivity index of non-phosphate phase NaCaSiO is greater than this of the crystalline apatitic phases, which are formed in the bioactive glass when exposed in protracted temperatures, while the release of phosphorus from the glassy matrix promotes the precipitation of carbonate hydroxyapatite from the solution at rates similar to them of the parent bioactive glass [90].

Following the dissolution, the condensation and the repolymerization of the surface silanol groups result in the formation of 2-, 3- and 4-silicate rings, which have been identified as active regions for the manifestation of bioactivity [96, 123, 124]. The 3-rings of silanol ($Si_3O_6H_6$) constitute the areas, which are primarily responsible for the nucleation and the growth of apatite crystals in an aqueous medium with approximately neutral pH [125]. The characteristic, which has been proposed as the responsible variable for the manifestation of bioactivity in areas of $Si_3O_6H_6$, is the arrangement of Si and O atoms [98]. The siliceous 4- and 2-open rings with cubic structure are relatively more stable than the 3-rings, fact that makes them secondary "candidates" for the manifestation of bioactivity [123]. In contrast, the 3- rings are relatively more easily destabilized because of the deformation energy -due to their structure- and as a result, they can be hydrolyzed within 24 hours [126, 127, 128]. Furthermore, the O atoms of silanol are arranged into an equilateral triangle structure; an angle of 60° is formed between each pair of them. This angle corresponds to the angle formed between the atoms of O which are associated with the Ca in the { 0 1 0 } surface of hydroxyapatite, providing an ideal stereochemical correlation between the two compounds, which favors the apatite nucleation at these areas [125]. Finally, beyond the mechanisms of structural and stereochemical correlation, the electrostatic interpretation of nucleation considers that the proton formation in the surface, which is in contact with aqueous solutions, and the formation of silanols result in the formation of apatite nuclei by forming a negatively charged surface; this surface electrostatically attracts the positively charged ions of Ca, forming an inner-Ca spherical cluster [129]. The surface of silica gel displays a high negative charge, since the pH value, in which the charge is set equal to zero, is pHPZC (pHPZC: pH of Point Zero Charge) which is equal to 3-4. Since the intensity of the negative charge increases with the increase of pH, the silica gel is extremely negative during nucleation, when pH is equal to 7 [123]. The significance of the negative charge of silica gel is confirmed by studies, which indicate that the negatively charged surfaces are always favorable for the heterogeneous nucleation of apatite crystals, while the positively charged ones inhibit it [130].

Here it should be noted that the extensively investigated *in vitro* behavior of the silicate bioactive glasses and glass-ceramics will always be hampered by the complexity of the *in vivo* environment. From the hydrolysis to the formation of the apatitic surface layer, the behavior of the silicate substrates under *in vivo* conditions is the trigger for a molecular sequence [131] that activates mechanisms in genetic and cellular level [132]. Although these mechanisms have not been fully investigated, it appears that the effect of the bioactive glasses and glas-ceramics in the regulation of cell cycle is achieved through the involvement of the aforementioned surface factors, such as the chemical composition and topography, along with factors related to the type and dissolution rate of specific ions, which act in the solution [133]. Specifically, the bioactive glasses and glass-ceramics develop an extracellular action; the negatively charged silica gel in the surface presents favorable sites for the chemical absorption of growth factors and develops a direct intracellular action; the release of Ca and Si regulates gene expression [134]. Xynos et al. (2000) in Hench's team have published a series of articles, defining the specific ratio and the minimum concentration of Si (15-30ppm) and Ca (60-90ppm) ions [135, 136] as the main stimulus for the activation of the cellular mechanism. Studying 1176 genes using *in vitro* cell models of osteoblasts and stem cells, they report that the Si and Ca ions activate/regulate seven families of genes: a. transcription factors and cell cycle regulators; b. proteins of DNA reconstruction; c. regulators of apoptosis; d. growth factors and cytokines; e. antigens and cell surface receptors; f. transduction molecules; and g. compounds of extracellular matrix [137]. In that way, the cell cycle, mitosis and differentiation is regulated towards fast tissue regeneration [138].

Specifically, it is mentioned that cell proliferation is achieved by reducing the transition time of the cell cycle from G1 and S phases to G2 phase; cells not being in the proper phase of the cell cycle, are led to apoptosis [139]. The aforementioned process, from the hydrolysis of the bioactive glass to the formation of the apatitic film on the surface, is completed in a framework of 5 stages, which lasts 24 hours. These stages are conducted simultaneously with the activation of wound healing, during which plasma proteins are absorbed directly from the surface and are recognized by the integrin receptors of most cells (step 6); clot formation and phagocytic activity begin after 24 hours (step 7) [140]. The adhesion of mesenchymal stem cells (step 8) begins after 24-48 hours followed by differentiation into mature cells (stage 9) These cells secrete the organic portion of the extracellular matrix (step 10) and crystallization takes place through the formation of apatite [141].

CONCLUSION

The natural biomineralization towards bone and dental tissues formation differs significantly from precipitation of apatite crystals conditioned by material substrates. Thus, pursuing clinically-relevant strategies in the biomedical engineering field of hard tissue regeneration is a challenging undertaking, hampered by the complexity of recreating microenvironments encompassing all the features of natural biomineralization. Even if a substantial body of evidence on apatite mineralization has been accumulating over the years, the underlying mechanism still remains largely obscure, holding back the designing and manufacturing of new functional bioactive self-healing biomedical products. This urges for a functional bridging the gap between the nature and the laboratory in order to create strategies that could be implemented in clinical reality.

REFERENCES

[1] Sarikaya, M., Tamerler, C., Jen, A.K., Schulten, K., Baneyx, F., (2003). Molecular biomimetics: nanotechnology through biology. *Nat. Mater.* 2, 577-585.

[2] Cranford, S., Buehler, M.J., (2010). Materiomics: biological protein materials, from nano to macro. *Nanotech. Sci. Appl.* 3, 127-148.

[3] Ackbarow, T., Buehler, M.J., (2008). Hierarchical Coexistence of Universality and Diversity Controls Robustness and Multi-Functionality in Protein Materials. *J. Comput. Theor. Nanosci.* 5, 1193-1204.

[4] Reis, R.L., (2003). Biomimetics. *Curr. Opinion. Solid. State. Mater. Sci.* 7, 263–264.

[5] Bar-Cohen, Y., (2005). Biomimetics: mimicking and inspired-by biology. *Proceedings of the SPIE Smart Structures Conference, San Diego, CA.SPIE. 5759-02, 1-8, DOI:* 10.1117/12.597436.

[6] Vincent, J.F.V., Bogatyreva, O.A., Bogatyrev, N.R., (2006). Biomimetics: its practice and theory. *J. R. Soc. Interface.* 3, 471-482.

[7] Vincent, B., (2009). Biomimetic Chemistry and Synthetic Biology. *Int. J. Phil. Chem.* 15, 31-46.

[8] Vincent, J.F.V., (2009). Biomimetics-a review. *Proc. Inst. Mech. Eng. H.* 223, 919-939.

[9] De Yoreo, J.J., Vekilov, P.G., (2003). Principles of Crystal Nucleation and Growth. *Rev. Miner. Geochem.* 54, 57-93.

[10] Gilbert, P.U.P.A., Abrecht, M., Frazer, B.H., (2005). The Organic-Mineral Interface in Biominerals. *Rev. Mineral. Geochem.* 59, 157-185.

[11] De Stasio, G., Schmitt, M.A., Gellman, S.H., (2005). Spectrpmicroscopy at the organic-inorganic interface. *Am. J. Sci.* 305, 673-686.

[12] Skinner, H.C.W., (2005). Biominerals. *Mineral. Magazine.* 69, 621-641.

[13] Barskov, I.S., (2008). Biomineralization and Evolution. Coevolution of Mineral and Biological Worlds. In Dobretsov N (Ed.), Bioesphere origins and evolution (pp. 211-218) NY : Springer.

[14] Bazylinski, D.A., Frankel, R.B., (2003). Biologically controlled mineralization in prokaryotes. *Rev. Mineral. Geochem.* 54, 217-247.

[15] Frankel, R.B., Bazylinski, D.A., (2003). Biologically induced mineralization by bacteria. *Rev. Mineral. Geochem.* 54, 95-114.

[16] Veis, A., (2003). Mineralization in organic matrix frameworks. *Rev. Mineral. Geochem.* 54, 249-289.

[17] Mount, A.S., Wheeler, A.P., Paradkar, R.P., Snider, D., (2004). Hemocyte-mediated shell mineralization in the eastern oyster. *Science.* 304, 297-300.

[18] Albeck, S., Addadi, L.,Weiner, S., (1996). Regulation of calcite crystal morphology by intracrystalline acidic proteins and glycoproteins. *Connect. Tissue Res.* 35, 419-424.

[19] Aizenberg, J., Muller, D.A., Grazul, J.L., Hamann, D.R., (2003). Direct fabrication of large micropatterned single crystals. *Science* 299, 1205-1208.

[20] Gotliv, B.A., Kessler, N., Sumerel, J.L., Morse, D.E., Tuross, N., Addadi, L., Weiner, S., (2005). Asprich: a novel aspartic acid-rich protein family from the prismatic shell matrix of the bivalve Atrina rigida. *Chem. Biochem.* 6, 304-314.

[21] Kwak, S.Y., Wiedemann-Bidlack, F.B., Beniash, E., Yamakoshi, Y., Simmer, J.P., Litman, A., Margolis, H.C., (2009). Role of 20-kDa amelogenin (P148) phosphorylation in calcium phosphate formation in vitro. *J. Biol. Chem.* 284, 18972-18979.

[22] Weiner, S., Addadi, L., (1991). Acidic macromolecules of mineralized tissues: The controllers of crystal formation. *Trends Biol. Sci.* 16, 252-256.

[23] Weiner, S., Addadi, L., (1997). Design strategies in mineralized biological materials. *J. Mater. Chem.* 7, 689-702.

[24] Cölfen, H., (2003). Precipitation of carbonates: recent progress in controlled production of complex shapes. *Curr. O. Coll. Interface Sci.* 8, 23-31.

[25] Beniash, E., Simmer, J.P., Margolis, H.C., (2005). The effect of recombinant mouse amelogenins on the formation and organization of hydroxyapatite crystals in vitro. *J. Struct. Biol.* 149, 182-90.

[26] Cai, Y.R., Tang, R.K., (2009). Towards understanding biomineralization: calcium phosphate in a biomimetic mineralization process. *Fron.t Mater. Sci. China.* 3, 124-131.

[27] Aizenberg, J., (2010). New Nanofabrication Strategies: Inspired by Biomineralization. *MRS. BULLETIN.* 35, 323-330.

[28] Weiner, S., Addadi, L., (2011). Crystallization Pathways in Biomineralization. *Ann. Rev. Mater. Res.* 41, 21-40.

[29] Frankel, R.B., Bazylinskib, D.A., Schülerb, D., (1998). Biomineralization of magnetic iron minerals in bacteria. *Supramol. Sci.* 5, 383-390.

[30] Vijayan, K.K., Diwan, A.D., (1996). Fluctuations in Ca, Mg and P Levels in the Hemolymph, Muscle, Midgut Gland and Exoskeleton During the Moult Cycle of the Indian White Prawn, Penaeus indicus (Decapoda: Penaeidae). *Camp. Biochem. Physiol.* 114A, 91-97.

[31] Fu, G., Valiyaveettil, S., Wopenka, B., Morse, D.E., (2005). CaCO3 Biomineralization: Acidic 8-kDa Proteins Isolated from Aragonitic Abalone Shell Nacre Can Specifically Modify Calcite crystal morphology. *Biomacromol.* 6, 1289-1298.

[32] Gago-Duport, L., Briones, M.J., Rodríguez, J.B., Covelo, B., (2008). Amorphous calcium carbonate biomineralization in the earthworm's calciferous gland: pathways to the formation of crystalline phases. *J. Struct. Biol.* 162, 422-435.

[33] Dorozhkin, S.V., Epple, M., (2002). Biological and Medical Significance of Calcium Phosphates. *Angewandte. Chem. Int. Edition.* 41, 3130-3146.

[34] Dorozhkin, S.V., (2007). Calcium Orthophosphates. *J. Mater. Sci.* 42, 1061-1095.

[35] Dorozhkin, S.V., (2009). Nanodimensional and Nanocrystalline Apatites and Other Calcium Orthophosphates in Biomedical Engineering, Biology and Medicine. *Materials.* 2, 1975-2045.

[36] Dorozhkin, S.V., (2009). Calcium Orthophosphates in Nature, Biology and Medicine. *Materials* 2, 399-498.

[37] Mehmel, M., (1930). On the structure of apatite. *I. Z. Kristallogr.* 75, 323-331.

[38] Wopenka, B., Pasteris, J.D., (2005). A mineralogical perspective on the apatite in bone. *Mater. Sci. Eng. C.* 25, 131-143.

[39] Frank-Kamenetskaya, O.V., (2008). *Structure, Chemistry and Synthesis of Carbonate Apatites-The Main Components of Dental and Bone Tissues. In: Krivovichev SV (Ed.), Minerals as Advanced Materials I (pp.241-252).* Berlin: Springer.

[40] Hughes, J.M., Rakovan, J., (2002). *The crystal structure of apatite, Ca5 (PO4)3(F,OH,Cl).* In: Rakovan J, Hughes JM, Kohn MJ (Ed.), Phosphates: Geochemical, Geobiological and Material Importance, Reviews in Mineralogy and Geochemistry (pp.1-12). Washington, DC: Mineralogical Society of America.

[41] Mathew, M., Takagi, S., (2001). Structures of Biological Minerals in Dental Research. *J. Res. Natl. Inst. Stand. Technol.* 106, 1035-1044.

[42] Eichert, D., Salomé, M., Banu, M., Susini, J., Re,y C., (2005). Preliminary characterization of calcium chemical environment in apatitic and non-apatitic calcium phosphates of biological interest by X-ray absorption spectroscopy. *Spectrochim. Acta. Part B: Atom. Spectrosc.* 60, 850-858.

[43] Rey, C., Combes, C., Drouet, C., Sfihii, H., Barroug, A., (2007). Physico-chemical properties of nanocrystalline apatites: Implications for biominerals and biomaterials. *Mater. Sci. Eng. C.* 27, 198-205.

[44] Pasteris, J.D., Wopenka, B., Valsami-Jones, E., (2008). Bone and Tooth Mineralization: Why Apatite? *Elements.* 4, 97-104.

[45] Rey, C., Combes, C., Drouet, C., Sfihi, H., (2006). Chemical diversity of apatites. *Adv. Sci. Tech.* 49, 27-36.

[46] Elliot, J.C., Holcomb, D.W., Young, R.A., (1985). Infrared determination of the degree of substitution of hydroxyl by carbonate ions in human dental enamel. *Calcif. Tissue Int.* 37, 372-375.

[47] Leventouri, Th., Antonakos, A., Kyriacou, A.,Venturelli, R., (2009). Crystal Structure Studies of Human Dental Apatite as a function of age. *Int. J. Biomater.* 698547, 1-6.

[48] Palmer, L.C., Newcomb, C.J., Kaltz, S.R., Spoerke, E.D., Stupp, S.I., (2008). Biomimetic systems for hydroxyapatite mineralization inspired by bone and enamel. *Chem. Rev.* 108, 4754-4783.

[49] Boskey, A.L., (2006). Assessment of bone mineral and matrix using backscatter electron imaging and FTIR imaging. *Curr. Osteoporos. Rep.* 4, 71-75.

[50] Elliott, J.C., (2002). *Calcium phosphate biominerals. In: Rakovan J, Hughes JM Kohn MJ (Ed.), Phosphates: Geochemical, Geobiological and Material Importance, Reviews in Mineralogy and Geochemistry (pp. 427-454).* Washington: Mineralogical Society of America.

[51] Gross, K.A., Berndt, C.C., (2002). *Biomedical application of apatites. In: Rakovan J, Hughes JM Kohn MJ (Ed.), Phosphates: Geochemical, Geobiological and Material Importance, Reviews in Mineralogy and Geochemistry (pp. 631-672).* Washington: Mineralogical Society of America

[52] Liao, S., Wataric, F., Xud, G., Ngiama, M., Ramakrishnaa, S., Chan, C.K., (2007). Morphological effects of variant carbonates in biomimetic hydroxyapatite. *Materials Letters.* 61, 3624-3628.

[53] Rey, C., Combes, C., Drouet, C., Glimcher, M.J., (2009). Bone mineral: update on chemical composition and structure. *Osteoporos. Int.* 20, 1013-1021.

[54] Wheeler, E.J., Lewis, D., (1977). X-Ray study of paracrystalline nature of bone apatite. *Calc. Tissue Res.* 24, 243-248.

[55] Wang, L., Nancollas, G.H., (2010). Dynamics of Biomineralization and Biodemineralization. *Met. Ions. Life. Sci.* 4, 413-456.

[56] Gower, L.B., (2008). Biomimetic Model Systems for Investigating the Amorphous Precursor Pathway and Its Role in Biomineralization. *Chem. Rev.* 108, 4551-4627.

[57] Dorvee, J.R., Veis, A., (2013). Water in the formation of biogenic minerals: peeling away the hydration layers. *J. Struct. Biol.* 183, 278-303.

[58] Navrotsky, A., (2004). Energetic clues to pathways to biomineralization:Precursors, clusters, and nanoparticles. *PNAS.* 101, 12096–12101.

[59] Weiner, S., (2006). Transient precursor strategy in mineral formation of bone. *Bone.* 39, 431-433.

[60] Grynpas, M.D., (2007). Transient precursor strategy or very small biological apatite crystals? *Bone.* 41, 162-164.

[61] Glimcher, M.J., Krane, S.M., (1968). *Biology of Collagen (pp.68-251).* London: Academic Press.

[62] Landis, W.J., Song, M.J., Leith, A., McEwen, L., McEwen, B.F, (1993). Mineral and organic matrix interaction in normally calcifying tendon visualized in 3 dimensions by high-voltage electron-microscopic tomography and graphic image-reconstruction. *J. Struct. Biol.* 110, 39–54.

[63] Landis, W.J., Hodgens, K.J., Son,g M.J., Arena, J., Kiyonaga, S., Marko, M., Owen, C., McEwen, B.F., (1996). Mineralization of collagen may occur on fibril surfaces: evidence from conventional and high-voltage electron microscopy and three-dimensional imaging. *J. Struct. Biol.* 117, 24-35.

[64] Crane, N.J., Popescu, V., Morris, M.D., Steenhuis, P., Ignelzı, M.A., (2006). Raman spectroscopic evidence for octacalcium phosphate and other transient mineral species deposited during intramembranous mineralization. *Bone.* 39, 434-442.

[65] Addadi, L., Raz, S., Weiner, S., (2003). Taking advantage of disorder: amorphous calcium carbonate and its roles in biomineralization. *Adv. Mater.* 15, 959-970.

[66] Lowenstam, H.A., Weiner, S., (1985). Transformation of amorphous calcium phosphate to crystalline dahillite in the radular teeth of chitons. *Science.* 227, 51-53.

[67] Weiss, I.M., Tuross, N., Addadi, L., Weiner, S., (2002). Mollusc larval shell formation: amorphous calcium carbonate is a precursor phase for aragonite. *J. Exp. Zool.* 293, 478-491.

[68] Polit, Y., Arad, T., Klein, E., Weiner, S., Addadi, L., (2004). Sea urchin spine calcite forms via a transient amorphous calcium carbonate phase. *Science* 306, 1161–1164.

[69] Politi, Y., Levi-Kalisman, Y., Raz, S., Wilt, F., Addadi, L., Weiner, S., Sagi, I., (2006). Structural Characterization of the Transient Amorphous Calcium Carbonate Precursor Phase in Sea Urchin Embryos. *Adv. Funct. Mater.* 16, 1289-1298.

[70] Politi, Y., Metzler, R.A., Abrecht, M., Gilbert, B., Wilt, F.H., Sagi, I., Addadi, L., Weiner, S., Gilbert, P.U.P.A., (2008). Transformation mechanism of amorphous calcium carbonate into calcite in the sea urchin larval spicule. *PNAS* 105, 17362-17366.

[71] Addadi, L., Vidavsky, N., Weiner, S., (2012). Transient precursor amorphous phases in biomineralization. In the footsteps of Heinz A. Lowenstam. *Z. Kristallogr.* 227, 711–717.

[72] Brown, W,E,, Eidelman, N., Tomazic, B., (1987). Octacalcium Phosphate as a Precursor in Biomineral Formation. *ADR.* 1, 306-307.

[73] Termine, J.D., Posner, S.A., (1967) Amorphous/crystalline interrelationships in bone mineral. *Calc. Tiss. Inter.* 1, 8-23.

[74] Posner, S.A., Betts, F., (1975). Synthetic amorphous calcium phosphate and its relation to bone mineral structure. *Acc. Chem. Res.* 8, 273-281.

[75] Mahamid, J., Sharir, A., Addadi, L., Weiner, S., (2008). Amorphous calcium phosphate is the major components of the forming fin bones of zebrafish: Indications for an amorphous precursor phase. *PNAS* 105, 12748-12753.

[76] Mahamida, J., Aichmayerb, B., Shimonic, E., Ziblata, R., Lib, C., Siegelb, S., Paris, O., Fratzl, P., Weiner, S., Addadi, L., (2010). Mapping amorphous calcium phosphate transformation into crystalline mineral from the cell to the bone in zebrafish fin rays. *PNAS* 107,6316-6321.

[77] Beniash, E., Metzler, R.A., Lam, R.S.K., Gilbert, P.U.P.A., (2009). Transient amorphous calcium phosphate in forming enamel. *J. Struct. Biol.* 166, 133-143.

[78] Okuda, M., Takeguchi, M., Tagaya, M., Tonegawa, T., Hashimoto, A., Hanagata, N., Ikoma, T., (2009). Elemental distribution analysis of type I collagen fibrils in tilapia fish scale with energy-filtered transmission electron microscope. *Micron.* 40, 665-668.

[79] Olszta, M.J., Cheng, X., Jee, S.S., Kumar, R., (2007). Bone structure and formation: A new perspective. *Mater. Sci. Engin. R.* 58, 77-116.

[80] Nudelman, F., Pieterse, K., George, A., Bomans, P.H., Friedrich, H., Brylka, L.J., Hilbers, P.A., de With, G., Sommerdijk, N.A., (2010). The role of collagen in bone apatite formation in the presence of hydroxyapatite nucleation inhibitors. *Nat. Mater.* 9, 1004-1009.

[81] Zeiger, D.N., Miles, W.C., Eidelman, N., Lin-Gibson, S., (2011). Cooperative calcium phosphate nucleation within collagen fibrils. *Langmuir* 27, 8263-8268.

[82] Nudelman, F., Lausch, A.J., Sommerdijk, N.A., Sone, E.D., (2013). In vitro models of collagen biomineralization. *J. Struct. Biol.* 183(2), 258-269.

[83] Beniash, E., (2011). Biominerals- hierarchical nanocomposites: the example of bone. *Wiley Interdiscip. Rev. Nanomed. Nanobiotechnol.* 3(1), 47–69.

[84] Nudelman, F., Bomans, P.H.H., George, A., de With, G., Sommerdijk, N.A.J.M., (2012). The role of the amorphous phase on the biomimetic mineralization of collagen. *Faraday Discuss.* 159, 357-370.

[85] Gu, L., Kim, Y.K., Liu, Y., Ryou, H., Wimmer, C.E., Dai, L., Arola, D.D., Looney, S.W., Pashley, D.H., Tay, F.R., (2011). Biomimetic analogs for collagen biomineralization. *J. Dent. Res.* 90, 82-87.

[86] Wang, J., Chen, Y., Li, L., Sun, J., Gu, X., Xu, X., Pan, H., Tang, R., (2013). Remineralization of dentin collagen by meta-stabilized amorphous calcium phosphate. *Cryst. Eng. Comm.* 15, 6151-6158.

[87] Lausch, A.J., Quan, B.D., Miklas, J.W., Sone, E.D., (2013). Extracellular matrix control of collagen mineralization in vitro. *Adv. Funct. Mater.* 23, 4906-4912.

[88] Thula, T.T., Rodriguez, D.E., Lee, M.H., Pendi, L., Podschun, J., Gower, L.B., (2011). In vitro mineralization of dense collagen substrates: a biomimetic approach toward the development of bone-graft materials. *Acta Biomater.* 7, 3158-3169.

[89] Clark, A.E., Pantano, C.G., Hench, L.L., (1976). Auger Spectroscopic Analysis of Bioglass Corrosion Films. *J. Am. Cer. Soc.* 59, 37-39.

[90] Peitl, O., Zanotto, E.D., Hench, L.L., (2001). Highly bioactive P2O5-Na2O-CaO-SiO2 glass–ceramics. *J. Non-Cryst. Solids.* 292, 115-126.

[91] Cerruti, M., Bianchi, C.L., Bonino, F., Damin, A., Perardi, A., Morterra, C., (2005). Surface modifications of bioglass immersed in TRIS-buffered solution. A multitechnical spectroscopic study. *J. Phys. Chem. B.* 109, 14496-14505.

[92] Sepulveda, P., Jones, J.R., Hench, L.L., (2002). In vitro dissolution of melt-derived 45S5 and sol-gel derived 58S5 bioactive glasses. *J. Biomed. Mater. Res.* 61, 301-311.

[93] Arcos, D., Greenspan, D.C., Vallet-Regi, M., (2003). A new quantitative method to evaluate the in vitro bioactivity of melt and sol-gel-derived silicate glasses. *J. Biomed. Mater. Res.* 65A, 344-351.

[94] Padilla, S., Roman, J., Carenas, A., Vallet-Regi, M., (2005). The influence of the phosphorus content on the bioactivity of sol-gel glass ceramics. *Biomater.* 26, 475-483.

[95] Hench, L.L., (1996). Biological application of bioactive glasses. *Life. Chem. Reports.* 13, 187-241.

[96] Sahai, N., (2005). Modeling apatite nucleation in the human body and in the geochemical environment. *Am. J. Sci.* 305, 661-672.

[97] Sahai, N., (2003). The effects of Mg2+ and H on apatite nucleation at silica surfaces. *Geochim. Cosmochim. Acta.*67, 1017-1030.

[98] Tilocca, A., (2010). Models of structure, dynamics and reactivity of bioglasses: a review. *J. Mater. Chem.* 20, 6848-6858.

[99] Kashchiev, D., (2000). *Nucleation: Basic Theory with Applications (pp.9-16).* Oxford : Butterworth-Heinemann.

[100] Markov, I.V., (1995). *Crystal growth for beginners: Fundamentals of Nucleation, crystal growth and epitaxy (pp.77-172).* Singapore: World Scientific.

[101] Mullin, J.W., (2001). *Crystallization (pp. 181-214).* Oxford : Butterworths-Heinemann Ltd.

[102] Weinberg, M.C., (1996). Glass-formation and crystallization kinetics. *Thermochimica Acta* 280/281, 63-71.

[103] Woodruff, D.P., (1973). *The solid-liquid interface (pp. 12-38).* NY: Cambridge University Press.

[104] Jiang, H., Liu, X.Y., (2004). Principles of Mimicking and Engineering the Self-organized Structure of Hard Tissues. *J. Biol. Chem.* 279, 41286-41293.

[105] Pereira, M.M., Hench, L.L., (1996). Mechanisms of hydroxyapatite formation on porous gel-silica substrates. *J. Sol-Gel Sci. Tech.* 7, 59-68.

[106] Liu XY., (2004). *From solid–fluid interfacial structure to nucleation kinetics: principles and strategies for micro/nanostructure engineering. In: De Yoreo JJ Liu XY (Ed.), Nanoscale structure and assemblyat solid-liquid interfaces (pp. 1-67).* Springer: Kluwer Academic Publishers.

[107] Liu, Y., Huang, S., Dan, X., Zhou, K., (2004). Growth of Hydroxyapatite Crystal in the Presence of Organic Film. *J. Mater. Sci. Technol.* 20, 223-226.

[108] Liu, X.Y., Lim, S.W., (2003). Templating and Supersaturation-Driven Anti-Templating: Principles of Biomineral Architecture. *J. Am. Chem Soc.* 125, 888-895.

[109] Benning, G., Waychunas, G., (2007). *Nucleation, growth, and aggregation of mineral phases: Mechanisms and kinetic controls. In: Kubicki, J.D., White, A.F., Brantley, S.L. (Ed.), Kinetics of Water-Rock Interaction (pp. 259-335).* NY: Springer.

[110] Jiang, H., Liu, X.Y., Zhang, G., Li, Y., (2005). Kinetics and Template Nucleation of Self-Assembled Hydroxyapatite Nanocrystallites by Chondroitin Sulfate. *J. Biol. Chem.* 280, 42061-42066.

[111] Tilocca, A., Cormack, A.N., (2011). The initial stages of bioglass dissolution: a Car-Parrinello molecular-dynamics study of the glass-water interface. *Proc. R. Soc. A.* 10.1098/rspa.2010.0519, 1-10.

[112] Hill R., (1996). An alternative view of the degradation of bioglass. *J. Matter. Sci. Letters.* 15, 1122-1125.

[113] Strnad Z., (1992). Role of Glass Phase in Bioactive Glass-Ceramics. *Biomater.* 13, 317-321.

[114] Serra, J., González, P., Liste, S., Chiussi, S., León, B., Pérez-Amor, M., Ylänen, H.O., Hupa, M., (2002). Influence of the non-bridging oxygen groups on the bioactivity of silicate glasses. *J. Mater. Sci. Mater. Med.* 13, 1221-1225.

[115] Helms, C.R., Poindexter, E.H., (1994). The silicon\–silicon dioxide system: Its microstructure and imperfections. *Rep. Prog. Phys.* 57, 791-851.

[116] Elgayar, I., Aliev, A.E., Boccaccini, A.R., Hill, R., (2005). Structural analysis of bioactive glasses. *J. Non-Cryst. Solids.*351, 173–183.

[117] Tilocca A., (2009). Structural models of bioactive glasses from molecular dynamics simulations. *Proc. R. Soc. A.* 465, 1003–1027.

[118] Marghussian, V.K., Mesgar, A., (2000). Effects of composition on crystallization behaviour and mechanical properties of bioactive glass-ceramics in the MgO-CaO-SiO2-P2O5 system. *Ceram. Int.* 26, 415-420.

[119] Nychka, J.A., Mazur, S.R.L., Kashyap, S., Li, D., Yang, F., (2009). Dissolution of bioactive glasses: The effects of crystallinity coupled with stress. *Biomed. Mater. Dev.* 61, 45-51.

[120] El-Ghannam, A., Hamazawy, E., Yehia, A., (2001). Effect of thermal treatment on bioactive glass microstructure, corrosion behavior, ζ potential, and protein adsorption. *J. Biomed. Mater. Res.* 55, 387-395.

[121] . Bretcanu, O., Chatzistavrou, X., Paraskeuopoulos, K., Conradt, R., Thompson, I., Boccaccini, A.R., (2009). Sintering and crystallisation of 45S5 Bioglass® powder. *J. Euro. Cer. Soc.* 29, 3299-3306.

[122] El Batal, H.A., Khalil, E.M.A., Monem, A.S., Hamdy, Y.M., (2003). Characterization of some bioglass-ceramics. *Mater. Chem. Phys.* 80, 599-609.

[123] Cho, S.B., Nakanishi, K., Kokubo, T., Soga, N., Ohtsuki, C., Nakamura, T., (1996). Apatite formation on silica gel in simulated body fluid: Its dependence on structures of silica gels prepared in different media. *J. Biomed. Mate.r Res.* 33, 145-151.

[124] Sahai, N., Tossell, J.A., (2000). Molecular orbital study of apatite (Ca5(PO4)3OH) nucleation at silica bioceramic surfaces. *J. Phys. Chem. B.* 104, 4322–4341.

[125] Sahai, N., Ansau, M., (2005). Cyclic silicate active site and stereochemical match for apatite nucleation on pseudowollastonite bioceramic–bone interfaces. *Biomater.* 26, 5763-5770.

[126] Brinker, C.J., Kirkpatrick, R.J., Tallant, D.R., Bunker, B.C., Montez, B., (1988). NMR confirnation of strained "defects" in amorphous silica. *J. Non-Cryst. Sol.* 99, 418-428.

[127] Wallace, S., West, J.K., Hench, L.L., (1993). Interactions of water with trisiloxane rings. I. Experimental analysis. *.J Non-Cryst. Sol.* 152, 101-108.

[128] West, J.K., Wallace, S., (1993). Interactions of water with trisiloxane rings. II. Theoretical analysis. *J. Non-Cryst. Sol.* 152, 109-117.

[129] Zhu, P., Masuda, Y., Koumoto, K., (2004). The effect of surface charge on hydroxyapatite nucleation. *Biomater.* 25, 3915–3921.

[130] Calvert, P., Mann, S., (1997). The negative side of crystal growth. *Nature* 386, 127-128.

[131] Hench, L.L., Splinter, R.J., Allen, W.C., Greenlee, T.K. (1971). Bonding mechanisms at the interface of ceramic prosthetic materials. *J. Biomed.Mate.r Res.* 5, 117-141.

[132] Valimaki, V.V., Aro, H.T. (2006). Molecular basis for action of bioactive glasses as bone graft substitute. *Scan. J. Surg.* 95, 95-102.

[133] Jell, G., Stevens, M.M. (2006). Gene activation by bioactive glasses. *J. Mater. Sci.: Mater. Med.* 17, 997-1002.

[134] Radin, S., Reilly, G., Leboy, P.S., Ducheyne, P. (2005). Osteogenic effects of bioactive glass on bone marrow stromal cells. *J. Biomed. Mater. Res.* 73A, 21-29.

[135] Xynos, I.D., Hukkanen, M.V.J., Batten, J.J., Buttery, I.D., Hench, L.L., Polak, J.M. (2001). BioglassR 45S5 stimulates osteoblast turnover and enhances bone formation in vitro: implications and applications for bone tissue engineering. *Calcif. Tiss. Int.* 67, 321-329.

[136] Xynos, I.D., Edgar, A.J., Buttery, L.D., Hench, L.L., Polak, J.M. (2000). Ionic dissolution products of bioactive glass increase proliferation of human osteoblasts and induce insulin-like growth factor II mRNA expression and protein synthesis. *Biochem. Biophys. Res. Commun.* 27,461-465.

[137] . Xynos, I.D., Edgar, A.J., Buttery, L.D.K., Hench, L.L., Polak, J.M. (2001). Gene expression profiling of human osteoblasts following treatment with the ionic dissolution products of BioglassR 45S5 dissolution. *J. Biomed. Mater. Res.* 55, 151–157.

[138] Gerhardt, L.C., Boccaccini, A.R. (2010). Bioactive glass and glass-ceramic scaffolds for bone tissue engineering. *Mater.* 3, 3867-3910.

[139] Sun, J.Y., Yang, Y.S., Zhong, J.P., Greenspan, D.C. (2007). The effect of ionic products of bioglass dissolution on human osteoblasts growth cycle in vitro. *J. Tissue. Eng. Regen. Med.* 1, 281-286.

[140] Horbett, T.A. (2004). *The role of adsorbed proteins in tissue response to biomaterials. In: Ratner, B.D. (eds), Biomaterials Science (pp. 237-245).* California: Elsevier Academic Press.

[141] Hench, L.L. (2009). Genetic design of bioactive glass. *J. Euro. Cer. Soc.* 29,1257-1265.

In: Apatite
Editors: M. Iafisco and J. M. Delgado-López
ISBN: 978-1-63321-500-9

Chapter 4

Distribution of Carbonate Ions in Biological Apatite and Excess Fluorine in Francolite

Michael E. Fleet*
Department of Earth Sciences, University of Western Ontario,
London, Ontario, Canada

Abstract

The out-of-plane bend (ν_2) infrared band areas for the channel (A) and phosphate (B) carbonate ions in synthetic sodium-bearing hydroxyapatite (CHAP) are in good agreement with site occupancies from X-ray structures, but the areas of the corresponding asymmetric stretch (ν_3) bands greatly overestimate the contribution of type B carbonate. This anomaly is attributed to overlap of A and B ν_3 doublet bands. The introduction of the Na^+ cation into the nearest-neighbor cation shell of the channel carbonate ion results in a downward shift of up to 50 cm^{-1} in the type A doublet band. This effect is well illustrated by a literature study of potassium-bearing CHAP. Wherever possible, the areas of ν_2 bands should be used to estimate the proportion of A and B carbonate ions in bone mineral and other biological apatites. Previous study of the location of the excess fluorine in francolite, the ore mineral of phosphorite, is supported by recent theoretical calculation, prompting further analysis of this interesting crystal chemical feature using the difference Fourier maps for synthetic carbonate fluorapatite.

Keywords: Carbonate hydroxyapatite; bone mineral; infrared spectra; francolite

Introduction

Biological apatite, a carbonate-bearing hydroxyapatite, is the dominant inorganic component in the hard tissue of vertebrates, accounting for up to about 65 % of cortical bone and 97 % of dental enamel [1]. The structural role of the carbonate ion in biological apatite

* E-mail: mfleet@uwo.ca.

has a critical bearing on the growth and strength of bone [2], bone physiology, and development of bone prostheses. The carbonate impurity also diminishes the hardness and chemical resistance of dental enamel. The common apatite minerals [3] are dominantly solid solutions of hydroxylapatite [HAP; ideally $Ca_{10}(PO_4)_6(OH)_2$], fluorapatite [FAP; $Ca_{10}(PO_4)_6F_2$] and chlorapatite [CLAP; $Ca_{10}(PO_4)_6Cl_2$]. The abbreviations HAP, FAP and CLAP presently refer to both ideal formulae and synthetic phases, and CHAP, CFAP and CCLAP are used similarly for the corresponding synthetic carbonate-bearing phases.

Natural calcium phosphate apatites all have hexagonal symmetry with space group $P6_3/m$ and one formula unit per unit cell [4]. In the basic apatite-type structure of FAP, isolated PO_4 tetrahedra centered on the mirror plane at a unit-cell height z = (1/4,3/4) are linked by Ca1 in ninefold (6+3) coordination and Ca2 in an irregular sevenfold (6+1) coordination (Figure 1a). A prominent feature of the structure is the large *c*-axis channel that accommodates the *X* anion component (F,OH,Cl) and is defined by triclusters of Ca2 cations at z = (1/4,3/4).

The structural role of carbonate in CHAP and CFAP has been investigated extensively by X-ray powder and single-crystal and neutron powder diffraction methods, as well as by infrared, Raman and nuclear magnetic resonance spectroscopy and theoretical simulations; all reviewed previously [5-7]. It has been established that the carbonate ion can be accommodated either in the *c*-axis structural channel (type A carbonate) or as a substituent for the phosphate group (type B carbonate). Biological apatite and CHAP generally contain detectable amounts of both A and B carbonate ions (type AB CHAP), because the carbonate ion fractionates between the two structural positions under most conditions of crystal growth.

However, the detailed crystal structure of biological apatite has remained elusive, definitive research being frustrated by a number of problems, including: the limited substitution of carbonate, especially of type B carbonate; small, nanoscale crystal size; fragility and reactivity of nanocrystals extracted from bone tissue; low degree of crystallinity; weak and overlapped electron density of carbonate atoms, and ambiguous chemical spectra. These limiting problems have been largely circumvented using the single-crystal X-ray structure method in combination with Fourier transform infrared (FTIR) spectroscopy and relatively large (50-200 μm) crystals of CHAP, CFAP and CCLAP of good diffraction quality synthesized from carbonate-rich calcium phosphate melts at high pressure and temperature [5, 8-14].

An important accomplishment of this extensive laboratory project was the demonstration that the areas of the A and B carbonate out-of-plane bend (ν_2) bands in infrared spectra were in linear proportion to the corresponding carbonate ion site occupancies from the X-ray structures (Figure 2; Table 1). This correlation permitted the proportions of A and B carbonate ions in the nanoscale biological apatite crystals to be estimated from the areas of the (ν_2) infrared bands. However, the proportion of channel (type A) carbonate ions obtained by this method for sodium-bearing carbonate apatites is generally in excess of that estimated from the corresponding asymmetric stretch (ν_3) band areas, challenging the prevailing view that the carbonate ions in biological apatites are dominantly substituents for the orthophosphate ions (type B). The present paper discusses this discrepancy and offers an explanation for the anomalous observed asymmetric stretch (ν_3) band areas, and concludes with a discussion of new developments on the structural location of the excess fluorine content of francolite, a variety of carbonate fluorapatite.

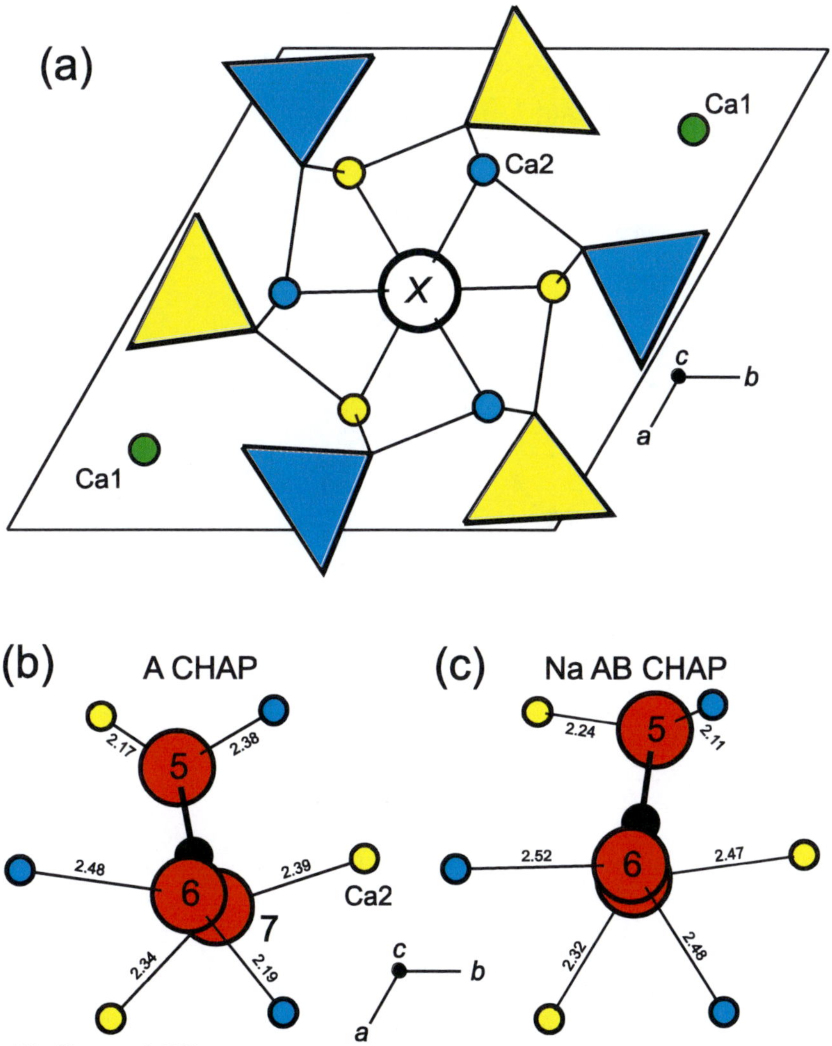

(after Fleet [6]; Fleet et al. [5]).

Figure 1. (a) Basic apatite-type structure in c-axis projection, showing location of X (= OH,F,Cl) anion in the c-axis structural channel defined by triclusters of Ca2 cations: PO_4 tetrahedra and Ca2 are colored yellow at height z = 1/4 and blue at z = 3/4; unit-cell outline is displaced by -0.5,-0.5,0.0. (b) and (c) show orientation of type A carbonate ion in c-axis structural channel of CHAP: (b) sodium-free A CHAP PC71; (c) sodium-bearing AB CHAP LM005; bond distances are Ångstroms.

X-RAY STRUCTURES

Apatite Compositions

The compositions of the crystals investigated by the single-crystal X-ray structure method correspond approximately to the substitution formula $Ca_{10-(y+z)}Na_y\square_z(PO_4)_{6-(y+2z)}(CO_3)_{y+2z}X_{2-2x}(CO_3)_x$, where $x \approx y$ up to 1.0 and $z \approx 0.0$ for CHAP, $x \approx y \approx 4z \approx 0.4$ for CCLAP, $x \approx y \approx 2z \approx 0.1$ for CFAP, and the open box symbol is a cation vacancy.

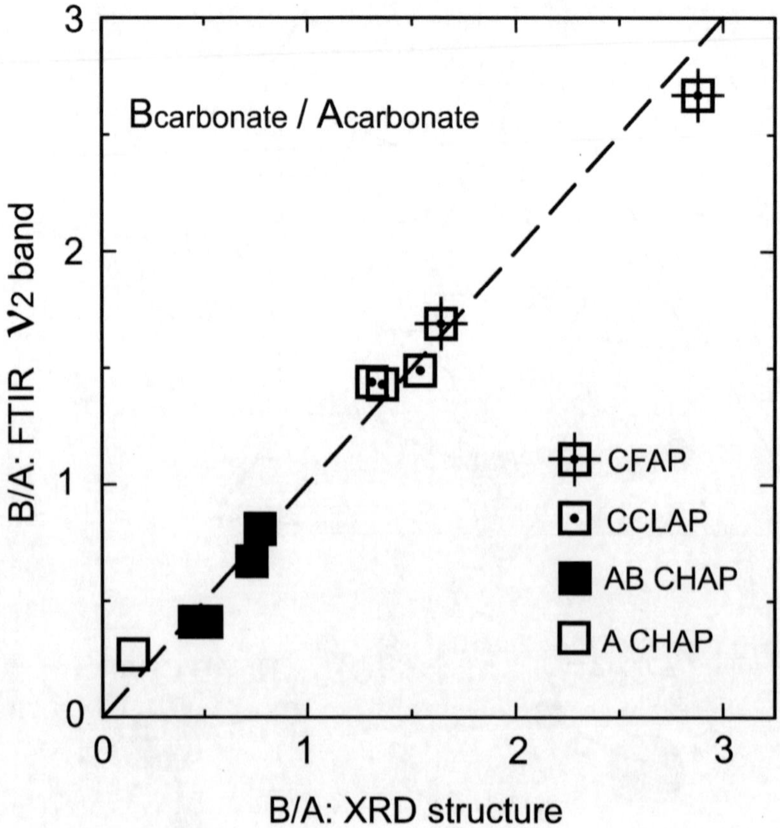

Figure 2. Ratio of formula amounts of B and A carbonate (B/A) in synthetic carbonate apatites, determined by FTIR and X-ray structure methods.

Table 1. Selected synthesis experiments and amounts of sodium, A and B carbonate (pfu[1])

	T	*P*	EPMA		X-ray structure			FTIR
Experiment[2]	(°C)	(GPa)	Na		A	B	B/A	B/A
Sodium-free A and AB CHAP								
PC71	1400	2.0	-		0.75	0.11(2)	0.15	0.27
PC55	1400	3.0	-		1.09(4)	0.56(2)	0.52	0.52
Sodium-bearing AB CHAP								
LM005	1200	0.5	0.87(3)		1.00(5)	0.77(3)	0.77	0.81
LM006	1200	1.0	0.35(4)		0.52(3)	0.38(2)	0.73	0.67
Sodium-bearing AB CCLAP								
LM173	1000	1.0	0.39(5)		0.42(3)	0.57(2)	1.36	1.43
Sodium-bearing AB CFAP								
LM142	1150	1.0	0.11(5)		0.14(3)	0.23(1)	1.64	1.69

[1] pfu is per formula unit [$Ca_{10}(PO_4)_6X_2$]; 2. Seventeen experiments yielded products investigated by X-ray structure analysis.

Two sodium-bearing CHAP experiments, LM006 and LM005, are particularly useful as analogue materials for biological apatite: LM006 has a similar content of CO_2 and sodium to cortical bone and dental enamel, and LM005 has a relatively high content of carbonate and yielded the most accurate structure (Table 1).

Type A Carbonate

The diffraction pattern of the type A CHAP from experiment PC71 (Table 1) is consistent with $P\bar{3}$ symmetry, a new space group for carbonate apatites [8]. Other space groups for type A CHAP have been proposed [15-18], but opinion does seem to be coalescing around $P\bar{3}$ [5]. The carbonate ion was located in the apatite channel from the difference electron density calculated after refinement of the basic apatite host structure and interpreted using model structures [5, 6].

The type A carbonate ion is oriented in the apatite channel with two oxygen atoms close to the *c*-axis, and the carbon atom and the third, off-axis, oxygen atom (O5) at a unit-cell height of $z \approx 0.5$. In type A CHAP (experiment PC71), the plane of the carbonate ion is rotated about 14° anticlockwise (Figure 1b) [5, 8], whereas in both sodium-free and sodium-bearing AB CHAP the rotation is clockwise (Figure 1c) [5], moving the O5 oxygen atom closer to the channel wall.

The type A structure of PC71, with $P\bar{3}$ symmetry, was the key to unraveling the more complex pattern of difference electron density that characterized all CHAP, CFAP and CCLAP structures with significant amounts of both A and B carbonate ions, where the space group is $P6_3/m$ and carbonate atom electron density features are replicated with a multiplicity of twelve.

Sodium-free AB CHAP crystals contained a second channel carbonate species, labeled A2 and oriented with only one oxygen atom close to the *c*-axis [5, 9]. Fleet et al. [5] suggested that this A2 carbonate was a high-pressure feature of sodium-free starting compositions and, therefore, not relevant to the present discussion.

Type B Carbonate

Substitution of the phosphate group by the carbonate ion requires the loss of one oxygen atom and charge balancing, resulting in some disruption of the phosphate matrix. Even though oxygen atoms of the type B carbonate ions were poorly resolved in the X-ray structures due to extensive overlap with the phosphate groups and weak difference electron densities, there were sufficient outlying density peaks to define the orientation and local environment of the type B carbonate ions in the various composition series investigated. Refinement of electron density features nearest to the phosphate groups showed that the type B carbonate ions were located close to a sloping face of the substituted phosphate tetrahedron, but tilted away from it in CCLAP and CHAP apatites, with the tilt angle (ϕ) increasing from -4° in CFAP to +8° in CCLAP and +18° in CHAP (Figure 3) [10, 12, 13].

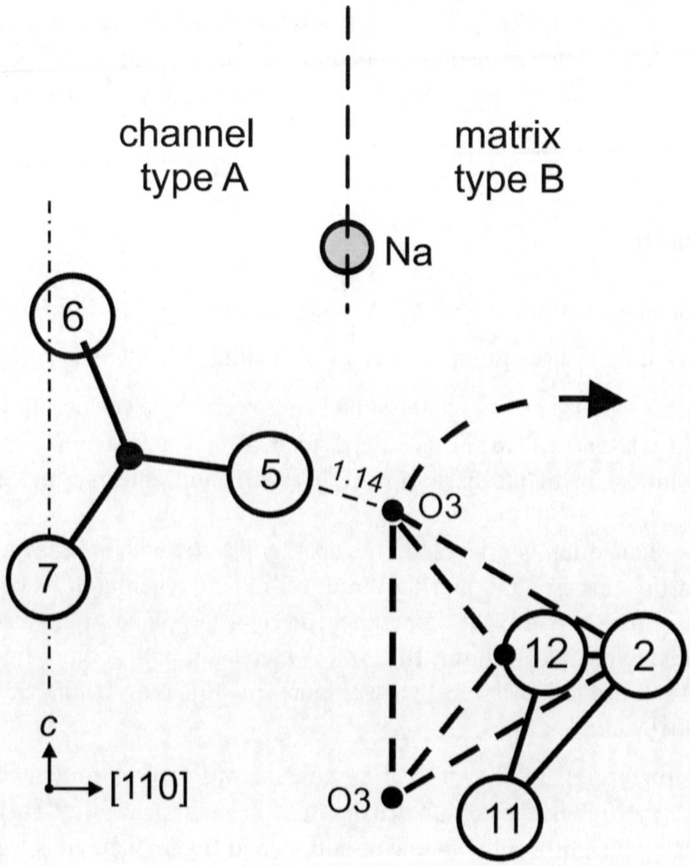

Figure 3. Fragment of the local structure of sodium-bearing AB CHAP LM005 illustrating coupling of A and B carbonate ions and sodium cations: large open circles are oxygen atoms of carbonate ions. Note that the prohibitively close O5-O3 distance is avoided in the local structure when the O3 oxygen site is eliminated (arrow).

Crystal Chemistry

Infrared Spectroscopy

Representative data for the positions of the carbonate bands in infrared spectra of carbonate apatites are given in Table 2 for both literature studies and the high-pressure-temperature synthesized crystals, and selected spectra are reproduced in Figures 4, 5, 6, and 7. The carbonate component of the infrared spectrum of a carbonate apatite with a single carbonate ion species consists of three bands; a strong symmetrical doublet band for the asymmetric stretch (ν_3) mode and a weak singlet band for the out-of-plane bend (ν_2) mode. All three bands are relatively sharp features and detectable at carbonate concentrations as low as about 0.1 wt.% CO_2. The asymmetric stretch mode (ν_3) is two-fold degenerate for the free carbonate ion but this constraint is lifted for site symmetries lower than trigonal.

Table 2. Infrared bands for carbonate in some apatites (cm^{-1})

Apatite	ν_3		ν_2	Type	Reference
Francolite	1453	1429	865	B	19
Dahllite	1455	1416	873	B	20
CHAP	1537	1451	874	A	21
A CHAP, PC71	1544	1461	878	A	8
A CHAP	1539	1467	879	A	18
B CHAP	1475	1419	873	B	22
AB CHAP, PC55	1569	1507	862	A2	9, 14
	1540	1470-1455[1]	879	A	
	1470-1455	1409	870	B	
CFAP	1456	1426	874	B	23
	-	-	866	B2	
Na AB CHAP, LM005	~1545[2]	1473-1449	878	A	10, 14
	1473-1449	1416	870	B	
Na CFAP, LM142	~1545[2]	1467-1452	881	A	12, 14
	1467-1452	1417	873	B	
	-	-	865	B2	
Dental enamel	1545	1450	880	A	19
	1465	1412	873	B	
Pig dental enamel	-	-	878	A	24
	-	-	871	B	
	-	-	866	L[3]	
Bone	1450	1409	877-865[4]		26

[1] Unassigned bands.

[2] Very weak, diffuse.

[3] L is labile.

[4] Broad bands not well resolved.

The site splitting is greater for type A carbonate than for type B, and there are rudimentary correlations with sample source. For spectra with resolved type A and type B doublets, the site splitting from literature spectra averages 76 cm^{-1} for type A CHAP, and increases to about 91 cm^{-1} for type A carbonate in heated dental enamel. For type B carbonate, it is 24-27 cm^{-1} in francolite, about 39 cm^{-1} in dahllite (the hydroxyl analogue of francolite), 40-56 cm^{-1} in type B CHAP, and 53 cm^{-1} in dental enamel. The magnitude of the site splitting in type AB carbonate apatites is generally obscured by band overlap in the 1475-1445 cm^{-1} region, involving the low-frequency limb of the type A doublet and the high-frequency limb of the type B doublet (Table 2). Band assignments in this complex region are considered to be arbitrary, unless supported by independent evidence.

In sodium-free A CHAP PC71, the type A carbonate ion is characterized by infrared bands at 1544, 1461 (ν_3; Figures 4a,5a) and 878 cm^{-1} (ν_2). In comparison, the characteristic infrared bands of the type A2 carbonate ion [9], with only one oxygen atom close to the *c*-axis, are shifted markedly to about 1569 and 1507 (ν_3) and 862 cm^{-1} (ν_2), respectively.

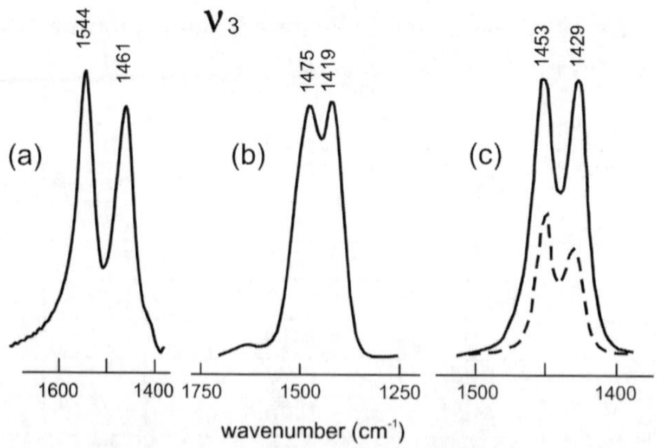

Figure 4. Carbonate asymmetric stretch (ν_3) infrared spectra: (a) type A carbonate in A CHAP PC71 [8]; (b) type B carbonate in sodium-bearing CHAP [22]; (c) polarized spectra for type B carbonate in francolite from Fowey Consols Mine, Cornwall, UK. [19].

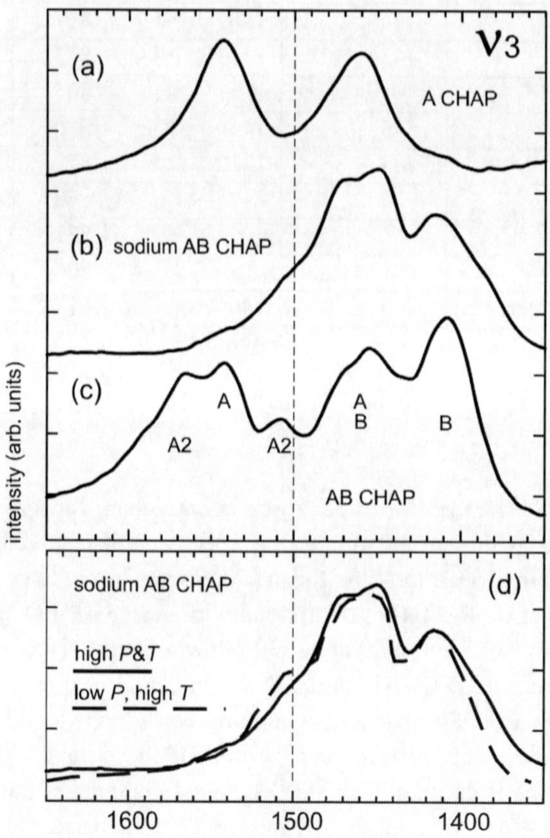

Figure 5. Comparison of infrared spectra in the asymmetric stretch (ν_3) region for CHAP synthesized at high pressure and temperature: (a) sodium-free A CHAP PC71; (b) sodium-bearing AB CHAP LM005; (c) sodium-free AB CHAP PC55. Note the near absence of absorption intensity in sodium-bearing AB CHAP beyond 1500 cm^{-1}. Part (d) shows that ν_3 spectrum for sodium-bearing AB CHAP synthesized at low pressure and high temperature in Nelson and Featherstone [29] is comparable to that of LM005.

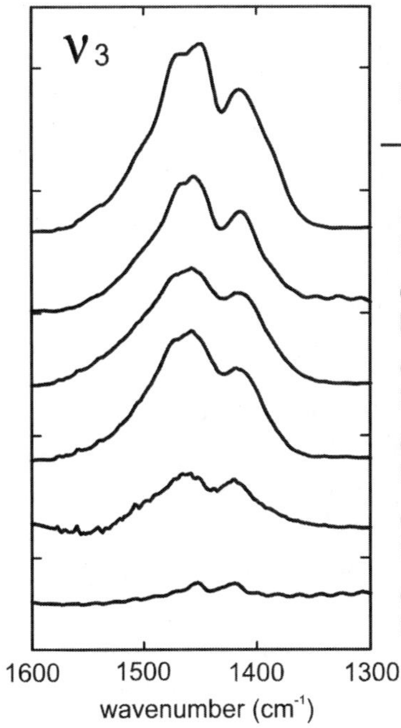

Expt	CO_2 wt.%	carbonate (pfu) A	B
CHAP LM005	8.2	1.00	0.77
CHAP LM006	3.5	0.52	0.38
CCLAP LM169	3.6	0.37	0.57
CCLAP LM171	2.9	0.46	0.58
CFAP LM142	1.9	0.14	0.23
CFAP LM134	0.3	-	-

Figure 6. Carbonate asymmetric stretch (ν_3) infrared spectra for sodium-bearing AB CHAP, AB CFAP, and AB CCLAP, synthesized at high pressure and temperature, showing that overall profile of absorption in the asymmetric stretch (ν_3) region is essentially independent of total carbonate content and X (= OH,F,Cl) composition series.

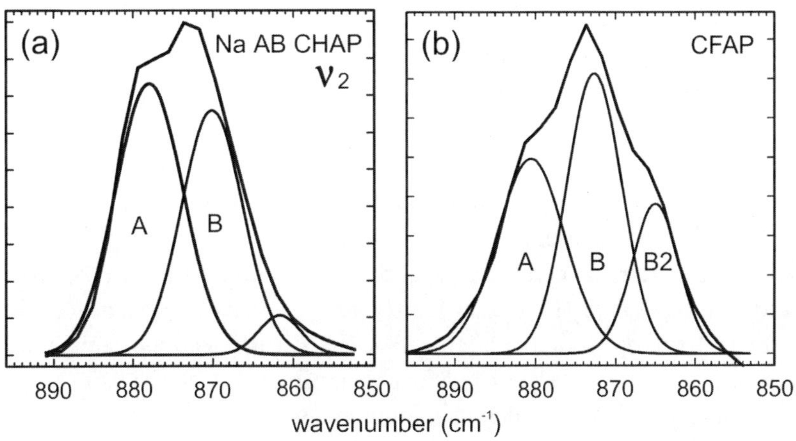

Figure 7. Out-of-plane bend (ν_2) infrared spectra fitted with Gaussian components for sodium-bearing carbonate apatite crystals grown at high pressure and temperature: (a) AB CHAP LM005; (b) CFAP LM142.

The type A spectrum of sodium-bearing AB carbonate apatites is anomalous in that the asymmetric stretch (ν_3) bands appear weaker and are shifted to lower frequency (Figure 5b). The type B carbonate ion has infrared bands at about 1470-1450 and 1409 (ν_3) and 870 cm^{-1} (ν_2) in sodium-free AB CHAP, 1475-1445 and 1416 (ν_3) and 870 cm^{-1} (ν_2) in sodium-bearing AB CHAP, and 1453 and 1429 (ν_3) and 865 cm^{-1} (ν_2) in francolite.

Clustering of Substituents

Fleet and Liu [10, 12, 13] observed that the sodium cation and A and B carbonate ion defects in sodium-bearing carbonate apatite must be locally coupled to facilitate charge compensation and minimize the effects of spatial accommodation. Coupling of these substituents is suggested from three independent lines of evidence: (1) the accommodation of the type B carbonate ion in the X-ray structures; (2) the approximate one-to-one correlation of the contents of sodium and A and B carbonate (Table 1); and (3) the common profile of infrared asymmetric stretch (ν_3) bands.

In the average structures of sodium-bearing AB carbonate apatites synthesized at high pressure and temperature, the off-axis oxygen atom O5 of a type A carbonate ion in the *c*-axis channel is prohibitively close, at 1.1-1.2 Å, to an O3 oxygen atom of a phosphate group (PO_4) (Figure 3). In the structure of sodium-free type A CHAP (PC71), the type A carbonate ion is centered in the channel with the carbon atom more-or-less on the *c*-axis, whereas in all of the type AB carbonate apatites presently investigated the type A carbonate ion is displaced towards the channel wall, to the extent that the other two oxygen atoms (O6 and O7) are now more-or-less on the *c*-axis and O5 is pushed into the channel wall opposite (Figure 1c). This close O5-O3 interaction is eliminated in the local structure when the phosphate group is replaced by a type B carbonate ion and the O3 oxygen atom in question is removed entirely from the structure, as indicated by the dashed arrow in Figure 3, thus coupling the type A and B carbonate ion defects.

The removal of the symmetry-related O3 atom in the structural fragment depicted in Figure 3 also provides an explanation for displacement of the type A carbonate ion towards the channel wall. The O5 oxygen atom is simply drawn into the space created by the removal of the O3 oxygen atom. One can view this as a collapse of the structure in the vicinity of the O3 oxygen vacancy. This distortion is just one of the numerous adjustments required to accommodate the replacement of a complex anion with four oxygen atoms by one with three.

It has been noted above that the asymmetric stretch (ν_3) bands of type A carbonate in sodium-bearing carbonate apatites are apparently shifted to lower frequency relative to equivalent spectra for sodium-free CHAP and extensively overlap the ν_3 spectrum of type B carbonate (Figure 5b). The resulting characteristic asymmetric stretch (ν_3) band profile is similar for all sodium-bearing carbonate apatite composition series (X = OH,F,Cl) and carbonate contents investigated (Figure 6) [6, 10, 12, 13]. This consistency shows that the chemical environments of the sodium cation and A and B carbonate ions are similar in all of these sodium-bearing apatites, and independently confirms that the substituents are likely coupled locally as a defect cluster, as depicted in Figure 3.

Proportion of A and B Carbonate Ions

In early research it was assumed that the carbonate content of bone mineral resided very largely in the calcium phosphate matrix; i.e., bone mineral was thought to be a type B carbonate apatite and the apatite channel had a largely passive role in bone physiology. This conclusion was based on the understanding that the typical two-peak asymmetric stretch (ν_3) region spectrum of bone mineral and other biological apatites represented type B carbonate to the exclusion of significant amounts of channel (type A) carbonate. However, Fleet [14] showed that the carbonate asymmetric stretch (ν_3) spectra give an ambiguous indication of the proportion of A and B carbonate ions, due to extensive overlap of the A and B carbonate ion doublet bands in sodium-bearing carbonate apatites (Table 2; Figure 5b). A better estimate of the true proportion of A and B carbonate ions is afforded by the singlet infrared out-of-plane bend (ν_2) bands. In apatite biomaterials the carbonate out-of-plane bend (ν_2) region is normally interpreted to be a composite of singlet bands at about 879 cm^{-1} (A carbonate), 871 cm^{-1} (B carbonate), and 866 cm^{-1} (labile carbonate; type L), using either deconvolution [24, 25] or peak fitting [14, 27, 28] procedures (Figure 7).

Following single-crystal X-ray structure study of high-pressure-temperature synthesized carbonate apatite crystals [9, 10, 12, 13], Fleet [14] demonstrated that the ratio of the fitted out-of-plane bend (ν_2) band areas for A and B carbonate ions (B/A) is in good agreement with B/A calculated from the corresponding X-ray structure site occupancies (Table 1; Figure 2). This agreement exists for all high-pressure-temperature synthesized crystals and is independent of the composition series investigated (i.e., X = OH,F,Cl) and of sodium content. The areas of corresponding A and B Gaussian components of ν_2 spectra are indeed in proportion to the amounts of carbonate present, and the effects of difference in molar extinction coefficient for A and B carbonate are essentially insignificant, for all carbonate apatites. Figure 2 establishes the FTIR out-of-plane bend (ν_2) band area as a viable independent method for estimating the proportion of carbonate species in CHAP precipitated from aqueous solutions, apatite biominerals and francolite, and reinforces earlier studies on ν_2 bands in infrared spectra of biological CHAP [24, 25].

Influence of Alkali Metals

The alkali metals are minor-trace components in compact bone; the contents in the mineral fraction of bovine cortical bone being about 1.0 wt.% for sodium and 0.07 wt.% for potassium. Hence, sodium and especially potassium could be ignored in laboratory studies on the inorganic chemistry of bone mineral were it not for the marked influence of sodium on the infrared spectra of AB CHAP, CCLAP and CFAP (Figures 5,6). The anomalous displacement to lower frequency of the asymmetric stretch (ν_3) bands for type A carbonate in high-pressure-temperature synthesized sodium-bearing AB carbonate apatites is now well established. This effect results in a significant underestimation of the contribution of type A carbonate to CHAP and, thus, may have relevance for the interpretation of the infrared spectra of biological apatites which in early studies were generally considered to be dominated by type B carbonate.

An important feature of the infrared spectra of sodium-free AB CHAP is that the band at about 1545 cm^{-1}, which is the high-frequency limb of the type A doublet, remains prominent for all samples having appreciable amounts of type A carbonate confirmed independently from either the out-of-plane bend (ν_2) spectra or X-ray structure analysis (Figure 5c). In marked contrast, although the type A carbonate sites are fully occupied in sodium-bearing AB CHAP (LM005), the 1545 cm^{-1} band is essentially absent (Figure 5b). The inference here is that the absorption associated with the majority of the type A carbonate ions has been shifted, presumably into the region of the composite 1475-1445 cm^{-1} band. This same anomalous behavior is presently noted for all sodium-bearing AB CHAP samples synthesized at high temperature [29, 30] (Figure 5d).

The synthesis of potassium-bearing AB CHAP by solid state reaction at high temperature in a nearly dry CO_2 atmosphere by Verbeeck et al. [31] adds a chemical dimension to this problem. CHAP compositions were controlled by the exchange relationship:

$$[K + CO_3 = Ca + PO_4] \quad (1)$$

Formula amounts of potassium and type A and type B carbonate ranged from 0.0, 0.70 and 0.06 (their sample 7) to 1.0, 0.90 and 1.03 (sample 6), respectively.

Sample 7 is a type A potassium-bearing CHAP, and sample 6 is effectively the potassium analogue of sodium-bearing AB CHAP LM005 (Figure 8). The infrared spectra for sample 6 are clearly in good agreement with LM005, and the visually estimated areas for type A and type B carbonate in the out-of-plane bend (ν_2) spectra for samples 6 and 7 are more-or-less in proportion to the amounts estimated to be present.

Note that the high-frequency limb of the type A doublet at about 1542 cm^{-1} in the asymmetric stretch (ν_3) spectra remains resolved, even at the maximum type B carbonate content investigated, and does not shift position with increase in total carbonate content. But the area of this high-frequency limb decreases progressively with increase in type B carbonate content from sample 7 to sample 6 (Figure 8a), giving the misleading impression that type B carbonate makes a far greater contribution to the total absorption in the asymmetric stretch (ν_3) region than type A carbonate.

These observations point to a second environment for type A carbonate in the *c*-axis channel of potassium-bearing AB CHAP, having asymmetric stretch (ν_3) doublet bands that are shifted to lower frequency and located entirely within the broad band of infrared absorption that extends from about 1500 to 1418 cm^{-1} (Figure 8a). This second type A environment is presently labeled type A(b) and suggested to represent the fraction of type A carbonate coupled to type B carbonate. The content of A(b) carbonate increases progressively with increase in the content of type B carbonate and potassium content, and attains a maximum value in sample 6. At low contents of potassium and type B carbonate (e.g., sample 7), the channel carbonate ions are mainly in the orientation established for sodium-free type A CHAP, with two oxygen atoms near the *c*-axis and the plane of the carbonate ion rotated 14° counterclockwise, as depicted in Figure 1b. This fraction of type A carbonate is presently labeled type A(a) and is equivalent to type A carbonate in sodium-free A CHAP (Figure 5a,c). The A(b) carbonate ion is similarly oriented with two oxygen atoms near the *c*-axis, but now with the O5 oxygen pushed into the channel wall and the plane of the carbonate ion rotated 13° clockwise.

In summary, the broad envelope of absorption in the asymmetric stretch spectrum of sample 6 (Figure 8a) represents six main bands, two for A(a), two for A(b) and two for B carbonate, whose frequencies are more-or-less independent of bulk composition. The progressive change in the ν_3 spectra for intermediate compositions (see Figure 2 of Verbeeck et al. [31]) represents only change in band intensity; i.e., diminution of A(a) and enhancement of A(b) and B with increase in potassium content.

The characteristic shift of the A(b) doublet to lower frequency is readily attributed to change in the nearest-neighbor Ca2 cation population of the channel carbonate ion.

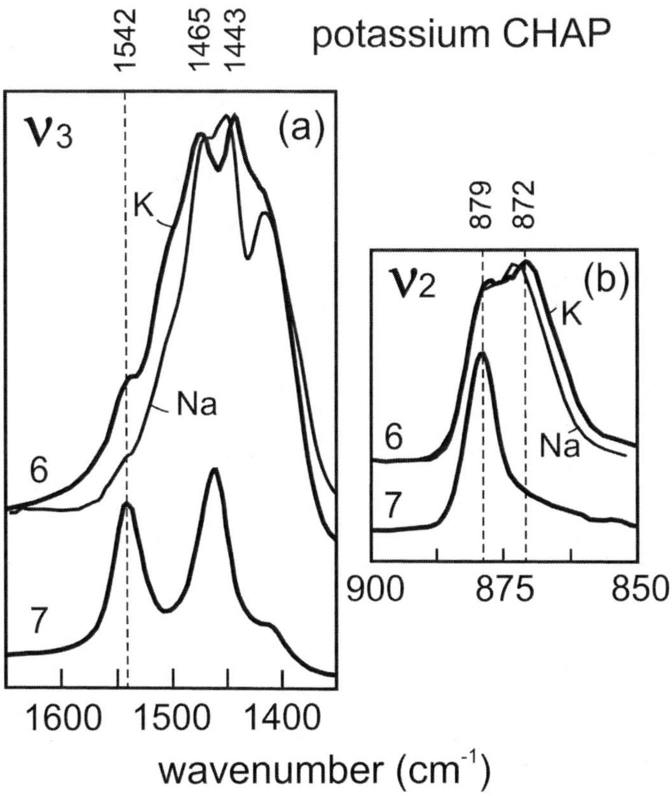

Figure 8. Infrared spectra for potassium-bearing AB CHAP at limiting low (experiment No. 7) and high (No. 6) potassium contents [31] (K), compared with spectra for sodium-bearing AB CHAP LM005 (Na).

The small fraction of sodium cations on Ca2 sites is expected to be concentrated in the channel wall in the vicinity of A(b)-B carbonate ion clusters, forming a coupled Na-A(b)-B carbonate cluster and weakening the bonding to the channel carbonate ion. Evidently, the high-frequency limb of the ν_3 doublet for the channel carbonate ion is sensitive to change in the nearest-neighbor cation charge field but relatively insensitive to other stereochemical details. As a result of the coupling of Na, A(b) and B substituents, the overall profile of the infrared absorption in the asymmetric stretch (ν_3) region is independent of the actual contents of sodium and A and B carbonate ions present, as long as these three components are present in broadly similar amounts. The characteristic two-band asymmetric stretch (ν_3) spectrum of sodium-bearing carbonate apatites synthesized at high temperature is unchanged even when as little as one percent of the large cation complement is sodium, as it is in the present samples of CFAP (Table 1; Figure 6).

Biological Apatites

The most relevant high-pressure-temperature products for understanding the crystal chemistry of biological apatites are those of the sodium-bearing AB CHAP experiments LM006 and LM005 (Table 1), which yielded approximate CHAP formulae of

$Ca_{9.5}Na_{0.5}(PO_4)_{5.5}(CO_3)_{0.5}(OH)(CO_3)_{0.5}$ for LM006 and $Ca_9Na(PO_4)_5(CO_3)(CO_3)$ for LM005. Although the crystal growth conditions for aqueous-precipitated CHAP are closer to those of the biological systems, it is not too surprising that the essential features of the structure of bone mineral and other biological apatites are reproduced by high-pressure-temperature synthesis, given the extensive pressure-temperature stability field of calcium phosphate apatites [2, 32]. There is no evidence of phase transition in this class of apatites up to the limits of pressure and temperature investigated for LM006 (1.0 GPa, 1200°C), and the principal changes on quenching to ambient conditions are likely to have been limited to relaxation of the effects of compression and thermal expansion.

The use of LM005 and LM006 as analogue materials for biological apatites is supported by close comparison of infrared spectra and X-ray powder diffraction patterns, taking into account that the spectra and powder patterns for the biological materials are degraded by low degree of crystallinity and nanocrystalline grain size [33], and the asymmetric stretch (ν_3) spectra of bone and dentin are largely obscured by bands from amide absorption. The overall similarity of the infrared spectra is especially reassuring, given that carbonate ion spectra are sensitive to slight change in the nearest-neighbor cation charge field. The good agreement between the asymmetric stretch (ν_3) and out-of-plane-bend (ν_2) spectra for human enamel [19] and LM005 CHAP (Figure 6 of Fleet and Liu [10]) is surely confirmation that the carbonate environments in these biological and synthetic carbonate apatites are similar. Other literature examples [34, 35] support the suggestion that displacement of the ν_3 doublet for type A carbonate to lower frequency is also a characteristic feature of biological apatites, and results in low estimates for the A carbonate content. On the other hand, the fitted out-of-plane bend (ν_2) spectra show that type A carbonate is invariably present in significant amounts in biological apatites [27, 28], as it is in sodium-bearing CHAP.

It has been known for some time that the carbonate content of bone is the reservoir for controlling excess acidity in the body [36-38]. In particular, in vivo experiments have shown that a considerable fraction of the carbonate associated with bone (about 30-40 %) [39, 40] is exchangeable with carbon species in body fluids. This exchangeable fraction is believed to be hydrogencarbonate (bicarbonate) located either in the hydration shell or surface layer of the carbonate hydroxyapatite nanocrystals, although these suggestions have not been confirmed by subsequent study. The presence of the hydrogencarbonate ion in carbonate hydroxyapatite was reported for the first time by Fleet and Liu [41], who quenched sodium-bearing type AB CHAP from high pressure and temperature under a high partial pressure of CO_2. Although the asymmetric stretch (ν_3) infrared spectrum of hydrogencarbonate is distinctive, it was not helpful in assigning the structural location of this new carbonate species in CHAP. However, a band at 837 cm^{-1} in the out-of-plane bend (ν_2) region is diagnostic of hydrogencarbonate and permits quantification of the amounts of the various carbonate ion species present [6].

It was appreciated in Fleet and Liu [41] that the infrared bands for hydrogencarbonate in CHAP crystals diminished in intensity with aging of the quenched high-pressure-temperature product for up to one year at low and room temperature. This change was quantified using the ν_2 band area method, showing that the sums of the contents of channel carbonate (i.e., type A, labile (L) and hydrogencarbonate) for three separate CHAP experiments are essentially unchanged with aging, as are the contents of type B carbonate [6]. The gain in type A and labile carbonate on aging is equal to the loss of hydrogencarbonate, within experimental error, whereas the amount of B carbonate replacing phosphate in the calcium phosphate matrix of the CHAP structure is unchanged. This agreement is confirmation that hydrogencarbonate and labile carbonate in these synthetic CHAP crystals are both restricted to the *c*-axis structural channel. Moreover, the carbonate component of the hydrogencarbonate ion does not appear to be mobile in the apatite channel at room temperature, in relatively large CHAP crystals. It is not lost from the crystal system in the form of gaseous CO_2. And the matrix carbonate (type B) is conserved as well.

The synthetic AB CHAP crystals show that the apatite channel is a second, and possibly alternative, location for hydrogencarbonate ions associated with bone mineral nanocrystals. The limited literature studies in the infrared out-of-plane bend (ν_2) region reveal appreciable amounts of type A and disordered labile (L) carbonate ions in apatite extracted from bone and dental enamel [1, 24, 25].

There are sufficient channel carbonate ions in biological apatite crystals to account for the fraction of exchangeable CO_2 reported by Neuman and Mulryan [39] and Poyart et al. [40, 42] especially if the labile carbonate ions are assumed to be channel species. The observations on change in the proportion of carbonate and hydrogencarbonate species on aging [6] point to a possible role for the apatite channel in mediating acid-base reactions in the body [43]. It is suggested that channel constituents may account for a considerable proportion of the carbonate ions available for acid-base homeostasis and mediation of mild acidosis. In addition, the channel of bone mineral crystals could contribute to the maintenance of acid-base homeostasis and mediation of mild acidosis by consuming and releasing protons through the simplified reaction

$$H^+ + CO_3^{2-} \rightleftharpoons HCO_3^-, \qquad (2)$$

where carbonate and hydrogencarbonate ions are bound species and the proton migrates in the channel from or to the extracellular fluid. For the forward reaction, the excess positive charge in the channel wall is dissipated in the surrounding structure and eventually neutralized by the release of a Na^+ counter ion into the extracellular fluid or hydration shell.

EXCESS FLUORINE IN FRANCOLITE

Francolite, a variety of carbonate fluorapatite, is the ore mineral of phosphorite deposits and the source of phosphorus for agricultural fertilizers and the manufacture of industrial chemicals. The mineral has proven to be very difficult to characterize because it occurs mainly as poorly crystallized, nanoscale grains intimately intermixed with gangue (non-ore) minerals. Infrared spectra show that the carbonate content is exclusively present as type B, substituting for phosphate [19], and chemical analyses are assumed to correspond approximately to the substitution formula $Ca_{10-x-y}Na_xMg_y(PO_4)_{6-z}(CO_3)_zF_{0.4z}F_2$, showing an excess of fluorine ideally equal to 0.4 times the B carbonate content [44,45]. There has been extensive discussion of the structural location and chemical state of the excess fluorine in francolite, and especially of the proposal that excess fluorine is present in a tetrahedral $CO_3{\cdot}F^{3-}$ complex anion; see recent article by Yi et al. [46]

Although excess fluorine was not anticipated in an earlier study of high-pressure-temperature synthesized CFAP crystals [12], difference Fourier maps for the X-ray structure of one experimental product LM142 revealed a weak difference electron density peak at x = 0.387, y = 0.213, z = 0.5 (equipoint 12i), which corresponded to a realistic structural position (labeled "Fx") for a fluoride anion. A second CFAP product also shows significant residual electron density at the location of Fx, but two other CFAP products did not. Thus, the Fx feature appears to be present close to the minimum detection limit of the Fourier map technique.

The excess fluoride anion is in near equilateral triangular coordination with one Ca1 cation, at a unit-cell height $z \approx 0.5$, and two Ca2 cations, at z = 1/4 and 3/4, leaving the triangular $FxCa_3$ cluster inclined at an angle of 56° to the c-axis (Figure 9). Interatomic (O-O) distances and bond angles of approximately 2.28, 2.42 and 2.32 Å and 115, 122 and 123°, respectively, are similar to those of the channel fluoride anion in CFAP (2.303 Å and 120°), which is a good indication that Fx does indeed represent minor-trace amounts of fluoride.

In the average structure, Fx is close (at 1.99 Å) to an O1 oxygen atom of the phosphate (PO_4) tetrahedron (Figure 9), but this Fx-O1 interatomic distance would be lengthened by local relaxation of the host structure in the vicinity of the Fx defect. Fx is located much closer (at 1.17 Å) to one of the O3 oxygen atoms of the phosphate tetrahedron in the average structure, but it is this feature that connects Fx with the content of type B carbonate ions. Following the earlier discussion on coupling of sodium and A and B carbonate ion substituents, this prohibitively short Fx-O3 interaction in the average structure is eliminated when the phosphate group is substituted locally by a type B carbonate ion, and the O3 oxygen atom in question is removed entirely from the structure.

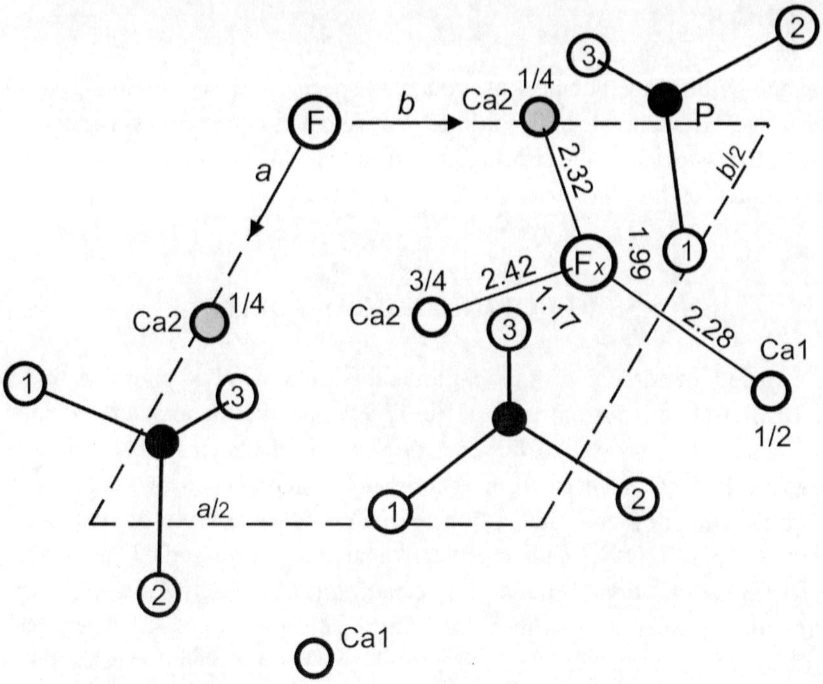

Figure 9. Location of excess fluoride anion (Fx) in the average structure of carbonate fluorapatite CFAP LM142 [12]: interatomic distances are Ångstroms; note that avoidance of the prohibitively close Fx-O3 interaction, in the local structure, is an explanation for the correlation of the contents of excess fluorine and type B carbonate; symmetry-equivalent Fx site is plotted.

The residual peak height for LM142 corresponds to an F*x* site occupancy of about 1 % and, since there are twelve F*x* sites per unit cell, to an excess fluorine content of about 0.13 per formula unit (pfu).

The B carbonate content of LM142 is 0.23 pfu (Table 1; Figure 6): therefore, only about half of the available F*x* sites are occupied.

If the carbon atom of the B carbonate ion is located at the geometric center of the tetrahedron, the F*x*-C interatomic distance would be 2.47 Å (Figure 9). However, the X-ray structure study of CFAP [12] indicates that the B carbonate ion is more likely to be closely parallel to the inclined tetrahedral face furthest from the F*x* site. This configuration would extend the F*x*-C distance to about 2.94 Å.

The new information on the structure of CFAP provides a ready crystal-chemical explanation for the association of excess fluorine with type B carbonate in francolite. The F*x* anion occupies an interstitial position in the structure of FAP, but an F*x* site cannot be occupied in the absence of an adjacent type B carbonate ion, and it is suggested that the content of B carbonate has to be significantly in excess of F*x*, as encountered in natural francolite compositions.

However, the present structure does not reveal evidence for a fixed ratio of F*x*:B, as required by the formula of McClellan [45]. Moreover, although the F*x* anion is located adjacent to a type B carbonate ion, it is not bonded to the carbon atom, as suggested in the proposed $CO_3{\cdot}F^{3-}$ complex ion. An F*x*-C interatomic distance of about 2.94 Å is too long for a bonded interaction, which is estimated from covalent radii [47] to be about 1.33 Å. The association of the F*x* anion with a carbonate ion is now readily attributed to the removal of the blocking O3 oxygen following introduction of the B carbonate ion. This is yet another example of the clustering of substituents in carbonate apatites.

The structural location of the excess fluorine content in francolite proposed in Fleet and Liu [12] and this study is supported by the recent theoretical simulation of Yi et al. [46], who used density functional theory calculations to propose a carbonate-fluoride tetrahedron defect model.

But the two structural models do differ in details. In particular, in the theoretical study, F*x* is displaced to $x = 0.353$, $y = 0.113$, $z = 0.5$, which locates it close to the blocking O3 oxygen atom. This displacement distorts the $FxCa_3$ cluster from the near-trigonal symmetry in Fleet and Liu [12], giving interatomic distances and bond angles of approximately 2.61, 1.98 and 2.65 Å and 118, 126 and 100°, respectively, relative to the Ca1 and Ca2 positions for the LM142 structure.

CONCLUSION

Asymmetric stretch (ν_3) infrared spectra give an ambiguous indication of the proportion of A and B carbonate in biological and sodium-bearing synthetic apatites due to the complex overlap of the various spectral contributions. However, the out-of-plane bend (ν_2) band method for the B/A ratio, which is based on well characterized synthetic apatites, can be extended to biological apatites. Wherever possible, the areas of ν_2 bands should be used to estimate the proportion of A and B carbonate ions in bone mineral and other biological apatites. The excess fluorine content of francolite is accommodated in an interstitial position of the carbonate fluorapatite structure, as an isolated fluoride anion.

ACKNOWLEDGMENT

I thank the natural Sciences and Engineering Research Council of Canada for financial support.

REFERENCES

[1] Elliott, J. C., (2002). Calcium phosphate biominerals. In Reviews in Mineralogy and Geochemistry: Vol. 48. Phosphates (Kohn, M. J., Rakovan, J., Hughes, J. M., edits.), Mineralogical Society of America, Washington, DC, pp 427-453.

[2] Liu, Xi, Shieh, S. R., Fleet, M. E., Zhang, L., He, Q., (2011). Equation of state of carbonated hydroxylapatite at ambient temperature up to 10 GPa: Significance of carbonate. *Amer. Mineral.*, 96, 74-80.

[3] Pasero, M., Kampf, A. R., Ferraris, C., Pekov, I. V., Rakovan, J. R., White, T. J., (2010). Nomenclature of the apatite supergroup minerals. *Eur. J. Mineral.*, 22, 163-179.

[4] Hughes, J. M., Cameron, M., Crowley, K. D., (1989). Structural variations in natural F, OH, and Cl apatites. *Amer. Mineral.*, 74, 870-876.

[5] Fleet, M. E., Liu, X., Liu, Xi, (2011). Orientation of channel carbonate ions in apatite: Effect of pressure and composition. *Amer. Mineral.*, 96, 1148-1157.

[6] Fleet, M. E., (2012). The carbonate ions in hydroxyapatite and biological apatite. In Calcium Phosphate: Structure, Synthesis, Properties, and Applications, (Heimann, R. B., edit.), Nova Science Publishers, New York, pp. 41-61.

[7] Fleet, M. E., (2013). The carbonate ion in hydroxyapatite: Recent X-ray structure and infrared results. *Frontiers in Bioscience,* (Elite Edit) 5 (January 1), pp. 643-652.

[8] Fleet, M. E., Liu, X., (2003). Carbonate apatite type A synthesized at high pressure: new space group ($P\bar{3}$) and orientation of channel carbonate ion. *J. Solid State Chem.*, 174, 412-417.

[9] Fleet, M. E., Liu, X., (2004). Location of type B carbonate ion in type A-B carbonate apatite synthesized at high pressure. *J. Solid State Chem.*, 177, 3174-3182.

[10] Fleet, M. E., Liu, Xi, (2007). Coupled substitution of type A and B carbonate in sodium-bearing apatite. *Biomaterials,* 28, 916-926.

[11] Fleet, M. E., Liu X., (2005). Local structure of channel ions in carbonate apatite, *Biomaterials,* 26, 7548-7554.

[12] Fleet, M. E., Liu, Xi, (2008). Accommodation of the carbonate ion in fluorapatite synthesized at high pressure. *Amer. Mineral.*, 93, 1460-1469.

[13] Fleet, M. E., Liu, Xi, (2008). Type A-B carbonate chlorapatite synthesized at high pressure. *J. Solid State Chem.*, 181, 2494-2500.

[14] Fleet, M. E., (2009). Infrared spectra of carbonate apatites: ν_2-region bands. *Biomaterials,* 30, 1473-1481.

[15] Suetsugu, Y., Shimoya, I., Tanaka, J., (1998). Configuration of carbonate ions in apatite structure determined by polarized infrared spectroscopy. *J. Amer. Ceram. Soc.*, 81, 746-748.

[16] Suetsugu, Y., Takahashi, Y., Okamura, F. P., Tanaka, J., (2000). Structure analysis of A-type carbonate apatite by a single-crystal X-ray diffraction method. *J. Solid State Chem.*, 155, 292-297.

[17] Ivanova, T. I., Frank-Kamenetskaya, O. V., Kol'tsov, A. B., Ugolkov, V. L., (2001). Crystal structure of calcium-deficient carbonated hydroxyapatite. Thermal decomposition. *J. Solid State Chem.*, 160, 340-349.

[18] Tonegawa, T., Ikoma, T., Yoshioka, T., Hanagata, N., Tanaka, J., (2010). Crystal structure refinement of A-type carbonate apatite by X-ray powder diffraction. *J. Mater. Sci.*, 45, 2419-2426.

[19] Elliott, J. C., (1994). Structure and Chemistry of the Apatites and Other Calcium Orthophosphates, Elsevier, Amsterdam.

[20] Carlström, D., (1968). Mineralogical carbonate-containing apatites. In *Proceedings of International Symposium on Structural Properties of Hydroxyapatite and Related Compounds*, (Brown, W.E., Young, R.A., edits.), Gaithersburg, Maryland, Chap 10.

[21] Bonel, G., (1972). Contribution à l'étude de la carbonatation des apatites. I. - Synthèse et étude des propriétés physico-chimiques des apatites carbonatées du type A. *Annales Chim.,* Paris 7, 65-88.

[22] Wilson, R. M., Elliott, J. C., Dowker, S. E. P., Smith, R. I., (2004). Rietveld structure refinement of precipitated carbonate apatite using neutron diffraction data. *Biomaterials,* 25, 2205-2213.

[23] Shimoda, S., Aoba, T., Moreno, E. C., Miake, Y., (1990). Effect of solution composition on morphological and structural features of carbonated calcium apatites. *J. Dent. Res.*, 69, 1731-1740.

[24] Rey, C., Renugopalakrishnan, V., Shimizu, M., Collins, B., Glimcher, M. J., (1991). A resolution-enhanced Fourier transform infrared spectroscopic study of the environment of the CO_3^{2-} ion in the mineral phase of enamel during its formation and maturation. *Calcif. Tissue Int.*, 49, 259-268.

[25] Rey, C., Collins, B., Goehl, T., Dickson, I. R., Glimcher, M. J., (1989). The carbonate environment in bone mineral: A resolution-enhanced Fourier transform infrared study, *Calcif. Tissue Int.*, 45, 157-164.

[26] Baxter, J. D., Biltz, R. M., Pellegrino, E. D., (1966), The physical state of bone carbonate: A comparative infra-red study in several mineralized tissues. *Yale J. Biol. Medicine,* 38, 456-470.

[27] Paschalis, E. P., DiCarlo, E., Betts, F., Sherman, P., Mendelsohn, R., Boskey, A. L., (1996). FTIR microspectroscopic analysis of human osteonal bone. *Calcif. Tissue Int.*, 59, 480-487.

[28] Verdelis, K., Lukashova, L., Wright, J. T., Mendelsohn, R., Peterson, M. G. E., Doty, S., Boskey, A. L., (2007). Maturational changes in dentin mineral properties. *Bone,* 40, 1399-1407.

[29] Nelson, D. G. A., Featherstone, J. D. B., (1982). Preparation, analysis, and characterization of carbonated apatites. *Calcif. Tissue Int.*, 34, S69-S81.

[30] Driessens, F. C. M., Verbeeck, R. M. H., Heijligers, H. J. M., (1983). Some physical properties of Na- and CO_3-containing apatites synthesized at high temperatures. *Inorg. Chim. Acta,* 80, 19-23.

[31] Verbeeck, R. M. H., De Maeyer, E. A. P., Driessens, F. C. M., (1995). Stoichiometry of potassium- and carbonate-containing apatites synthesized by solid state reactions. *Inorg. Chem.*, 34, 2084-2088.

[32] Fleet, M. E., Liu, Xi, Shieh, S. R., (2010). Structural change in lead fluorapatite at high pressure. *Phys. Chem. Minerals,* 37, 1-9.

[33] LeGeros, R. Z., (1991). Calcium Phosphates in Oral Biology and Medicine, Karger, Basel.

[34] Arends, J., Davidson, C. L., (1975). HPO_4^{2-} content in enamel and artificial carious lesions. *Calcif. Tissue Res.*, 18, 65-79.

[35] Farlay, D., Panzcer, G., Rey, C., Delmas, P. D., Boivin, G., (2010). Mineral maturity and crystallinity index are distinct characteristics of bone mineral. *J. Bone Miner. Metab.*, 28, 433-445.

[36] Rey, C., Combes, C., Drouet, C., Glimcher, M. J., (2009), Bone mineral: update on chemical composition and structure. *Osteoporos Int.*, 20, 1013-1021.

[37] Green, J., Kleeman, C. R., (1991). Role of bone in regulation of systemic acid-base balance. *Kidney Int.*, 39, 9-26.

[38] Bettice, J. A., (1984). Skeletal carbon dioxide stores during metabolic acidosis. *Amer. J. Physiol.*, 247, F326-F330.

[39] Neuman, W. F., Mulryan, B. J., (1967). Synthetic hydroxyapatite crystals. III. The carbonate system. *Calcif. Tissue Res.*, 1, 94-104.

[40] Poyart, C. F., Bursaux, E., Fréminet, A., (1975). The bone CO_2 compartment: evidence for a bicarbonate pool. *Respiration Physiol.,* 25, 89-99.

[41] Fleet, M. E., Liu, Xi, (2007). Hydrogen-carbonate ion in synthetic high-pressure apatite. *Amer. Mineral.*, 92, 1764-1767.

[42] Poyart, C. F., Fréminet, A., Bursaux, E., (1975), The exchange of bone CO_2 *in vivo.* *Respiration Physiol.*, 25, 101-107.

[43] Bushinsky, D. A., Smith, S. B., Gavrilov, K. L., Gavrilov, L. F., Li, J., Levi-Setti, R., (2002). Acute acidosis-induced alteration in bone bicarbonate and phosphate. *Amer. J. Physiol- Renal Physiol.*, 283, F1091-F1097.

[44] McClellan, G. H., Lehr, J. R., (1969). Crystal chemical investigation of natural apatites. *Amer. Mineral.*, 54, 1374-1391.

[45] McClellan, G. H., (1980). Mineralogy of carbonate fluorapatites. *J. Geol. Soc. London,* 137, 675-681.

[46] Yi, H., Balan, E., Gervais, C., Segalen, L., Fayon, F., Roche, D., Person, A., Morin, G., Guillaumet, M., Blanchard, M., Lazzeri, M., Babonneau, F., (2013). A carbonate-fluoride defect model for carbonate-rich fluorapatite. *Amer. Mineral.*, 98, 1066-1069.

[47] Cordero, B., Gomez, V., Platero-Prats, A. E., Reves, M., Echeverria, J., Cremades, E., Barragan, F., Alvarez, S., (2008). Covalent radii revisited. *Dalton Trans.*, 2008, 2832-2838.

In: Apatite
Editors: M. Iafisco and J. M. Delgado-López
ISBN: 978-1-63321-500-9

Chapter 5

DIAGNOSIS OF APATITE DEPOSITS IN SKIN CALCIFIED TISSUES: BIOMEDICAL APPLICATIONS OF VIBRATIONAL SPECTROSCOPY

Shan-Yang Lin*
Laboratory of Pharmaceutics and Biopharmaceutics,
Department of Biotechnology and Pharmaceutical Technology,
Yuanpei University, Hsin Chu, Taiwan, ROC

ABSTRACT

Skin calcification is an uncommon disorder for the deposition of insoluble calcium salts in cutaneous tissues. Until recently, clinical diagnosis of the calcified deposits occurred in or under the skin is usually achieved via time-consuming histopathological or immunohistochemical evaluation with several complicated procedures, which is always based on the empirical evidence and clinical experience of physicians. The use of FT-IR and Raman vibrational spectroscopy has been recently shown to have great potential as a diagnostic tool for different diseased tissues, as compared with other methods. The purpose of this chapter is to introduce the pathological calcification and apatite formation in the skin, and review the biomedical diagnosis of the chemical composition and carbonate apatite in human skin calcified tissues using advanced vibrational spectroscopy. The increased biomedical applications of these diagnostic tools suggest that both FT-IR and Raman vibrational spectroscopic techniques will soon become important clinical tools to complement the standard methods currently used in various fields of clinical diagnosis.

Keywords: Skin, calcification, diagnosis, FT-IR, Raman, hydroxyapatite, carbonate apatite

* Corresponding author: Prof. Shan-Yang Lin, Ph.D., Yuanpei University, No. 306, Yuanpei Street, Hsin Chu, 30015, Taiwan, ROC. Email: sylin@mail.ypu.edu.tw.

INTRODUCTION

The progressive movement of calcium and other minerals from the bone and into the soft tissues of the body is one of the more crucial symptoms of the aging process [Boskey 2003; Ehrlich et al., 2008; 2009; Mayer 2006]. The former is mineral loss (demineralization), the latter is mineral uptake (mineralization/calcification). With the increase of aging, the deficient absorption of minerals may be often suffered and cause a progressive mineral deficiency. In order to meet many regulatory demands in human body, the body robs its own resources, particularly from bones, leading to demineralization. On the other hand, much of the calcium leached from bones may ends up gradually accumulating in other human soft tissues, leading to soft tissue calcification [Goto 2012; Peacock 2010]. Excess calcium is usually dissolved in the bloodstream for excretion in the urine, but a certain amount may be re-absorbed into an area of the body tissues; this re-absorpted calcium then calcify the tissue. Over time, the calcification can more exacerbate and disrupt normal bodily processes. Although the real mechanism of calcification in human body is still unclear, a variety of factors such as aging, or the results of an injury, infection, inflammation, disorders of calcium metabolism, or genetic or autoimmune disorders had been proposed to be correlated to the calcification [Kumar et al., 2010; Fernandez-Flores 2011].

With age, the human body also loses its ability to regulate calcium balance and subsequently undergoes systemic calcification. During this process, calcium deposits tend to accumulate in tissues throughout the body. Physiological calcification (mineralization) usually occurs within hard tissues, whereas pathological calcification occurs in soft tissues. Both calcifications are a complex multifactorial process. Physiological calcification is restricted to skeletal tissues and teeth, as well as the hypertrophic zone of growth plate cartilage, in which this highly complex process plays critical functions during development of these tissues [Kirsch 2007]. Pathologic calcification is a common process in a wide variety of disease states; it implies the abnormal deposition of calcium salts, together with smaller amounts of iron, magnesium, and other minerals [Kumar et al., 2010]. Therefore, it is recognized as a passive process resulting from injury, disease, or old age [Orimo, 2010; Bazin et al., 2012; 2013]. Many chronic diseases have been shown to be associated with calcification, including arterial stiffening, kidney/bladder/prostate stones, arthritis, cataracts, aortic valve stenosis, bone fractures, skin calcinosis, and coronary atherosclerosis [Kart et al., 1992].

Pathological Calcification and Apatite Formation

Pathological calcification of soft tissues occurs when calcium and other mineral salts are deposited in a tissue or in a passage [Black and Kanat, 1985; Golub, 2011; Chander and Gordon, 2012; Kirsch, 2012]. Calcium salt deposits in soft tissues other than bones and teeth gradually grow and aggregate into clusters, causing loss of flexibility and pliability of soft tissues, subsequently leading to sclerosis, thickening, and/or deterioration [Kirsch, 2006; Peacock, 2010]. Abnormal calcification within soft tissues is an important structural or functional factor that causes many chronic diseases due to the induction of inflammation by mineral crystals [Kart et al., 1992; Kirsch, 2006; Peacock, 2010; Golub, 2011]. Calcification becomes a problem when its location, shape or size interferes with organ function. Almost all tissues are easily calcified. Of these, the skin, kidney, tendons, and cardiovascular tissues appear to be particularly prone to mineralization [Giachelli, 1999; Bonucci, 2006; Kawasaki et al., 2009; Golub, 2011; Lin, 2014].

Calcium phosphates are the principal inorganic constituent of both normal and pathological calcifications in biological systems. Almost all hard tissues in human body are made of phosphate minerals, and among these, hydroxyapatite (HA) predominates. Most pathologic calcifications throughout the body contain mixtures of carbonate-substituted HA and octacalcium phosphate [Ciftçioğlu and McKay, 2010]. HA is a naturally occurring mineral form of calcium apatite with the formula $Ca_{10}(PO_4)_6(OH)_2$. The apatite structure allows cationic (for Ca^{2+}) and anionic (for PO_4^{3-} and/or OH^- groups) substitutions [Dorozhkin, 2009; 2010; Sader et al., 2013]. Such substitutions affect the crystallography (crystal size, lattice parameters), physicochemical properties (e.g. solubility), and cell and tissue response [Pan and Darvell, 2010]. The effect of CO_3^{2-} substitution on the apatite lattice may be directly related to the dissolution-precipitation cycles that occur during bone tissue regeneration [Porter et al., 2005; Dorozhkin, 2011]. Biological apatite is usually referred to as carbonate apatite [De Aza et al., 2005]. Carbonate can be incorporated into the apatite lattice ($Ca_{10}[PO_4]_6[OH]_2$) in the following ways: A-type carbonate apatite, in which the OH^- group is substituted by CO_3^{2-} ; B-type carbonate apatite, in which the PO_4^{3-} group is substituted by CO_3^{2-} ; and L-type (labile) carbonate apatite, in which the carbonate is located within the surface hydration layer of the apatite crystals [Elliott et al., 1985; LeGeros, 1991; Wopenka and Pasteris, 2005; Peroos et al., 2006; Kannan et al., 2011].

The incorporation of CO_3^{2-} in the apatite lattice has been reported to cause a decrease in the crystal shape, resulting in significant differences in the specific area which contributed to the release of Ca^{2+} in the dissolution kinetics [Sader et al., 2013]. In addition, carbonate apatites present higher solubility when compared to carbonate-free apatite due to the fact that Ca-PO_4 bonds are stronger than Ca-CO_3 bonds, making the carbonate apatite more susceptible to acid dissolution [Wopenka and Pasteris, 2005]. The representative configurations of HA lattice and the apatite lattices with A-type and B-type substitution are shown in Figure 1 [Peroos et al., 2006]. The value of the A-type : B-type ratio in carbonate apatite is reported to increase with age in humans. A-type carbonate apatite is mostly found in old tissue, while B-type is found in young tissue.

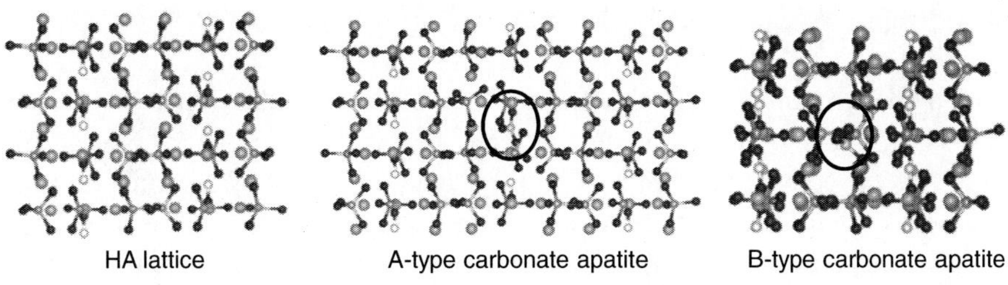

Figure 1. The representative configurations of HA lattice and the apatite lattices with A-type and B-type substitution. The location of the carbonate substitution is indicated with a circle. (modified from *Peroos* et al., 2006).

Skin Calcification

Calcinosis is a condition of unknown cause in which calcium salts build up in soft tissues. When this condition occurs in the skin and subcutaneous tissue, it is known as calcinosis cutis (CC) [Walsh and Fairley, 1995; Fernandez-Flores, 2011; Reiter et al., 2011]. It may be focal or widespread, symptomatic or asymptomatic and a variety of factors can result in this condition. The signs and symptoms of CC vary according to the underlying cause. In many cases the lesions gradually develop and are often symptomless. The lesions consist of firm or stony, flesh-colored or white-yellowish papules, plaques or nodules on the surface of the skin. A solitary lesion may develop, although multiple lesions are more common.

Over time, lesions may become tender and ulcerate, discharging chalk-like creamy material consisting of varying forms of calcium. This calcification may follow cutaneous injury; it may be secondary to connective tissue or metabolic disorders.

Based on the pathogenesis of the deposition, CC can be classified as metastatic, dystrophic, idiopathic, and iatrogenic calcifications. Dystrophic calcification appears as a result of local tissue damage in the presence of normal serum calcium and phosphate levels. Metastatic calcification is characterized by abnormal calcium and/or phosphate metabolism, leading to the precipitation of calcium within the cutaneous and subcutaneous tissues. Idiopathic CC is cutaneous calcification of unknown cause without any underlying tissue damage or metabolic disorder. Skin calcification in iatrogenic CC is a common side effect of therapy.

Clinical diagnosis of CC is usually achieved via time-consuming routine histopathologic or immunohistochemical evaluation using several complicated procedures. Moreover, the conventional visual examination is only be qualitatively identified by experienced practitioners. The von Kossa stain is a well-known and widely used method to quantify mineralization in cell cultures and tissue sections [Meloan and Puchtler, 1985; Rungby et al., 1993]. However, von Kossa staining is not specific to detecting apatitic minerals and is subject to interference from any residual buffer phosphate or halide (e.g., chloride, fluoride) present in a specimen [Bonewald et al., 2003; Morris and Finney, 2004]. The problem is especially severe in cell cultures, where small, or even large, amounts of inorganic phosphates may be misidentified by staining [Stewart et al., 2002].

Vibrational Spectroscopy and Its Biomedical Applications

Vibrational spectroscopy is an analytical tool used in the fields of analytical chemistry and measurement science that incorporates infrared (IR) and Raman spectroscopy. Both of these types of spectroscopy have been used to provide complementary information about the molecular structure of different samples for a long time. The use of these techniques has increased substantially over the past 20 years because of advances in computational and optical instrumentation technology [Colthup et al., 1990; Petrich, 2001; Mantsch et al., 2002; Thygesen et al., 2003; Sathyanarayana, 2007]. The IR and Raman spectroscopy measure the vibrational energies of molecules and depend on different selection rules, i.e., an absorption process and a scattering effect. IR spectroscopy is based on the absorption of electromagnetic radiation, whereas Raman spectroscopy relies upon inelastic scattering of electromagnetic radiation. The sample is radiated by IR light in IR spectroscopy, and the vibrations induced by electrical dipole moment are detected. In Raman spectroscopy, the sample is illuminated by a monochromatic visible or near IR light from a laser source and its vibrations during the electrical polarisability changes are determined [Adar et al., 2003; Pistorius 1995; Smith 2011]. Thus both vibrational techniques may provide complementary information, since the shapes and positions of both spectral bands on the energy scale are closely located for the specific functional groups of the molecule (Figure 2) [Lin et al., 2007].

Many vibrations are observed in both spectra to serve as a valuable confirmatory evidence of samples. Therefore, the combination of using IR and Raman spectroscopy can offer a more potential approach for analyzing intact sample and provides more detailed chemical information.

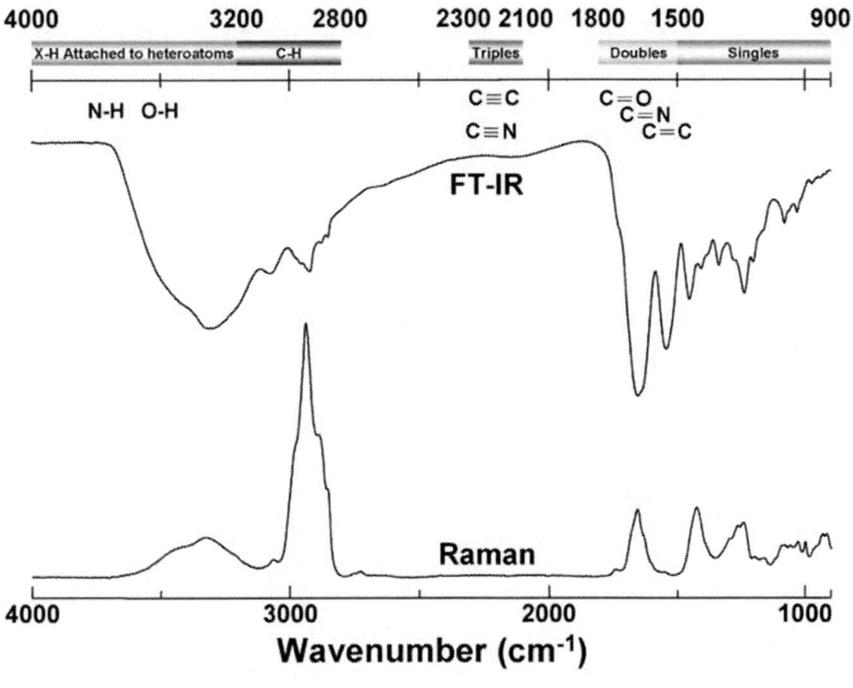

Figure 2. Typical FT-IR and Raman spectra of collagen (adapted from *Lin* et al., *2007*).

IR and Raman spectroscopy provide information on the chemical composition of all materials and can be used in chemical fingerprinting, particularly when they are combined with the Fourier transform (FT) system. When such FT instruments are also equipped with a computer system, weak signals can be measured faster and more sensitively than when using older dispersive instruments. A computerized FT instrument may also save the FT-IR or Raman spectrum of a sample for subsequent data processing. Thus, the use of these advanced techniques to investigate the intact structural components of bio-samples is of interest to a wide range of researchers in the fields of biochemistry, biomedicine, biophysics, and biomaterials [Gremlich and Yan, 2001; Crupi et al., 2004; Krafft, 2004; Lin et al., 2007; Siebert and Hildebrandt, 2008; Severcan and Haris, 2012; Lin, 2014]. More recently, FT-IR and Raman spectroscopy, used in conjunction with microscopes and computer systems, have been shown to be capable of microanalysis of minute samples. These advanced FT-IR or

Raman vibrational microspectroscopic techniques can provide further molecular information about samples at the microscopic level and with a high resolution [Wetzel and LeVine, 1998; Ellis and Goodacre, 2006; Krafft and Sergo, 2006; Walsh et al., 2007; Lin et al., 2007; Singh et al., 2012; Lin, 2014].

Both types of vibrational microspectroscopy have significant benefits compared with other diagnostic techniques for investigating the chemical components of ultrastructual tissues during various disease states. Compared with conventional histopathologic diagnosis, microspectroscopic techniques have been found to be more objective and rapid. Because these two vibrational microspectroscopic methods can distinguish qualitative and quantitative differences in the spectral characteristics of many molecules, molecular and submolecular spectral profiles can be used to define and differentiate between diseased and healthy human tissues [Clark and Hester, 1996; Kalasinsky et al., 2002; Crupi et al., 2004; Krafft and Sergo, 2006; Lin et al., 2007; Singh et al., 2012; Lin, 2014]. Specifically, mineralized tissues found in vertebrates, whether they developed via normal physiologic or pathologic processes or not, contain an analogue of the naturally occurring mineral HA or other matrices. These components are quite distinct and vibrational spectroscopy can easily be used to identify and study them [Carden and Morris, 2000; Adele and Richard, 2005; Boskey and Mendelsohn, 2005].

Diagnosis of Human Skin Calcification Using Vibrational Spectroscopy

In order to prevent the possible misidentified problem of staining [Stewart et al., 2002], FT-IR and Raman spectroscopy have been considerably applied to overcome these problems [Eikje et al., 2005; Ellis and Goodacre, 2006; Krafft et al., 2009; Chiou et al., 2010; Kendall et al., 2011; Diem et al., 2012; Singh et al., 2012; Tosi et al., 2012; Krafft and Bird, 2013; Severcan et al., 2013]. Because FT-IR and Raman spectroscopy are considered to be complementary techniques, in some cases that the combined application of both techniques has proved to be an invaluable method in the investigation of the exact information of molecular-structural fingerprints of the calcified samples. In this chapter, the applications of both advanced vibrational spectroscopic techniques to directly identify and diagnose the chemical composition of the calcified deposits in the skin of patients with dystrophic and idiopathic calcifications are reviewed.

1) Dystrophic calcification

Baldet et al. examined subcutaneous calcification in a human case of CRST (calcinosis cutis, Raynaud's phenomenon, sclerodactyly, and telangiectasia) syndrome using electron microscopy, X-ray diffraction (XRD) and IR spectroscopy [Baldet et al., 1981]. The results of XRD, IR spectroscopy, and differential thermal analysis (DTA) showed that the deposits consisted of B-type carbonate apatite. On the other hand, Daculsi et al. also used electron microscopic microanalysis and IR spectroscopy to investigate a subcutaneous heterotopic calcification from a patient with Thibierge-Weissenbach's syndrome (a type of creeping scleroderma included in CRST syndrome) [Daculsi et al., 1983]. It was demonstrated that these calcified materials were highly heterogeneous and appeared in a nonstoichiometric, carbonated, calcium ion-deficient apatitic solid phase. Three typical IR spectral peaks at 870, 1430, and 1455 cm^{-1}, corresponding to a B-type carbonate apatite, were defined. Bettoli et al.

found that the IR absorption bands at 870 and 1455 to 1430 cm^{-1} in patient suffered with porphyria cutanea tarda were observed in the IR spectra of these deposits, suggesting that a B-type carbonate apatite had occurred [Bettoli et al., 1992].

Konno et al., analyzed the calcified materials from ten human cases of CC [Konno et al., 2003]. All of the calcified materials were identified as HA by XRD examination. However, FT-IR and Raman spectral analyses of these specimens showed that each CC contained not only calcium phosphate but also calcium carbonate, a finding that was confirmed by total carbon analysis. Thus, they concluded that the components of these calcified materials were mainly crystalline HA and amorphous calcium carbonate, rather than HA alone. This indicates that vibrational spectroscopy may have benefits over XRD analysis in the determination of the composition of samples.

Cheng et al., used Raman microspectroscopy to analyze the changes in the structural conformation and chemical composition of human skin pilomatrixoma (PMX) masses [Cheng et al., 2005]. Samples were taken from two patients diagnosed with typical PMX, in which one specimen was a soft mass and the other was a hard mass with somewhat calcified deposits (Figure 3-A and B). The differences between Raman peak intensity in the normal skin dermis and soft or hard PMX masses were obvious at 1622 to 1558, 1400 to 1230, 1128, 1000 to 850, 749, and 509 cm^{-1}, as shown in Figure 3-C.

The peak at 1665 cm^{-1} corresponding to the amide I band was shifted to 1655 cm^{-1} and the peak at 1246 cm^{-1} corresponding to the amide III band was reduced in its intensity in the hard PMX mass. The peak intensities of the Raman spectrum at 1622, 1588, 1398, 1339, 1309, 1128, 960, 749, and 509 cm^{-1} were stronger for soft or hard PMX mass. The higher content of tryptophan (749 cm^{-1}) in the soft PMX mass and higher S-S disulfide formation (509 cm^{-1}) in the hard PMX mass stood out. In particular, two weak peaks at 1081 and 960 cm^{-1} assigned to the stretching vibration of the PO_4^{3-} in HA also appeared in the Raman spectra of both soft and hard PMX masses, suggesting the occurrence of cutaneous calcification of HA in the PMX tissue, particularly in the hard PMX mass. These results indicate that micro-Raman spectroscopy may provide a highly sensitive and specific method for identifying the normal skin dermis and how it differs in chemical composition from different PMX tissues.

Gniadecka et al., used non-invasive Raman spectroscopy instead of the XRD method and routine histology to examine skin calcinosis in the intradermal nodules of three patients [Gniadecka et al., 2001].

They detected several new Raman spectral peaks at 1445 and 1420 cm^{-1} (C=N stretching), 1205 and 1060 cm^{-1} (C-O stretch and O-H bending), and 1010 cm^{-1} (ring vibration) in the Raman spectra of the nodular deposits in all fingertips of a patient with progressive kidney failure and in the nodular lesions on the back of the hands of another patient who had undergone renal transplantation. These unique Raman peaks were consistent with the Raman spectra of needle aspirate from these patients, attributed to the characteristic peaks of monosodium urate, as shown in Figure 4-a~d. On the other hand, an in vivo Raman spectrum obtained from the nodular lesions of the fingertips of a third patient was dominated by a peak at 960 cm^{-1}, indicating a phosphate group in the HA deposit (Figure 4-e). In vitro examination of material expressed from the finger lesion showed an identical Raman pattern (Figure 4-f).

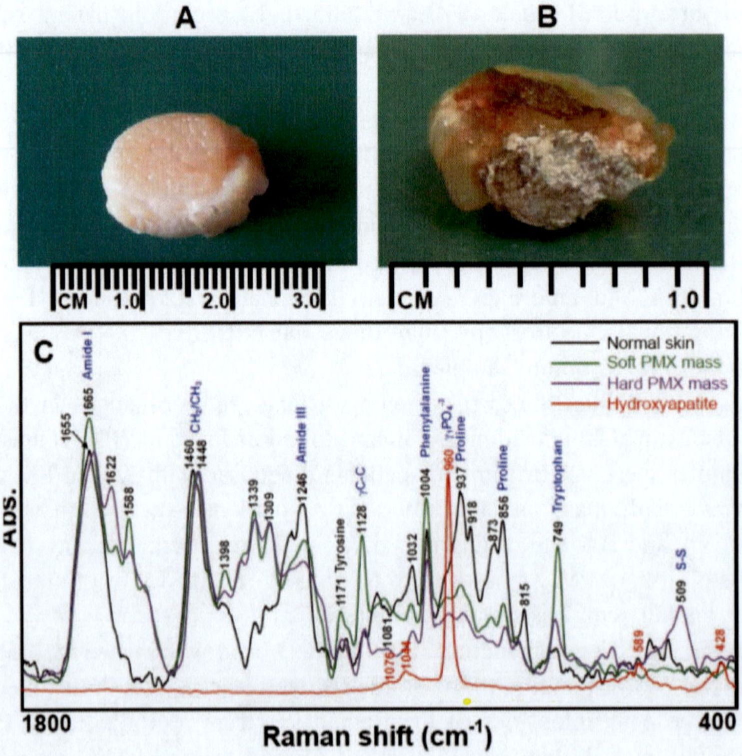

Figure 3. Photographs of the excised soft mass (A) and hard mass (B) of PMX from two patients and their Raman spectra as compared with normal skin dermis and HA after normalizing the peak intensity within the region of 1460-1448 cm^{-1} (modified from *Cheng et al., 2005*).

2) Idiopathic calcification

Juvenile dermatomyositis (juvenile DM) is an autoimmune disease affecting younger children. Juvenile DM is often associated with CC, cutaneous ulcerations, and vasculopathy characterized by intimal proliferation of small blood vessels and infarctions [Rider and Miller, 1997]. The cause of the disease is unknown but it appears to begin spontaneously [Feldman et al., 2008].

The crystal structure and mineral content of the calcific deposits from five children with juvenile DM (two boys and three girls) were examined by Pachman et al., using XRD, FT-IR spectroscopy, and imaging [Pachman et al., 2006]. They found that the value of 1660 cm^{-1}/1690 cm^{-1} (an indicator of collagen maturity), was slightly higher in the deposit than in bone and the maximum values were observed in the center of the deposit rather than on the periphery. Additionally, the area ratio of carbonate (840–890 cm^{-1}) : phosphate (916–1180 cm^{-1}) (an index of carbonate substitution) in the deposit mineral was comparable to that of bone. These results indicate that chronic cutaneous inflammation in children with juvenile DM might contribute to the formation of HA.

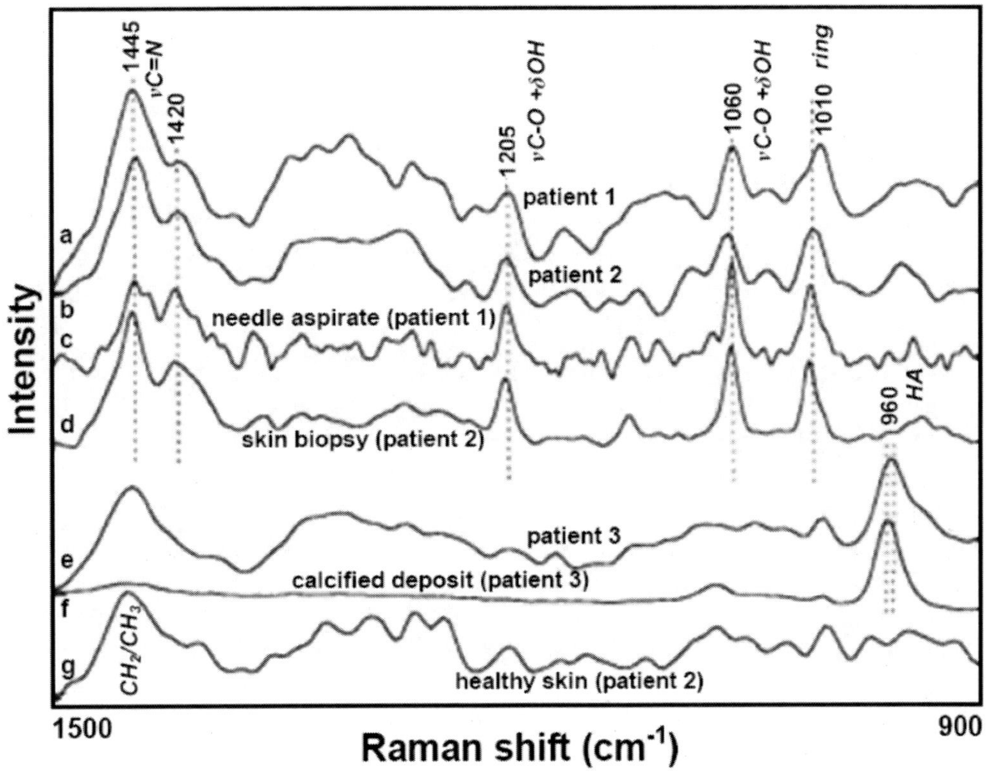

Figure 4. Raman spectra of the nodular lesions detected from patients 1-3 (a, b and e) and healthy skin from patient 2 (g). Spectra were also performed on a needle aspirate in patient 1 (c), a skin biopsy in patient 2 (d) and a calcified deposit in patient 3 (f) (modified from *Gniadecka et al., 2001*).

Eidelman et al., first found that the density of minerals present in the skin calcified deposits (SCD) of juvenile DM patients was most similar to that of enamel, but was clearly different to that of bone [Eidelman et al., 2009]. Calcium and phosphate presented in SCD might have been precipitated as carbonate apatite determined by FT-IR spectroscopy with reflectance mode. The apatite was much more crystallized than bone and dentin and its calcium:phosphate ratio was closer to that of stoichiometric HA. The deposit also contained magnesium and sodium.

Liu et al., used both FT-IR and Raman microspectroscopic analyses to detect the chemical composition of a SCD excised from a protruding subepidermal calcified nodule on the left buttock of a female patient [Liu et al., 2005]. The result indicated that the whole SCD in the skin lesion was a well-developed, mature, and hard mass. Several FT-IR absorption bands at 873 (870), 961, and 1031 cm^{-1} (the stretching modes of carbonate and phosphate in HA, and 1547 and 1658 cm^{-1} (the amide II and I bands of collagen) were detected in the IR spectrum of SCD (Figure 5-A). While the Raman spectral bands at 1450 and 1665 cm^{-1} (collagen); 1519 and 1156 cm^{-1} (β-carotene); and 958, 1072 and 1450cm^{-1} (HA) were also obtained (Figure 5-B). Information gained via both techniques may be used to elucidate all of the components of the SCD deposit in this patient. As shown in Figure 5-A, the appearance of a 1031 cm^{-1} peak might explain the mature, stoichiometric, and crystalline properties of HA in the SCD sample.

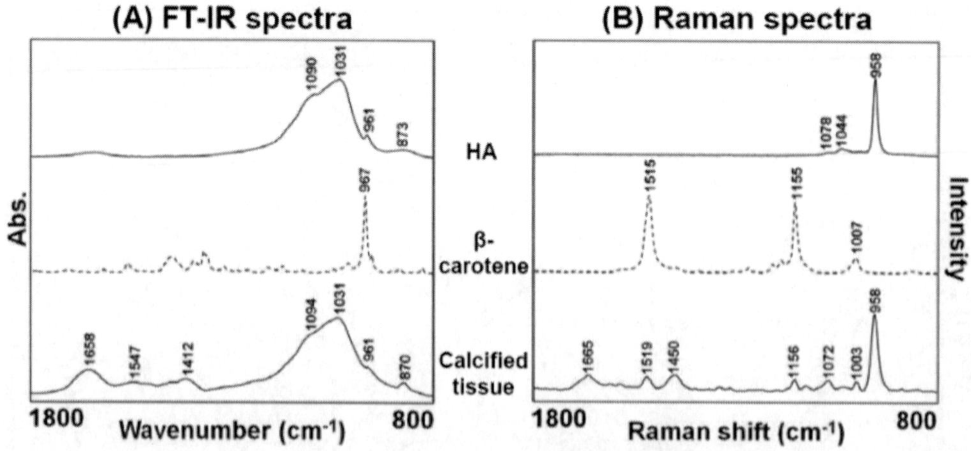

Figure 5. FT-IR (A) and Raman (B) spectra of the skin calcified deposit from a femal patient (modified from *Liu et al., 2005*).

In addition, the IR spectra of the carbonate apatite exhibited three components of the carbonate bands in the spectral region of 860 to 890 cm^{-1}: 878 cm^{-1}, A-type carbonate apatite; 871 cm^{-1}, B-type carbonate apatite; and 866 cm^{-1}, an unstable carbonate apatite, respectively [Rey et al., 1989 and 1991; Fleet, 2013]. Liu et al., also used a curve-fitting technique to investigate the carbonate substitutions within the HA structure [Liu et al., 2005]. As shown in Figure 6, three compositions in the IR spectrum of the standard HA reference sample were deduced using a curve-fitting analysis: 882 (15.1%), 875 (65.1%), and 868 (19.8%) cm^{-1}. However, three compositions at 881 (12.0%), 875 (60.0%), and 868 (28.0%) cm^{-1} were obtained for the curve-fitted IR spectrum of SCD. Obviously, the strong peak at 875 cm^{-1} with a higher composition for the curve-fitted IR spectrum of SCD suggests that B-type carbonate apatite was a predominant mature component in this SCD.

Niu et al., found a rare idiopathic CC in a healthy five-year-old girl with multiple nodules on her left knee [Niu et al., 2011]. They used both FT-IR and Raman microscopic techniques to detect and identify the components of the calcified deposits. The FT-IR spectral results indicated several unique absorption bands at 876, 961, 1033, 1240, 1453, 1549, 1656, 1746, 2853, and 2925 cm^{-1} in the IR spectrum of the idiopathic calcified deposit (Figure 7-A). Here, the peaks at 876, 961, and 1033 cm^{-1} were probably due to the carbonate and phosphate stretching modes of HA, but the IR peaks at 1240, 1453, 1549, 1656, 1746, 2853, and 2925 cm^{-1} were due to the stretching vibrations of lipids and proteins similar to the IR spectrum of non-calcified tissue [Haris and Chapman, 1994; Jackson and Mantsch, 1995]. This strongly suggests that apatite particles interspersed within the subcutaneous tissue were the predominant component in the calcified deposit for this patient with idiopathic CC.

On the other hand, the Raman spectrum of this calcified deposit of idiopathic CC was more clearly manifested by the presence of β-carotene, except in presence of HA (Figure 7-B). Two unique peaks at 961 and 587 cm^{-1} were found in the Raman spectrum of the calcified deposit, which were mainly due to ν_1 phosphate stretching of HA.

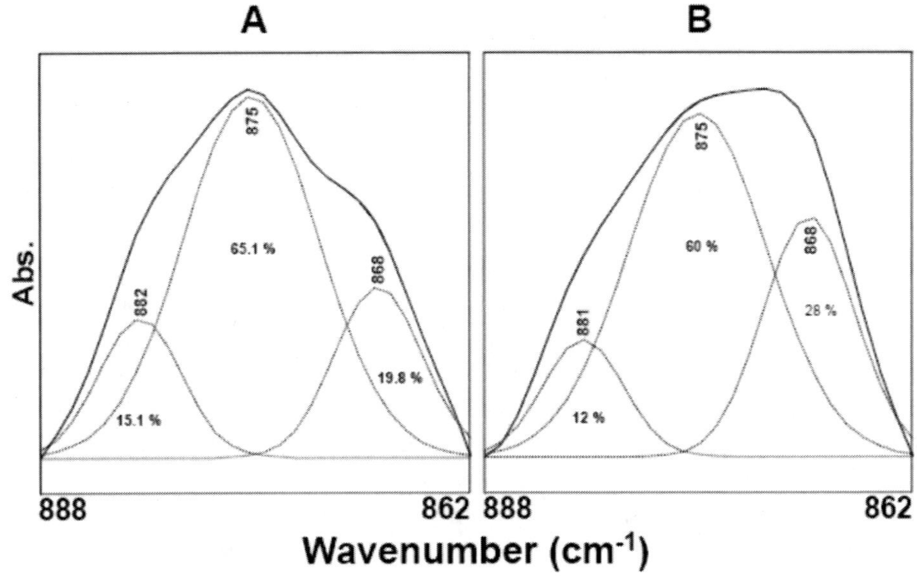

Figure 6. The assignment and composition of the curve-fitted original IR spectra of standard HA reference sample (A) and skin calcified deposit (B) (adopted from *Liu et al., 2005*).

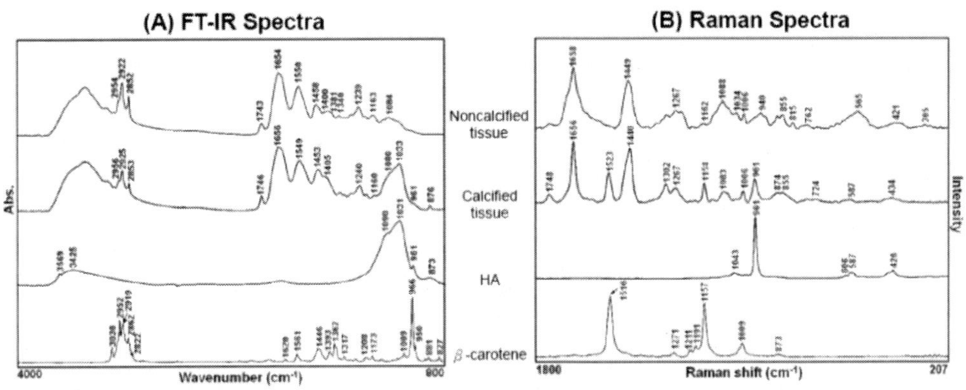

Figure 7. FT-IR and Raman spectra of noncalcified tissue and calcified deposit of idiopathic patient, HA and β-carotene (modified from *Niu et al., 2011*).

Two strong peaks at 1516 (1523) and 1157 (1158) cm^{-1} corresponded to C=C and C-C stretching vibrations and a shoulder at 1009 (1006) cm^{-1} assigned to the rocking motion of the methyl groups, were observed in both the Raman spectra of β-carotene and the calcified deposit [Ermakov et al., 2004; Hammond and Wooten, 2005]. This implies that β-carotene was also present in the calcified idiopathic CC deposit. Two major components of B-type carbonate apatite and β-carotene interspersed within the subcutaneous tissue were clearly shown to make up the calcified deposits in this idiopathic CC sample. The combination of FT-IR and Raman microscopic techniques was very useful for the simultaneous assessment of the intact components of the calcified deposits in idiopathic CC.

Conclusion

This chapter summarizes the advances in vibrational spectroscopy and their impact on the diagnosis of calcified deposits in various skin diseases. Because the FT-IR and Raman spectra of the mineral components are quite distinct, both vibrational techniques have been applied as structural probes to study all of these mineralized tissues, providing information on the nature of the mineral phases, quantitative information on the changes in mineral and matrix composition during mineralization, and the nature and amounts of substituents in the mineral [Carden and Morris, 2000; Adele and Richard, 2005; Boskey and Mendelsohn, 2005]. In particular, the non-invasive technique can be used in population screening, early diagnosis, prognosis, monitoring of therapy and subsequent aspects of treatment in the future [Gniadecka et al., 2001; Ermakov et al., 2004; Stone et al., 2007; Saha et al., 2011; Kendall et al., 2011; Surmacki et al., 2013]. The recent advent of the use of vibrational imaging to investigate mineralized tissues is particularly exciting. This new method holds great promise for generating high-definition chemical-state images that may prove invaluable for future studies of mineralized tissue physiology and even in clinical diagnostics [Carden and Morris, 2000; Bhargava, 2012; Li et al., 2013; Surmacki et al., 2013].

Acknowledgments

The author would like to thank Drs. Liu MT, Liu HN, Niu DM, Pan CC, Yang DM, and Lin CC for their supporting the human samples and kindly comments. I would also like to acknowledge Misses Cheng WT and Li MJ technical assistances in much sample preparations and data collection. The spectral studies have been partly supported by National Science Council, Taipei, Taiwan, Republic of China.

References

Adele, L.B., Richard, M., (2005). Infrared spectroscopic characterization of mineralized tissues. *Vib. Spectrosc.* 38, 107-114.

Baldet, P., Pernot, F., Blotman, F., Bonnel, F., Simon, L., (1981). CRST syndrome. Ultrastructural and physico-chemical studies of calcifications. *Ann. Pathol.* 1, 259-269.

Bazin, D., Daudon, M., Combes, C., Rey, C., (2012). Characterization and some physicochemical aspects of pathological microcalcifications. *Chem. Rev.* 112, 5092-5120.

Bazin, D., Haymann, J.P., Letavernier, E., Rode. J., Daudon, M., (2013). Pathological calcifications: A medical diagnosis based on their physicochemical properties. *Presse. Med.* S0755-4982(13)00658-1.

Bettoli, V., Bigi, A., Cojazzi, G., Roveri, N., Strumia, R., (1992). Structural and chemical characterization of a cutaneous calcification. *J. Therm. Anal.* 38, 2719-2728.

Bhargava, R., (2012). Infrared spectroscopic imaging: the next generation. *Appl. Spectrosc.* 66, 1091-1120.

Black, A.S., Kanat, I.O., (1985). A review of soft tissue calcifications. *J. Foot Surg.* 24, 243-250.

Bonewald, L.F., Harris, S.E., Rosser, J., Dallas, M.R., Dallas, S.L., Camacho, N.P., Boyan, B., Boskey, A., (2003). Von Kossa staining alone is not sufficient to confirm that mineralization in vitro represents bone formation. *Calcif. Tissue Int.* 72, 537-547.

Bonucci, E., (2006). Biological Calcification: Normal and Pathological Processes in the Early Stages. Springer Verlag: Heidelberg, Germany.

Boskey, A.L., (2003). Biomineralization: an overview. *Connect. Tissue Res.* 44 Suppl 1, 5-9.

Boskey, A.L., Mendelsohn, R., (2005). Infrared spectroscopic characterization of mineralized tissues. *Vib. Spectrosc.* 38, 107-114.

Carden, A., Morris, M.D., (2000). Application of vibrational spectroscopy to the study of mineralized tissues. *J. Biomed. Opt.* 5, 259-268.

Chander, S., Gordon, P., (2012). Soft tissue and subcutaneous calcification in connective tissue diseases. *Curr. Opin. Rheumatol.* 24, 158-164.

Cheng, W.T., Liu, M.T., Liu, H.N., Lin, S.Y., (2005). Micro-Raman spectroscopy used to identify and grade human skin pilomatrixoma. *Microsc. Res. Tech.* 68, 75-79.

Chiou, H.J., Hung, S.C., Lin, S.Y., Wei, Y.S., Li, M.J., (2010). Correlations among mineral components, progressive calcification process and clinical symptoms of calcific tendonitis. *Rheumatology.* 49, 548-555.

Ciftçioğlu, N., McKay, D.S., (2010). Pathological calcification and replicating calcifying-nanoparticles: general approach and correlation. *Pediatr. Res.* 67, 490-499.

Clark, R.J.H., Hester, R.E., (1996). Biomedical Applications of Spectroscopy. Wiley: Chichester, UK.

Colthup, N.B., Daly, L.H., Wiberely, S.E., (1990). Introduction to IR and Raman Spectroscopy, 3rd ed, Academic Press Inc., New York, USA.

Crupi, V., Venuti, V., Majolino, D., (2004). FT-IR Spectroscopy: An Advanced Tool for Studying Biomedical Problems. *Spectroscopy.* 17 (7), 22-30.

Daculsi, G., Faure, G., Kerebel, B., (1983). Electron microscopy and microanalysis of a subcutaneous heterotopic calcification. *Calcif. Tissue Int.* 35, 723-727.

De Aza, P.N., De Aza, A.H., De Aza, S., (2005). Crystalline Bioceramic Materials. *Bol. Soc. Esp. Ceram.* 44, 135-145.

Diem, M., Miljković, M., Bird, B., Chernenko, T., Schubert, J., Marcsisin, E., Mazur, A., Kingston, E., Zuser, E., Papamarkakis, K., Laver, N., (2012). Applications of infrared and Raman microspectroscopy of cells and tissue in medical diagnostics: present status and future promises. *Spectroscopy: AIJ.* 27, 463-496.

Dorozhkin, S.V., (2009). Calcium orthophosphates in nature, biology and medicine. *Materials.* 2, 399-498.

Dorozhkin, S.V., (2010). Bioceramics of calcium orthophosphates. *Biomaterials.* 31, 1465-1485.

Dorozhkin, S.V., 2011. Calcium orthophosphates: Occurrence, properties, biomineralization, pathological calcification and biomimetic applications. *Biomatter.* 1-2, 121-164.

Ehrlich, H., Koutsoukos. P.G., Demadis, K.D., Pokrovsky, O.S., (2008). Principles of demineralization: modern strategies for the isolation of organic frameworks. Part I. Common definitions and history. *Micron.* 39, 1062-1091.

Ehrlich, H., Koutsoukos. P.G., Demadis, K.D., Pokrovsky, O.S., (2009). Principles of demineralization: modern strategies for the isolation of organic frameworks. Part II. Decalcification. *Micron.* 40, 169-193.

Eidelman, N., Boyde, A., Bushby, A.J., Howell, P.G., Sun, J., Newbury, D.E., Miller, F,W., Robey, P.G., Rider, L.G., (2009). Microstructure and mineral composition of dystrophic calcification associated with the idiopathic inflammatory myopathies. *Arthritis Res. Ther.* 11 (5), R159.

Eikje, N.S., Aizawa, K., Ozaki, Y., (2005). Vibrational spectroscopy for molecular characterisation and diagnosis of benign, premalignant and malignant skin tumours. *Biotechnol. Annu. Rev.* 11, 191-225.

Elliott, J.C., Holcomb, D.W., Young, R.A., (1985). Infrared determination of the degree of substitution of hydroxyl by carbonate ions in human dental enamel. *Calcif. Tissue Int.* 37, 372-375.

Ellis, D.I., Goodacre, R., (2006). Metabolic fingerprinting in disease diagnosis: biomedical applications of infrared and Raman spectroscopy. *Analyst.* 131, 875-885.

Ermakov, I.V., Ermakova, M.R., Gellermann, W., Lademann, J., (2004). Noninvasive selective detection of lycopene and beta-carotene in human skin using Raman spectroscopy. *J. Biomed. Opt.* 9, 332-338.

Feldman, B.M., Rider, L.G., Reed, A.M., Pachman, L.M., (2008). Juvenile dermatomyositis and other idiopathic inflammatory myopathies of childhood. *Lancet.* 371, 2201-2212.

Fernandez-Flores, A., (2011). Calcinosis cutis: critical review. *Acta Dermatovenerol Croat.* 19, 43-50.

Fleet ME., (2013). The carbonate ion in hydroxyapatite: recent X-ray and infrared results. *Front Biosci (Elite Ed).* 5, 643-652.

Giachelli, C.M., (1999). Ectopic calcification: Gathering hard facts about soft tissue mineralization. *Am. J. Pathol.* 154, 671-675.

Gniadecka, M., Wulf, H.C., Johansson,C.K., Ullman, S., Halberg, P., Rossen, K., (2001). Cutaneous tophi and calcinosis diagnosed in vivo by Raman spectroscopy. *Br. J. Dermatol.* 145, 672-674.

Golub, E.E., (2011). Biomineralization and matrix vesicles in biology and pathology. *Semin. Immunopathol.* 33, 409-417.

Goto, M., (2009). Emerging concept of ageing: Inflammageing. *Inflamm. Regen.* 29, 249-257.

Gremlich, H.U., Yan, B., (2001). Infrared and Raman Spectroscopy of Biological Materials, Marcel Dekker, Inc., New York, USA.

Hammond, B.R., Wooten, B.R. (2005). Resonance Raman spectroscopic measurement of carotenoids in the skin and retina. *J. Biomed. Opt.* 10 (5), 054002.

Haris, P.I., Chapman, D., (1994). Analysis of polypeptide and protein structures using Fourier transform infrared spectroscopy. *Meth. Mol. Biol.* 22, 183-202.

Jackson, M., Mantsch, H.H., (1995). The use and misuse of FTIR spectroscopy in the determination of protein structure. *Crit. Rev. Biochem. Mol. Biol.* 30, 95-120.

Kalasinsky, V.F., Marie Jenkins, H., Johnson, F.B., 2002. Applications of vibrational microspectroscopy to pathology specimens. *Vib. Spectrosc.* 28, 199-207.

Kannan, S., Vieira, S.I., Olhero, S.M., Torres, P.M.C., Pina, S., Da Cruz e Silva, O.A.B., Ferreira, J.M., (2011). Synthesis, mechanical and biological characterization of ionic doped carbonated hydroxyapatite/β-tricalcium phosphate mixtures. *Acta Biomater.* 7, 1835-1843.

Kart, C.S., Metress, E.K., Metress, S.P., (1992). Human Aging and Chronic Disease. Jones and Bartlett: Boston, USA.

Kawasaki, K., Buchanan, A.V., Weiss, K.M., (2009). Biomineralization in humans: making the hard choices in life. *Annu. Rev. Genet.* 43, 119-142.

Kendall, C., Hutchings, J., Barr, H., Shepherd, N., Stone, N., (2011). Exploiting the diagnostic potential of biomolecular fingerprinting with vibrational spectroscopy. *Faraday Discuss.* 149, 279-290.

Kirsch, T., (2006). Determinants of pathological mineralization. *Curr. Opin. Rheumatol.* 18, 174-180.

Kirsch, T., (2007). Physiological and pathological mineralization: a complex multifactorial process. *Curr Opin Orthop.* 18, 425-427.

Kirsch, T., (2012). Biomineralization--an active or passive process? *Connect. Tissue Res.* 53, 438-445.

Konno, M., Kikyo, S., Ando, K., Alta, S., Muraishi, S., Inoue, T., (2003). Analysis of the calcified materials from calcinosis cutis. *Jap. J. Clin. Chem.* 32, 249-254.

Krafft, C., (2004). Bioanalytical applications of Raman spectroscopy. *Anal. Bioanal. Chem.* 378, 60-62.

Krafft, C., Bird, B., (2013). Special Issue: Vibrational Spectroscopy in Medicine. *J. Biophotonics.* 6, 1-121.

Krafft, C., Sergo, V., (2006). Biomedical applications of Raman and infrared spectroscopy to diagnose tissues. *Spectroscopy: AIJ.* 20, 195-218.

Krafft, C., Steiner, G., Beleites, C., Salzer, R., (2009). Disease recognition by infrared and Raman spectroscopy. *J. Biophotonics.* 2, 13-28.

Kumar, V., Abbas, A.K., Fausto, N., Aster, J.C., (2010). Chapter 1. Cellular responses to stress and toxic insults: Adaptation, injury, and death. In Robbins and Cotran Pathologic Basis of Disease. 8 ed. Saunders, Elsevier Inc., PA, USA, pp. 3-42.

LeGeros, R.Z., (1991). Calcium phosphates in oral biology and medicine. *Monogr. Oral Sci.* 15, 1-201.

Li, Q., He, X., Wang, Y., Liu, H., Xu, D., Guo, F., (2013). Review of spectral imaging technology in biomedical engineering: achievements and challenges. *J. Biomed. Opt.* 18 (10), 100901.

Lin, S.Y., Li, M.J., Cheng W.T., (2007). FT-IR and Raman vibrational microspectroscopies used for spectral biodiagnosis of human tissues. *Spectroscopy: AIJ.* 21, 1-30.

Lin, S.Y., (2014). Vibrational spectral biodiagnosis of ocular calcification. *Appl. Spectrosc. Rev.* 49, 11-63.

Liu, M.T., Cheng, W.T., Li, M.J., Liu, H.N., Yang, D.M., Lin, S.Y., (2005). Identification of chemical compositions of skin calcified deposit by vibrational microspectroscopies. *Arch. Dermatol. Res.* 297, 231-234.

Mantsch, H.H., Choo-Smith, L.P., Shaw, R.A., (2002). Vibrational spectroscopy and medicine: an alliance in the making. *Vib. Spectrosc.* 30, 31-41.

Mayer, M., (2006). Calcification: The aging factor, how to defuse the calcium bomb. Lulu. Com., Raleigh, NC, USA

Meloan, S.N., Puchtler, H., (1985). Chemical mechanisms of staining methods: von Kossa's technique. What von Kossa really wrote and a modified reaction for selective demonstration of inorganic phosphate. *J. Histotechol.* 8, 11-13

Morris, M.D., Finney, W.F., (2004). Recent developments in Raman and infrared spectroscopy and imaging of bone tissue. *Spectroscopy: AIJ.* 18, 155-159.

Niu, D.M., Lin, S.Y., Li, M.J., Cheng, W.T., Pan, C.C., Lin, C.C., (2011). Idiopathic calcinosis cutis in a child: chemical composition of the calcified deposits. *Dermatology.* 222, 201-205.

Orimo, H., (2010). The mechanism of mineralization and the role of alkaline phosphatase in health and disease. *J. Nihon Med. Sch.* 77, 4-12.

Pachman, L.M., Veis, A., Stock, S., Abbott, K., Vicari, F., Patel, P., Giczewski, D., Webb, C., Spevak, L., Boskey, A.L., (2006). Composition of calcifications in children with juvenile dermatomyositis: association with chronic cutaneous inflammation. *Arthritis Rheum.* 54, 3345-3350.

Pan, H., Darvell, B.W. (2010). Effect of carbonate on hydroxyapatite solubility. *Crystal. Growth & Des.* 10, 845-850.

Peacock, M., (2010). Calcium metabolism in health and disease. *Clin. J. Am. Soc. Nephrol.* 5, S23-30.

Peroos, S., Du, Z., De Leeuw, N.H., (2006). A computer modelling study of the uptake, structure and distribution of carbonate defects in hydroxyapatite. *Biomaterials.* 27, 2150-2161.

Petrich, W., (2001). Mid-infrared and Raman spectroscopy for medical diagnostics. *Appl. Spectrosc. Rev.* 36, 181-237.

Porter, A., Patel, N., Brooks, R., Best, S., Rushton, N., Bonfield, W., (2005). Effect of carbonate substitution on the ultrastructural characteristics of hydroxyapatite implants. *J. Mater. Sci. Mater. Med.* 6, 899-907.

Reiter, N., El-Shabrawi, L., Leinweber, B., Berghold, A., Aberer, E., (2011). Calcinosis cutis. Part I. Diagnostic pathway. *J. Am. Acad. Dermatol.* 65:1-12;quiz 13-4.

Rey, C., Collins, B., Goehl, T., Dickson, I.R., Glimcher, M., (1989). The carbonate environment in bone mineral: a resolution-enhanced Fourier transform infrared spectroscopy study. *Calcif. Tissue Int.* 45, 157-164.

Rey, C., Renugopalakrishnan, V., Collins, B., Glimcher, M.J., (1991). Fourier transform infrared spectroscopic study of the carbonate ions in bone mineral during aging. *Calcif. Tissue Int.* 49, 251-258.

Rider, L.G., Miller, F.W., (1997). Classification and treatment of the juvenile idiopathic inflammatory myopathies. *Rheum. Dis. Clin. North Am.* 23, 619-655.

Rungby, J., Kassem, M., Eriksen, E.F., Danscher, G., (1993). The von Kossa reaction for calcium deposits: silver lactate staining increases sensitivity and reduces background. *Histochem. J.* 25, 446-451.

Sader, M.S., Lewis, K., Soares, G.A., LeGeros R.Z., (2013). Simultaneous incorporation of magnesium and carbonate in apatite: effect on physico-chemical properties. *Mater. Res.* 16, 779-784.

Saha, A., Barman, I., Dingari, N.C., McGee, S., Volynskaya, Z., Galindo, L.H., Liu, W., Plecha, D., Klein, N., Dasari, R.R., Fitzmaurice, M., (2011). Raman spectroscopy: a realtime tool for identifying microcalcifications during stereotactic breast core needle biopsies. *Biomed. Opt. Exp.* 2, 2792-2803.

Sathyanarayana, D.N., (2007). Vibrational Spectroscopy: Theory and Applications. New Age Int., New Delhi, India.

Severcan, F., Haris, P.I., Aksoy, C., Ozek, N.S., (2013). Progress in vibrational spectroscopy in diagnosis and screening. *Biomed. Spectrosc. Imaging.* 2, 73-81.

Severcan, F., Haris, P.I., (2012). Vibrational Spectroscopy in Diagnosis and Screening. IOS Press, Amsterdam, Netherlands.

Siebert, F., Hildebrandt, P., (2008). Vibrational Spectroscopy in Life Science. Wiley-VCH, Berlin, Germany.

Singh, B., Gautam, R., Kumar, S., Vinay Kumar, B.N., Nongthomba, U., Nandi, D., Mukherjee, G., Santosh, V., Somasundaram, K., Umapathy, S., (2012). Application of vibrational microspectroscopy to biology and medicine. *Curr. Sci.* 102, 232-244.

Stewart, S., Shea, D.A., Tarnowski, C.P., Morris, M.D., Wang, D., Franceschi, R., Lin, D.L., Keller, E., (2002). Trends in early mineralization of murine calvarial osteoblastic cultures: A Raman microscopic study. *J. Raman Spectrosc.* 33, 536-543.

Stone, N., Baker, R., Rogers, K., Parker, A.W., Matousek, P., (2007). Subsurface probing of calcifications with spatially offset Raman spectroscopy (SORS): future possibilities for the diagnosis of breast cancer. *Analyst.*132, 899-905.

Surmacki, J., Musial, J., Kordek, R., Abramczyk, H., (2013). Raman imaging at biological interfaces: applications in breast cancer diagnosis. *Mol Cancer.* 12, 48.

Thygesen, L.G., Løkke, M.M., Micklander, E., Engelsen, S.B., (2003). Vibrational microspectroscopy of food. Raman vs. FT-IR. *Trends Food Sci. Technol.* 14, 50-57.

Tosi, G., Giorgini, E., Rubini, C., Sabbatini. S., Librando, V., Alò, F., (2012). Vibrational spectroscopy as a supporting technique in clinical diagnosis and prognosis of atherosclerotic carotid plaques: a review. *Anal. Quant. Cytol. Histol.* 34, 214-232.

Walsh, M.J., German, M.J., Singh, M., Pollock, H.M., Hammiche, A., Kyrgiou, M., Stringfellow, H.F., Paraskevaidis, E., Martin-Hirsch, P.L., Martin, F.L., (2007). IR microspectroscopy: potential applications in cervical cancer screening, *Cancer Lett.* 246, 1-11.

Wetzel, D.L.B., LeVine, S.M., (1998). Fourier Transform Infrared (FTIR) Microspectroscopy: A new molecular dimension for tissue or cellular imaging and in situ chemical analysis. *Cell Mol. Biol.* 44, 1-272.

Wopenka, B., Pasteris, J.D., (2005). A mineralogical perspective on the apatite in bone. *Mater. Sci. Engineer.* C 25, 131-143.

In: Apatite
Editors: M. Iafisco and J. M. Delgado-López
ISBN: 978-1-63321-500-9

Chapter 6

THE DISSOLUTION MECHANISM OF CALCIUM APATITES IN ACIDS

Sergey V. Dorozhkin*
Moscow, Russia

ABSTRACT

Eight dissolution models of calcium apatites (both fluorapatite (FA) and hydroxyapatite (HA)) in acids were drawn from the published literature, analyzed and discussed. Major limitations and drawbacks of the models were conversed in details. The models were shown to deal with different aspects of apatite dissolution phenomenon and none of them was able to describe the dissolution process in general. Therefore, an attempt to combine the findings obtained by different researchers was performed which resulted in creation of the general description of apatite dissolution in acids. For this purpose, eight dissolution models were assumed to complement each other and provide the correct description of the specific aspects of apatite dissolution. The general description considers all possible dissolution stages involved and points out to some missing and unclear phenomena to be experimentally studied and verified in future. This creates a new methodological approach to investigate reaction mechanisms based on sets of affine data, obtained by various research groups under dissimilar experimental conditions.

Keywords: Calcium orthophosphates, fluorapatite, hydroxyapatite, dissolution, model, mechanism

INTRODUCTION

Calcium apatites (hydroxyapatite, HA: $Ca_{10}(PO_4)_6(OH)_2$ and fluorapatite, FA: $Ca_{10}(PO_4)_6F_2$) appear to be of a special interest to the mankind, because natural apatites are

* E-mail: sedorozhkin@yandex.ru.

the main source of inorganic phosphorus in nature [1], while calcified tissues (bones and teeth) of mammals consist of ion-substituted calcium apatites of biological origin [2-4]. Therefore, HA and FA are widely used in various aspects of human being, starting from the major source of phosphorus-containing fertilizer production and ending with various biomedical applications, for example, using as artificial bone grafts. The latter resulted in a great number of research papers published on calcium apatites and related calcium orthophosphates (extensively reviewed in Refs. [5-7]). Since the chemical nature of natural phosphate ores and the inorganic phase of bones and teeth is almost the same, such different processes as fertilizer production, dental caries, osteoporosis, as well as both *in vitro* and *in vivo* biodegradation of artificial bone grafts might be simulated by dissolution of chemically pure calcium apatites (HA and FA) in acids. Therefore, establishing of the true dissolution mechanism of calcium apatites in acids appears to be of the paramount importance for our species.

In chemistry, a reaction mechanism is the step by step sequence of elementary reactions by which the overall chemical change occurs [8]. Regarding the dissolution process of calcium apatites in acids, eight models have been already proposed to explain the processes involved. These models already provided the important information about the factors associated with solution (pH, composition, saturation and hydrodynamics), bulk solid (chemical composition, solubility, particle sizes) and surface (defects, adsorbed ions, "history", phase transformation) of the apatite crystals. These findings cannot be underestimated because they have already made a significant impact on understanding of the medical aspects of both dental caries and osteoporosis, as well as they created the necessary fundamentals for fertilizer production industry. Nevertheless, there are still a lot of elements missing in the overall explanation for the processes involved.

To clarify possible directions of further investigations on apatite dissolution, a critical analysis of currently available models has been made in this review followed by an attempt to combine the data available into a general dissolution mechanism. This creates a new methodological approach to investigate reaction mechanisms based on sets of affine data, obtained by various research groups under dissimilar experimental conditions.

Critical Analysis of the Dissolution Models of Calcium Apatites

First, a clear distinction must be made between *solubility*, which is a reversible process measured under the equilibrium conditions, and *dissolution*, which is an irreversible process measured under the non-equilibrium conditions. Only the latter process will be considered, analyzed and discussed in this review.

Diffusion (or Transport) and Kinetically (or Surface) Controlled Models

A diffusion (or transport) and a kinetically (or surface) controlled [9-31] dissolution models, as well as the two-site model [15, 32-34], are devoted to comparison of transport rates of chemical reagents (H^+ and anions of acid) from bulk solution to the surface of apatite crystals and those of products (Ca^{2+}, F^- and orthophosphate anions) obtained backwards with the kinetics of chemical interactions on the crystal surface. According to the diffusion-controlled and two-site models, a transport of either the chemical reagents and/or the products of chemical reactions mainly determine the overall dissolution rate, whereas the kinetically controlled model emphasizes chemical transformations on the surface as the limiting factor (Figure 1) [35]. An important consequence of the two types of rate-controlling mechanism is that the solution immediately next to the crystal surface is undersaturated with respect to apatite in the surface controlled process and this solution is saturated in the diffusion-controlled process [36]. Both models usually operate with a so-called "driving force" which means either a concentration gradient within the Nernst diffusion layer (the diffusion controlled model) or a gradient of ionic chemical potentials between the apatite crystal surface and bulk solution (the kinetically controlled model).

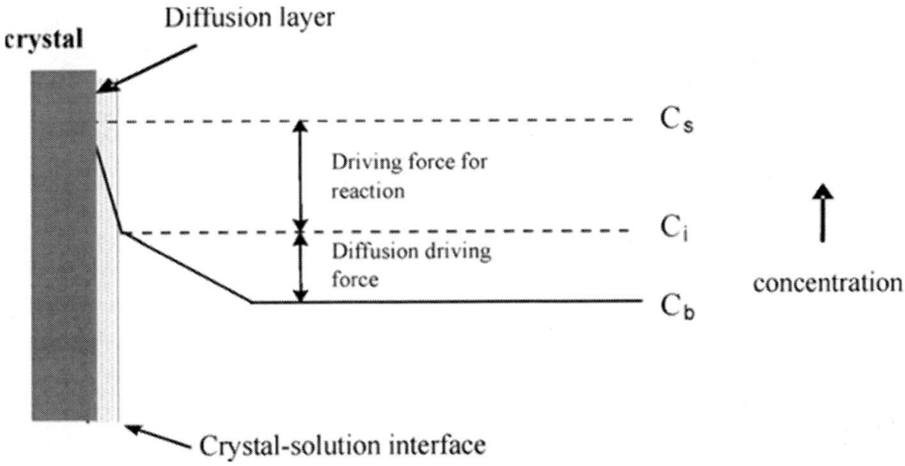

Figure 1. A dissolution process according to the diffusion and kinetically controlled models. Here: C_s – solute concentration on the surface, C_i – solute concentration on the interface, C_b – solute concentration in bulk. Reprinted from Ref. [35] with permission.

To investigate whether a dissolution process follows either of these models, one should study the influence of hydrodynamics, concentration of reagents, ionic strength, solution undersaturation, pH, temperature, crystal dimensions and the presence of chemical modifiers on the dissolution kinetics. A rotary disk [10, 13, 14], constant composition [9, 16, 18, 19, 20, 37-39], dual constant composition [22, 40, 41] or stirred suspension techniques are used for the experimental investigations. The results obtained are usually plotted as an uptake of H^+ ions (e.g., as a titrant volume added) and/or a release of calcium, orthophosphate and fluoride ions versus the dissolution time. Afterwards, calculations of the numeric values for activation energy, rate constants, effective reaction order, diffusion layer thickness, characteristic adsorption impedance, diffusive jump distance, *etc.* are performed [9-42]. The conclusion on whether a dissolution process is kinetically or diffusion controlled is made based on the numeric values calculated. For example, when analysis of the rate data gave a numeric value of the effective reaction order $n = 6$, a surface controlled dissolution mechanism was suggested [22].

Basic thermodynamic principles predict that dissolution rates should increase with increasing driving force or chemical potential; however, the experimental studies show that this dependence is complex. Namely, dissolution of apatite in some cases was found to be diffusion controlled [19, 20, 23, 25, 31], in some other cases – kinetically controlled [21, 22] and even intermediate (*i.e.* both kinetically and diffusion controlled) in still other cases [14]. Furthermore, the dissolution process might be composed of two stages: the first one was found to be both surface and diffusion controlled, while the second one was an exactly diffusion-controlled dissolution [24]. Therefore, the results obtained for these models are valid within the experimental conditions studied only; no extrapolation can be made outside the ranges studied. For example, after, say, an agitation decreasing or temperature increasing, an initially kinetically controlled dissolution might be controlled by diffusion [30].

Furthermore, in some cases, dissolution curves (a titrant volume against time) reach plateaus prior to complete dissolution of HA, thus indicating a creation of metastable states in which the reaction is effectively terminated even though apatite crystals remain in contact with the undersaturated solutions [43-56]. There is an opinion [57] that this effect resulted from a false solubility product value for HA. Thus, a high sensitivity to the experimental conditions appears to be the main drawback of these models.

Mono- and Polynuclear Models

The surface dissolution may be described by three different models shown schematically in Figure 2. As may be seen in Figure 2A it is possible to start dissolution from one center from which originates the units detachment leading to step disintegration. This is a mononuclear dissolution. It is also possible that detachment takes place simultaneously from multiple centers (Figures 2B and 2C) which may be developed in one or more steps on the crystal surface. This is a polynuclear dissolution. Alternatively, it has been suggested that the crystal units' detachment takes place along spirals originating from an active site on the crystal surface following an Archimedean spiral (Figure 2D). The unit detachment along the spirals takes place at constant velocity [35].

A polynuclear model [58-67] was created from the study on apatite dissolution and growth kinetics under the constant composition conditions. Polydispersed samples of apatite were put into a stirred undersaturated (for dissolution experiments) or supersaturated (for those on crystal growth) solutions and the solution pH and amount of chemicals added (an acid for the dissolution experiments and a base for those on crystal growth) were permanently recorded as functions of time. The results obtained were plotted versus either undersaturation or supersaturation values: straight lines were obtained in the specific logarithmic coordinates typical for this model [68]. According to this model, dissolution nuclei, *i.e.* collections of vacant sites for Ca^{2+}, PO_4^{3-} and OH^- ions, are formed on the crystal surface of apatite and spread over the surface with a definite lateral rate [59, p. 30].

In polynuclear model, the nucleation rate is assumed as a function of the mean ion activity. A lateral growth rate of the nuclei is assumed proportional to the difference between the total concentration of calcium ions in the saturated solution and in a solution, while the rate constant is related to the frequency for calcium ions to make a diffusion jump into a kink and, simultaneously, partly dehydrate [65, 66]. Investigations on apatite dissolution and crystal growth using atomic force microscopy revealed that the rate-determining step was not the diffusion but two-dimensional surface nucleation [69, 70]. These observations provided a valuable support for the polynuclear model.

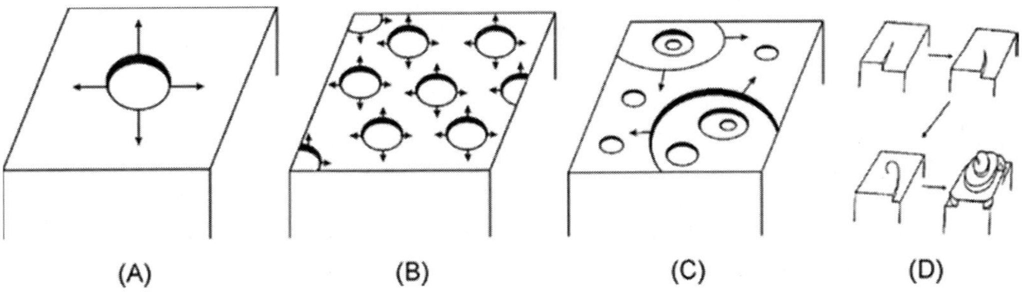

Figure 2. (A) Mononuclear model. (B) Polynuclear model in one crystal step. (C) Polynuclear model in multiple steps: birth and spread. (D) Spiral model. Reprinted from Ref. [35] with permission.

The experimental data on dissolution and growth rates versus undersaturation ($S < 1$) and supersaturation ($S > 1$) values, respectively, were obtained. Numeric values of surface tension (40 mJ/m^2 for apatite dissolution and 100 mJ/m^2 for apatite growth [65, 66]), ionic frequencies to enter or leave a kink ($1.6 \times 10^5/s$ [64]), the mean distance between kinks in a surface nucleus (approximately 3.07 Å [63]) and the critical nuclei for both HA [58-62, 65, 66] and FA [63, 64] were calculated. Furthermore, within $5.0 < pH < 7.2$ the dissolution kinetics was found to follow the polynuclear model, while that of growth appeared to follow the polynuclear model within $5.5 < pH < 6.5$ and the combined mono-polynuclear model at $pH = 5.0$ [63, 64].

A new idea about hydroxide ions formation at the hydroxide sites of HA from water molecules trapped under the crystal surface appeared in the most recent papers of this series [65, 66]. Probably, this might be a reason for the above changing in the growth model found for HA at pH = 5.0. To conclude, the polynuclear model seems to be better elaborated among others because it is valid for apatite dissolution within solution pH of 5.0 – 7.2, saturation degrees of $0.1 < S < 7$, and Ca/P molar ratios of 0.1 – 20 [66]. However, nothing has been published on a validity of this model in more acidic solutions.

Self-Inhibition (Calcium-Rich Layer Formation) Model

A model with a self-inhibition layer formation describes a formation of a calcium-rich layer on the surface of apatite during dissolution [19, 51-55]. This model was created from the results on dissolution kinetics of apatite powders in acidic buffers at solution pH within 3.7 – 6.9 (the constant composition technique was used). Following concentration of calcium and solution pH, the authors found that during the initial period (2 – 5 min) of dissolution release of calcium into the bulk solution was always less when compared with consumption of H^+. When dissolution progressed, rates of calcium release permanently increased and hydrogen uptake decreased until the equivalence (consumption of two hydrogen resulted in release of one calcium) was reached, while the over-all rate of apatite dissolution permanently decreased [19, 51-55]. Similar decreasing of apatite dissolution rates were obtained by other research groups [43-50, 56, 71].

According to this model, apatite is dissolved by ionic detachment of calcium and orthophosphate ions from the surface to a solution. When an initial portion of apatite has been dissolved, some amount of calcium cations (presumably, in connection with anionic counter ions) is returned from the solution and adsorbed back onto the surface of apatite. The latter results in formation of a semi-permeable ionic membrane consisting of a positively charged layer containing strongly adsorbed calcium ions, *i.e.* a calcium-rich layer is formed [19, 51-55]. An indirect confirmation of the presence of a calcium-rich layer was obtained in the most recent paper of this series: for initial strong calcium undersaturation in solutions, attainment of a pseudo-steady state was very long [55]. The areas per adsorbed single PO_4^{3-} and Ca^{2+} ions on HA surface at the point of zero charge (pH = 8.5) were estimated to be 33.1 ± 2.7 and 23.0 ± 2.1 $Å^2$, respectively [72].

The authors of calcium-rich layer formation model suggested that the rate control of apatite dissolution was diffusion of calcium and/or orthophosphate ions through an interfacial surface layer formed during the initial portion of the dissolution process. This interface would be composed of the Nernst layer that would be adjacent to the calcium-rich layer. This last one would present low permeability making the dissolution process slower [19, 51-55]. It is important to notice that these conclusions were made from the experiments on following calcium concentration and solution pH only; no measurements of the surface composition of apatites were performed.

As said by this model, ionic diffusion in solution results in a situation when ionic concentration near the surface layer of apatite is always higher than that in the bulk solution. Calcium concentration in the solution is also inhomogeneous: it decreases when the distance from the liquid/solid interfaces increases [19, 51-55]. Therefore, the calcium-rich layer appears to be undoubtedly rich in calcium when compared with the bulk solution only. However, when compared with the bulk of solid apatite (Ca/P = 1.67) this layer appears to be calcium-rich (Ca/P > 1.67) when the counter anions A^{n-} (it is assumed, that apatite is dissolved in acid H_nA, where A^{n-} is an anion) adsorbed from the solution are not taken into account. If the counter anions are calculated, a numeric value for Ca/(P+A) molar ratio on the surface will be around 1.67 in the case of zero charge on the surface.

On the other hand, ions of calcium might be adsorbed without the counter anions. This results in an electrical double layer formation on the solid/liquid interface with positive charge on apatite. Indeed, apatite is charged positively in acidic solutions. This effect was explained by adsorption of either protons only [73] or protons and calcium cations from the solution [74, 75]. Certainly, the presence of a positive charge on apatite surface is in favor for the calcium-rich layer model but one should not forget that this charge might be caused by adsorption of protons only [73]. The dissolution rate reduction found by the authors and explained by the state of the interface (adsorption of calcium [58]) might also be explained by adsorption of some impurities those always present in the solution. Other explanations are also available [43-50]. Thus, unless the direct measurements of the calcium content on the surface are made, this model cannot be considered as experimentally proven.

A contradiction between the calcium-rich layer formation and polynuclear models can be outlined: sorption of calcium is likely to happen on the dissolution nuclei. In this case, some amount of the nuclei should disappear, as a result. No suggestions how to overcome this contradiction were found in the references.

To conclude this model, one should notice that due to the formation of a calcium-rich layer, a non-stoichiometric (incongruent) dissolution (see the next model) is suggested by default.

Stoichiometric/Non-Stoichiometric (Congruent/Incongruent) Dissolution

A congruent/incongruent (or stoichiometric/non-stoichiometric) dissolution [11, 17, 19, 21, 41, 42, 52, 56, 71, 76-95] is based on the direct measurements of either the ionic concentrations in a solution [19, 21, 41, 42, 56, 71, 77-79, 83, 86, 87, 90, 91, 94] or the surface composition of apatites during dissolution [17, 80, 84, 93, 95]. According to the definitions, ions of a solid might be dissolved either simultaneously with the dissolution rates proportional to their molar concentrations (a congruent dissolution) or non-simultaneously with different values of the dissolution rates for each ion (an incongruent dissolution). The latter case always results in a situation when a surface layer with a chemical composition different from that of the bulk of solid is formed.

For acidic dissolution of FA, ions of F^- were found to dissolve faster (or in a greater ionic proportion) when compared to ions of calcium and orthophosphate [17, 71, 80, 86, 91, 92]. A similar phenomenon of faster (or in a greater ionic proportion) dissolution of calcium when compared to that of orthophosphate was also found [42, 78, 79, 82-85, 93-95]. In the latter case, a surface layer of acidic calcium orthophosphates, presence of HPO_4^{2-} ions and/or Ca-depleting of the surface were either suggested [54, 64, 88] or found [84, 89, 91-93, 95]. Thus, if incongruent dissolution occurs from a stoichiometric HA surface, the surface Ca/P ratio will become lower than for the bulk and a Ca-deficient HA (CDHA) must then be present, even if there is no phase change [57, 93, 95]. However, there are contradictory results. For example, when FA was dissolved, the solution concentrations of calcium and fluoride ions were non-stoichiometrically lower with respect to that of orthophosphate ions due to a possibility that some amount of sparingly soluble calcium fluoride was precipitated [42, 89, 91, 92, 94, 96-99]. Furthermore, one should mention on the studies, in which release of calcium ions into the bulk solution was always less, when compared with release of orthophosphate ions [56, 71]. On the other hand, having followed the release kinetics of calcium and orthophosphate ions into solution, apatites were found to dissolve stoichiometricaly (congruently) [21, 76, 100].

Interestingly, but apatites might be dissolved stoichiometrically or non-stoichiometrically depending on the experimental conditions. For example, CDHA was found to dissolve congruently at the solution $pH < 4.5$, while an incongruent dissolution with Ca/P ratio in solution attaining values as high as ~ 25 was detected in solutions within $4.5 < pH < 8.2$. Furthermore, almost negligible dissolution with another type of a non-stoichiometry with Ca/P ratio close to zero in solutions was found at solution $pH > 8.2$ [90]. For $4.5 < pH < 8.2$ conditions, the authors explained the incongruent dissolution by re-adsorption of already dissolved orthophosphate anions from the solution onto the positively charged CDHA surface. While in more basic solutions, the surface of CDHA was charged negatively and, thus, the authors suggested a possibility of both re-adsorption of already dissolved calcium cations onto CDHA surface and desorption of orthophosphate ions from the CDHA surface. However, it has to be realized that at solution $pH > 8.2$ the uncertainties in the chemical analysis of Ca (in particular) and P were relatively high, making any quantitative interpretation of this re-adsorption/desorption difficult [90]. The same effect was discovered in another study [92]. Furthermore, dissolution of both FA and natural sedimentary carbonated FA was found to be incongruent in the early stages but tended toward congruency as the dissolution reaction proceeded to steady state [42, 71, 101]. According to the authors, this initial incongruency was likely a result of the exchange of solution H^+ ion for mineral bound Ca [42]. Similar transition from the initial incongruency to further congruency was also observed for dissolution of HA, $CaHPO_4 \cdot 2H_2O$ and a calcium orthophosphate cement in aqueous acidic (pH = 3) solutions [41]. Besides, a presence of simple inorganic salts (Na_2SO_4, CH_3COONa or NaCl) in solution appeared to have an influence on the equilibrium concentrations of orthophosphate and calcium ions released from apatite: the order of salts increasing orthophosphate in solution was Na_2SO_4 > CH_3COONa > NaCl, while that increasing calcium was opposite [87]. This effect was explained by a specific affinity of some ions (e.g., sulfate) for apatite and might be a reason for transition of congruent to incongruent dissolution in some cases [97-99, 102]. Similarly, presence of other compounds might influence the dissolution congruence [103]. Other reasonable explanations might be due to either experimental difficulties in precise following of Ca/P ratio both in the solution and on the crystal surface or differences in chemical and structural compositions of apatites studied by different researchers (both FA and HA are known to be often non-stoichiometric and calcium-deficient, especially on the surface [2-4, 90-92, 104]).

The idea on a great influence of a surface layer is supported by the data on dissolution of stoichiometric HA (Ca/P = 1.67) in aqueous solutions within pH 4.90 – 9.94 [79]. The authors found that after removal of an irreversible surface, which generated excessive calcium and orthophosphate solution concentrations, further behavior of HA became reproducible. A surface of one atom-layer thickness or less was believed to have formed during the preparation of the solid [79, p. 697]. On the other hand, dissolution of HA in aqueous medium appeared to be always non-stoichiometric at the beginning, but when the solid was successively equilibrated at any given pH, the solution Ca/P ratio approached a limiting value of 1.67. Once this value was reached, the solid only maintained this solution ratio by dissolving stoichiometricaly [79, p. 700]. The authors explained these phenomena by a previous history of the samples studied. A similar conclusion was also made in another paper [77]. Thus, the experimental techniques of apatite preparation might have a great influence to stoichiometry/non-stoichiometry of dissolution. This is a serious drawback.

Finally, the results on theoretical analysis of surface layer formation on HA should be reported [88]. The main point of that paper is an idea that both the stoichiometric HA (Ca/P = 1.67) and any non-stoichiometric CDHA (Ca/P ratio within 1.5 – 1.67) might be described as the same substance. More precisely, there is a continuous amount of intermediate compounds within $1.5 \leq Ca/P \leq 1.67$ having the same crystal structure which is drawn as a line on the solubility diagram CaO – P_2O_5 – H_2O [88]. Furthermore, according to the authors, at the steady state, both HA and CDHA are dissolved congruently only when both the solution and solid have the same Ca/P ratio. Therefore, any sample of HA and CDHA is dissolved incongruently except the only one solution composition having the similar Ca/P ratio with that particular sample. In all other cases, a surface layer having a Ca/P ratio less than 1.67, which is often used as the indication of incongruent dissolution of apatite, will form as a result (this, however, depends on the accuracy with which the analyses can be made) [88]. Thus, whether apatites are dissolved congruently or incongruently might also depend on both their initial chemical composition and the solution properties.

To conclude, one should notice that all the aforementioned cases of incongruent dissolution of both FA and HA describe the formation of a surface layer of a CDHA and/or acidic calcium orthophosphates, which is calcium-depleted when compared to the bulk apatite. Thus, there is a disagreement between the incongruent dissolution and calcium-rich layer formation model. Perhaps, this might be due to the differences in experimental conditions.

Chemical Model

A chemical dissolution model was proposed as well [105-107]. This model was developed from a self-evident supposition that it would be highly unlikely if apatite were dissolved by detachment of "single molecules" equal to the unit cells and consisting of 18 ions. Moreover, in the crystal lattice all ions are shared with neighboring unit cells and often cannot be attributed to the given "single molecule". Therefore, based on the experimental results obtained on the one hand and an analysis of the data found in references on the other hand, a sequence of four successive chemical reactions:

$$Ca_5(PO_4)_3(F, OH) + H_2O + H^+ = Ca_5(PO_4)_3(H_2O)^+ + HF, H_2O \quad (1)$$

$$2Ca_5(PO_4)_3(H_2O)^+ = 3Ca_3(PO_4)_2 + Ca^{2+} + 2H_2O(2)$$

$$Ca_3(PO_4)_2 + 2H^+ = Ca^{2+} + 2CaHPO_4 \quad (3)$$

$$CaHPO_4 + H^+ = Ca^{2+} + H_2PO_4^- \quad (4)$$

was proposed to describe the chemical processes of apatite dissolution [105-107]. According to this model, equations (1) – (4) should be used instead of the well-known net-reactions (5) and (6):

$$Ca_5(PO_4)_3(F, OH) + 7H^+ = 5Ca^{2+} + 3H_2PO_4^- + HF, H_2O \quad (5)$$

$$Ca_5(PO_4)_3(F, OH) \rightarrow 5Ca^{2+} + 3PO_4^{3-} + F^-, OH^- \quad (6)$$

Chemical model (1) – (4) was initially created for description of apatite dissolution in strong acidic media [106] and appeared to be a further development of some ideas proposed earlier. For example, already in 1960-s reactions (5) and (6) had been suggested "not to give the mechanism but merely express net-reactions only" [108, p. 333]. Later, Pearce [109] and Chow [36] discussed which of net-reactions (5) or (6) described the dissolution process better. For acidic dissolution of apatites, net-reaction (5) represents the direct chemical interaction between the surface of solid apatite and an acidic solution. However, Pearce suggested that, in principle, the dissolution process might also happen according to net-reaction (6) followed by a chemical interaction in the solution among the ions of apatite and acid *near* the crystal surface [109]:

$$5Ca^{2+} + 3PO_4^{3-} + F^-, OH^- + 7H^+ = 5Ca^{2+} + 3H_2PO_4^- + HF, H_2O \quad (7)$$

In spite of a lack of the experimental techniques, able to follow the detachment of single ions, a reasonable ionic description of apatite dissolution as a system of chemical equations (1) – (4) was created from the indirect measurements [105-107]. According to this model, chemical dissolution of apatite occurs step by step on the surface via intermediate formation of a mono-atomic layer of some virtual compounds, like $Ca_3(PO_4)_2$ and $CaHPO_4$ [105-107]. Both latter substances are known as precursors of apatite formation during chemical crystallization [3, 110]. Moreover, based on the sequence of ionic detachment (first fluorine for FA or hydroxide for HA, followed by calcium and afterwards by orthophosphate) it is suggested that dissolution of apatites is always incongruent (non-stoichiometric) at the atomic (ionic) level. However, a lack of the precise experimental verification transforms the chemical model into a reasonable hypothesis only.

To conclude this model, one may assume that some new relations between the kinetically controlled and chemical models might be established in future. Before the chemical model was created, chemistry of apatite dissolution had been always described by equations (5) or (6) + (7). For this reason, the kinetically controlled model was also adapted for these equations. For example, the overall rate constants for net-reactions (5), (6) and (7) are still used for calculations, whereas they are probably equal to either the lowest of the rate constants of (1) – (4) (the rate determining step) or some combination.

Etch Pit Formation

A process of etch pit formation describes acidic dissolution of apatite crystals containing structural defects (dislocations and inclusions) [43-50, 95, 101, 111-120]. A radius of dislocations outlets (hollow cores) on apatite was calculated to be within 8.3 – 20 Å for screw dislocations and 18 – 43 Å for edge ones [114]. The presence of dislocations accelerates dissolution, because dislocations give rise to continuous steps on the surface (Figure 2D) and the strain energy they cause in crystals favors etch pit formation. Thus, the dissolution process of solids is induced by formation of pits (Figure 3) and continues with spreading of their stepwaves [121, 122]. As a result, the surface of apatite becomes rough and the total edge length and edge free energy increases [43-50]. These pits provide dissolution sites and the entire reaction proceeds via nucleation and growth of the pits accompanying step flow. The pits appear at the dislocation outlets, usually they are 0.1 – 10 μm in size (their dimensions depend on dissolution kinetics and dissolution time: they increase when the dissolution progresses; furthermore, the crystals must be sufficiently big to provide enough room for large pits to form). The walls and bottom of etch pits consist of crystal faces possessing the highest dissolution rates, which allow them to grow. However, the growth kinetics of etch pits (therefore, the dissolution process of apatites in general) might be suppressed by addition of very simple chemicals, such as NaCl. This phenomenon might be attributed to a competition for surface protonation sites between Na^+ and H^+ ions [124].

For dissolution of apatites, the pits usually have a hexagonal shape (Figure 3) according to the crystal symmetry $P6_3/m$ of pure HA and FA [110-116, 123-128]. However, their shape might be different (e.g., triangular) when specific ions (e.g., fluoride) are present in a solution [123, 126]. It is important to mention that the growing process of etch pits might be inhomogeneous: movement of the dissolution steps on one side of any single pit might be faster or slower their counterparts on the other side, suggesting the formation of asymmetric pits, as schematically shown in Figs. 3a and 3d. Concerning height of the steps, it appears to depend on both crystal faces and dissolution conditions. For example, for an HA (100) surface exposed to deionized water, the average height of each step was found to be 0.84 ±

0.1 nm, which is close to the interlayer distance of the (100) surface (d(100) ≈ 0.82 nm). Furthermore, it remained constant during dissolution [125].

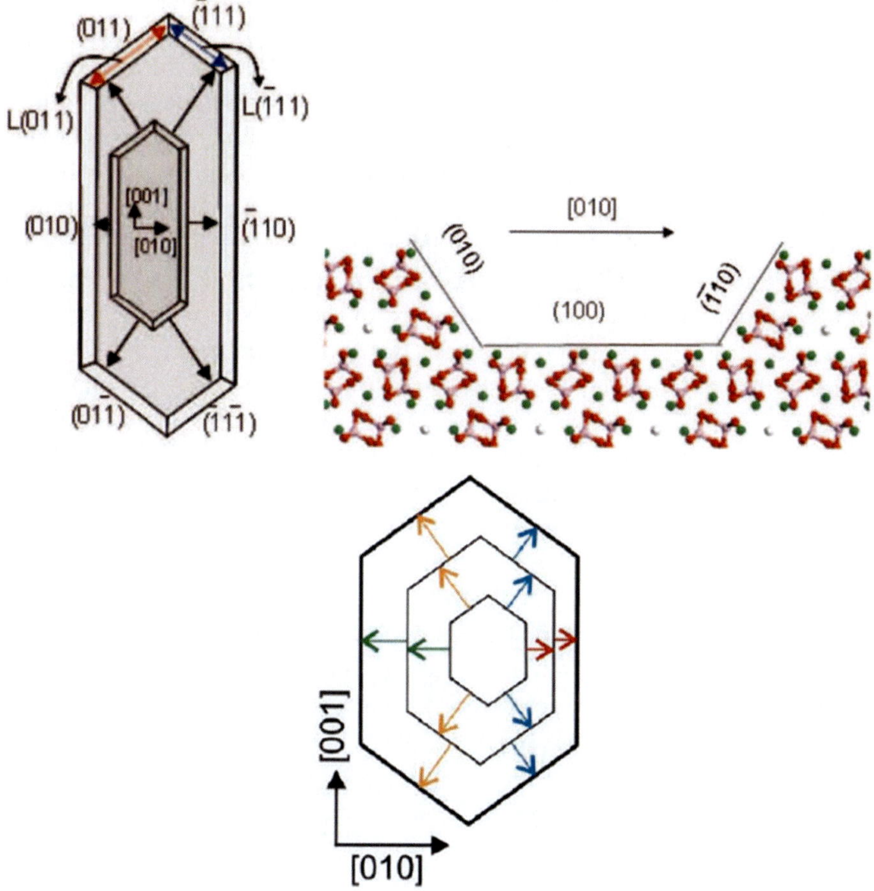

Figure 3. A typical etch pit on the surface of calcium apatites. (a) – Top view and a model of its evolution during dissolution. Arrows represent relative step velocities displaying different facets. (b) – View through the [001] direction in the (100) surface. The exposed step faces can be seen to be unequivalent owing to the opposite orientations of the orthophosphate groups. Reprinted from Ref. [123] with permission. (c) – Schematic of the evolution of hexagonal etch pits. Four differently colored arrows represent relative step velocities. Reprinted from Ref. [125] with permission.

The formation of etch pits on the crystal faces is the driving force for dissolution. However, in the case of nearly physiological conditions and sufficiently low solution undersaturation, a free energy barrier becomes too high for vacancy nucleation to occur on a time scale that is competitive with other processes. Thus, the dissolution process of apatites becomes spontaneous only when etch pits of critical sizes (determined by the Gibbs-Thomson effect, a well-known thermodynamic principle) are reached [43-50]. At this critical size, the free energy change goes through a maximum that defines an energy barrier to a pit formation. Therefore, only relatively large pits (of sizes greater than a critical value) appear to be active, with stepwaves contributing to dissolution, while the spreading velocities are also dependent on the pit sizes, decreasing with pit size decreasing. Moreover, during dissolution, the crystals become smaller and the average lengths of dissolution steps decrease (which leads to a decrease in dissolution rates) and approach the critical value. When dimensions of dissolving crystals are sufficiently reduced, in some cases dissolution is thought to be dynamically stabilized (dissolution suppression) due to a lack of space for active pits/defects formation on the very small crystallite surfaces [43-50]. This phenomenon involves particle-size-dependent critical conditions of energetic control at the molecular level. Although, the authors claim this effect as “a new dissolution model incorporating particle size considerations” [43-50], in fact, they just have introduced some boundary conditions (the minimal pit sizes and crystal dimensions) to the etch pit formation process. One should remind, that a similar effect of the dissolution self-inhibition was detected by the authors of the calcium-rich layer formation model [19, 51-55], although the interpretation of this effect was quite different (see above).

Different acids were used in experimental investigations but, unless the conditions of surface coating creation [125], no specific influence of the acids to etch pit formation and growth was found [111-120]. A theoretical basis for the pit formation and growth at the surface outlets of dislocations includes detachment of single ions from the kink sites. The bulk rate associated with dissolution stepwaves arises quite naturally from the equations describing the spreading of step trains from structure defects and densities [45]. However, until recently, the experimental investigations on pits formation and growth on apatites were made when their dimensions became at least 0.3 – 0.5 μm and even larger. Smaller dimensions of pits were out of the range of old microscopes. That is why, the initial stages of etch pits formation are not quite clear yet.

The investigations on apatite dissolution using atomic force microscopy provided new data on this point [43-50, 69, 70, 123-127]. For example, the growth process of a single pit on apatite was followed since its dimensions were 50 nm [69, 70]. Therefore, one may expect new results on even smaller dimensions of pits to appear soon.

Ion Exchange Model

An ion exchange dissolution model is based on a supposition about adsorption of protons [42] and anions of acid (e.g., citrate anions [130-132]) from a solution onto the surface of apatite and removing ions of calcium and orthophosphate into the solution instead. An uptake of citric acid [130] and citrate ions [131] from dilute (4 – 12.5 mM for citric acid [130] and 0.05 – 1 mM for sodium citrate [131]) aqueous solutions by a synthetic HA was found to be time independent and follow Langmuir isotherms rather good. However, it was not the case when solutions of higher concentrations were used. In more concentrated solutions (13 – 100 mM of citrate), this interaction was “reactive” and not “adsorptive” [130, p. 1421].

Adsorption was found to take place by ionic exchange of orthophosphate by citrate ions at the solid-solution interface, caused by a higher affinity of citrate than orthophosphate species for the Ca-sites on the surface. Citrate ions were proposed to interact in two ways: $C_6H_8O_7^{3-}$ interacted weakly in a bi-dentate manner (1 citrate per 2 calcium sites), whereas $C_6H_9O_7^{2-}$ interacted strongly in a mono-dentate manner (1 citrate per 1 calcium site) [131]. Recent computer simulations revealed that this process is followed by hydrogen-bonded interactions of citric acid to surface oxygen ions of apatites [133]. Furthermore, the adsorbed citrate molecules can use their free carboxylate and hydroxyl groups to self-assemble on the HA surface via intermolecular hydrogen bonding [134].

Later this model was updated by investigation of apatite dissolution in other acids and an adhesion-decalcification concept was introduced [135, 136] According to this concept, the mechanism of acidic interaction with HA was found to involve two phases. In the first phase, an acid bonds to surface calcium of HA with formation of calcium-acid complexes. Depending on both a solubility and a diffusion rate of the calcium-acid complexes from the HA surface to a solution, the acid will in the second phase either remain attached to the HA surface with only limited decalcification involved or the calcium-acid complexes will detach, resulting in a substantial decalcification effect. Rather similar data, in which carboxylic groups of polyalkenoic acid were adsorbed on the surface of an HA substrate, replaced orthophosphate ions and made ionic bonds with calcium ions of HA, were obtained in another study [137].

Recently, *ab initio* total energy methods were used to study the atomic structure and surface chemistry of HA [138]. All surfaces of HA were found to react strongly with water and a loss of Ca from the surface in exchange for two H appeared to be very favorable. Since these calculations were performed for the case of water adsorption on HA, it remained unclear whether they can be applied to the acidic solutions. Nevertheless, other researchers found this effect experimentally [89, 91]. Furthermore, another type of the exchange was proposed for HA, in which one calcium and one hydroxyl were replaced by one proton [90].

To conclude this model, the time independent ionic exchange of orthophosphate by citrate was detected in dilute solutions only. In more concentrated solutions it transferred into chemical reaction (5) followed by time dependent precipitation of calcium citrate when its solubility was exceeded [130, 132]. Application of this model to HA dissolution in other acids revealed its dependency on both the solubility and the diffusion rate of the calcium-acid complexes [135, 136]. Thus, new experimental data are necessary, to elaborate the ion exchange model further.

Hydrogen Catalytic Model

A hydrogen catalytic model is based on a reasonable suggestion about adsorption of protons onto negatively charged oxygen ions of the surface orthophosphate groups of apatites [139]. Sorption of protons results in partial transformation of surface PO_4^{3-} groups into HPO_4^{2-} and catalyses the dissolution process [9, 10, 15, 18, 20, 55, 57-64, 92]. A schematic representation of this process at different solution pH is shown in Figure 4 [124]. Besides, recent computer simulations revealed that both OH^- (for HA) and F^- (for FA) located on the surface are protonated as well and, of them, fluoride ions exhibited much lower proton affinities (pK_B = 10.8, a weak base) compared to hydroxide (pK_B = 0, a strong base) [140]. Furthermore, the surface protonation of apatites is intimately related to hydrogen bonding. The details on the latter process are available in literature [141].

One can notice, that the authors of different dissolution models (the diffusion and kinetically controlled [9, 10, 15, 18, 20] and polynuclear [59-64] models) have suggested a similar possibility of a catalytic effect of the adsorbed protons. At present, indirect data only are available to support this hypothesis. Namely, at neutral pH, dissociation of ions is penalized by more than 150 kJ/mol giving rise to very stable apatite-water interfaces [140]. This picture changes drastically with decreasing pH, as the protonation of orthophosphate and hydroxide ions lowers the free energy of calcium ions dissociation. An obvious trend in apatite dissolution is that the dissolution rate is increased as the solution pH decreases, which is a general phenomenon observed in dissolution of other minerals. This indicates that dissolution proceeds by surface protonation, which promotes breaking of Ca–O bonds.

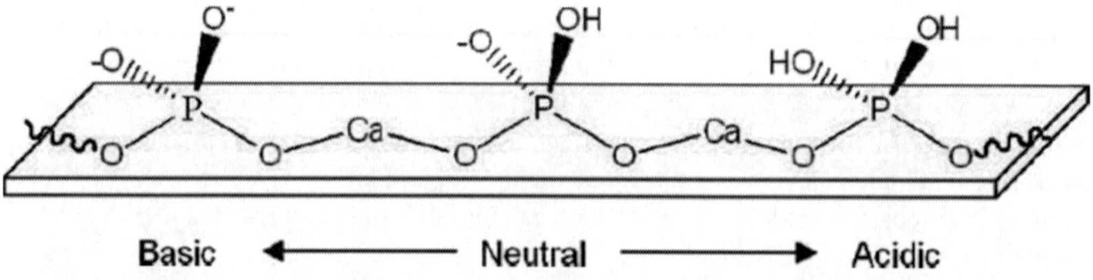

Figure 4. A schematic representation of surface protonation of apatites at different solution pH. Reprinted from Ref. [124] with permission.

In the course of proton uptake, the charge of the surface anions is reduced and the Coulomb attraction to the crystal surface is lowered considerably, which might be considered as “catalysis”. Currently, no precise information about a catalytic activity of adsorbed protons on the dissolution of apatites is available. However, it does not mean that this effect does not exist. Up to now, nobody has ever succeeded to reveal a catalytic effect unambiguously due to a lack of an experimental technique able to provide the experimental data necessary. Obviously, the general idea on a catalytic activity of the adsorbed protons seems reasonable and, perhaps, this effect will be better elucidated in future.

Summary on the Dissolution Models

There is an old tale on six blind men, who wanted to determine what an elephant looked like. An elephant was delivered and the men were allowed to feel different parts of its body. The blind man who felt a leg said the elephant was like a pillar; the one who felt the tail said the elephant was like a rope; the one who felt the trunk said the elephant was like a tree branch; the one who felt an ear said the elephant was like a hand fan; the one who felt the belly said the elephant was like a wall; and the one who felt the tusk said the elephant was like a solid pipe. Thus, six blind men appeared to be in complete disagreement. Then, a wise man explained to them: “All of you are right. The reason every one of you is telling it differently is because each one of you touched the different part of the elephant. So, actually the elephant has all the features you mentioned.” The story is widely used to indicate that the reality may be viewed differently depending upon one’s perspective, suggesting that what seems the absolute truth may be relative due to a deceptive nature of half-truths.

As the complete physical picture that explains widely observed variations in dissolution behavior is still lacking, a similar approach must be adapted to the aforementioned dissolution models of calcium apatites. Undoubtedly, each from the aforementioned models appears to have limitations and drawbacks of its own and none of them is able to describe the apatite dissolution in general. Moreover, most of the models were elaborated for apatite dissolution in either slightly acidic or nearly neutral ($4 < pH < 8$) aqueous solutions, relatively small values of solution undersaturation and temperatures of 25 and 37 °C only. Nothing is known about the validity of these models for apatite dissolution in strong inorganic acids (HCl,

HNO_3, H_2SO_4, H_3PO_4) for solution pH < 2 and temperatures above 70 °C. Moreover, in such "tough" experimental conditions, the numerical value of solution undersaturation by HA and FA becomes uncertain due to impossibility to determine their solubility [1-4]. The latter cases serious problems for application of current versions of calcium-rich layer formation and polynuclear models to apatite dissolution in strong acids, because both models operate with the solution supersaturation/undersaturation phenomena. The abovementioned "tough" dissolution conditions are common in industry of fertilizer production [142] and have to be taken into account by the final version of any dissolution model.

However, there are no doubts that each of the models discussed is correct for the specific experimental conditions studied. The different models are just devoted to diverse aspects of the same process of apatite dissolution and are likely complement each other. Undoubtedly, the dissolution process of calcium apatites has all the features elaborated within eight dissolution models. Unfortunately, it is still impossible to obtain the complete understanding for all the processes involved because experimental techniques able to provide direct following after diffusion, sorption, detachment and transformation of the single atoms, ions and molecules currently are lacking. Therefore, researchers are still forced to make conclusions about the events happening at the atomic (ionic) level based on indirect measurements and/or calculations only. That is why, combining the experimental and theoretical findings, obtained by different investigators (often under hardly comparable experimental conditions), currently, seems to be the only way to create a general depiction of the dissolution mechanism of calcium apatites in acids.

A Reasonable Classification of the Dissolution Models

At the fundamental level reactions between solids and liquids involve a coupled sequence of mass transport, adsorption/desorption phenomena, heterogeneous reaction, chemical transformations of intermediates, *etc.*, whose identification, separation and kinetic quantification are all necessary if the mechanism of the process is to be fully understood and described [143]. It is also generally agreed that during dissolution lattice ions undergo the following processes: (1) detachment from a kink site, (2) surface diffusion away from the crystal steps, (3) desorption from the surface and (4) diffusion into the bulk solution [144]. Therefore, the steady-state conditions of apatite dissolution in aqueous acidic media include the following simultaneous processes or steps:

- diffusion of chemical reagents (H^+ and anions of acid A^{n-}) from bulk solution to the solid/liquid interface;
- adsorption of the chemicals onto the surface of apatite;
- chemical transformations on the surface;
- desorption of products (ions of fluoride, calcium and orthophosphate) from the crystal surface;
- their diffusion into the bulk solution.

All the steps mentioned above are likely to be much more complicated. For example, processes 1 and 5 include chemical transformations happen with the ions during diffusion, because solution pH is known to depend on the distance from the solid/liquid interface (in acidic solutions it is higher near the surface of apatite and decreases when the distance increases) [145, 146]. Processes 2 and 4 include ionic diffusion along the surface both to (step 2a) and away from (step 4a) the crystal steps, respectively, as well as a diffusive jump (steps 2b and 4b) [144]. Finally, process 3 consists of several successive chemical transformations (step 3a – chemical equations (1) – (4)) [105-107] and includes ionic detachment from the kink sites (step 3b) [144]. Moreover, for adsorption (step 2) to take place, the following intermediate steps are necessary: (i) dehydration of a surface site, (ii) partial dehydration of ions and (iii) a diffusive jump toward the surface [144, p. 1692]. One can also expect three similar intermediate steps (diffusion jump from the surface, hydration both the ions and the surface site) for desorption (step 4).

As shown above, different models describe just diverse aspects of the same phenomenon of apatite dissolution. Commonly, the main distinction is made between the transport control and surface-reaction control. For example, in the diffusion controlled, two-site and polynuclear models steps 1 and 5 limit a reaction rate. On the contrary, in the kinetically controlled, chemical and hydrogen catalytic models a reaction rate is limited by step 3. In the cases of the ion exchange and self-inhibition (or calcium-rich layer formation) models, a reaction rate is likely to be controlled by both ionic diffusion (steps 1 and 5) and kinetics of the adsorption/desorption phenomena (steps 2 and 4). Processes of etch pit formation and dissolution stoichiometry – are the experimental facts, not models. Thus, they may be controlled by factors pertaining to any step from 1 – 5 and even all of them.

Brief Information on Apatite Structure

Briefly, crystals of HA and FA consist of individual orthophosphate tetrahedrically linked by calcium cations. The orthophosphate groups forms the skeleton having two types of calcium sites. The first (Ca(1) site) is occupied by four calcium atoms that have a nine-fold coordination, while the second (Ca(2) site) is occupied by six calcium atoms having a seven-fold coordination. Fluoride (for FA) or hydroxide (for HA) anions are located within the same channels as the Ca(2) sites and have a high mobility [1-4, 147, 148]. Thus, the apatite structure contains Ca–F (for FA), O–H (for HA), Ca–O and P–O bonds; consideration of the relative electrostatic strength of these bonds suggests that their relative destruction rate is consistent with: Ca–F ≈ Ca–OH > Ca(1)–O > Ca(2)–O > P–O. Considering its structure, once all Ca–F and Ca–O bonds are broken, apatite is destroyed; while breaking of P–O bonds is unnecessary for apatite dissolution [101].

Necessary Assumptions and Limitations

When apatite comes into a contact with an acidic aqueous solution, dissolution steps 1 – 5 (see section 4) begin. Before giving the detailed description of the processes involved, the following assumptions and limitations must be made:

The stoichiometric apatite (Ca/P = 1.67) is dissolved and neither non-stoichiometric layers nor other ions except of calcium, orthophosphate, hydroxide (for HA) or fluoride (for

FA) are initially present in the crystal lattice, whereas volume and surface defects (dislocations and dislocation outlets respectively) might be present and, if present, they are distributed randomly.

Except of the presence of dislocation outlets, the initial surface of apatite is perfect (molecularly sooth). Otherwise, each imperfection might act as a possible dissolution nucleus.

In spite of the limitations and drawbacks discussed above, all the afore-discussed dissolution models are correct and complementary to each other.

In some cases, anions of acid might have an influence on apatite dissolution due to a specific affinity (e.g., citrate) or by means of formation of insoluble compounds (e.g., sulfate, fluoride). That is why, anions are specified neither here nor below. It is just assumed that dissolution of apatite occurs in acid H_nA, where A^{n-} is an anion.

Only soluble compounds are formed during the apatite dissolution in acids. No precipitates are obtained.

Hydration effects for all the ions and molecules involved as well as those for crystal surface of apatite [104, 149-153] are omitted for simplicity.

All crystal faces of a solid apatite are equal and have a similar ionic arrangement. Although, it was found that HA {001} facets could be more resistant against acidic dissolution than {100} ones [154], no specific influence of different crystal faces on the dissolution process is considered.

Since the afore-described dissolution models have been elaborated for micron-sized apatite crystals, while the nanodimensional calcium orthophosphates possess the special properties [155], to eliminate the size effects, dissolution of large crystals is described. No specific influence of crystal dimensions on the dissolution mechanism is considered.

Based on these assumptions and limitations, an attempt to create the general description of the apatite dissolution mechanism at the atomic (ionic) level is given below.

Creation of the General Dissolution Mechanism

Atomic (Ionic) Description for the Perfect Crystals

Immediately after being placed to acidic aqueous solutions, the surface of apatite crystals becomes in contact with water molecules and ions of the acid, which results in sorption of all these chemicals with simultaneous formation of a solid/liquid interface and various adjacent strata, such as a Nernst diffusion layer, an electrical double layer, *etc.* Transport of the chemicals (H^+ and A^{n-}) to the solid/liquid interface always happens by ionic diffusion through the Nernst diffusion layer (Figure 1). A thickness of this layer depends on both solution hydrodynamics and dimensions of the crystals dissolving: it decreases when agitation intensity increases and crystal sizes decrease. However, in no cases the thickness becomes equal to the zero [156]. Thus, a concentration gradient is always present near the solid/liquid interface, the diffusion laws always remain valid and all findings of the diffusion-controlled model (e.g., an adsorption resistance and a diffusion jump) remain applicable for any case of apatite dissolution. A similar approach is correct for the surface phenomena: when dissolution occurs, adsorption and desorption of ions, as well as chemical transformations (1) – (4) always take place on the surface regardless the experimental conditions chosen.

After being delivered by diffusion to the solid/liquid interface, ions of H^+ and A^{n-} are adsorbed onto the surface of apatite. A great number of various surface complexes might be formed as a result [89-92, 157-159]. According to the diffusion-controlled model, there is an adsorption resistance for ions to be adsorbed onto the surface and in order to overcome the resistance the ions make a diffusive jump toward the surface to a distance corresponding to their size [144, 160]. Being charged positively, protons are adsorbed onto oxygen ions of orthophosphate groups [61] as well as onto ions of fluoride (in the case of FA) and hydroxide (in the case of HA). After the model by Wu *et al.*, the surface protonation of apatites proceeds via formation of ≡POH surface groups ("≡" stands for the surface) at $5 < pH < 7$, while apatite surfaces become fully protonated at $pH < 5$ [157]. One may expect to find some differences in the adsorption kinetics of protons onto fluoride, hydroxide and orthophosphate ions. Namely, due to a higher electro negativity of fluoride (when compared to oxygen ions of orthophosphate groups) and a higher basicity of hydroxide (when compared to orthophosphate), adsorption of protons might happen faster (or previously) onto these ions when compared with orthophosphate. However, recent results of computer simulations indicated a possibility of OH^- protonation by the nearest HPO_4^{2-} ion [140], indicating to the fact, that orthophosphate ions might be protonated faster (or previously) if compared to that for hydroxide ions. Obviously, this topic needs to be clarified in future.

Anyway, negatively charged anions A^{n-} may be adsorbed onto calcium cations only. According to the ion exchange model, the exchange process shows an adsorption of about one anion per unit-cell of apatite [130-132]. On the other hand, the surface of apatites is charged positively in aqueous acidic media and negatively in basic solutions (the point of zero charge is at solution pH within 6.8 – 8.5) with an electric double layer formation at the solid/liquid interface [73-75, 81, 89-91, 157-162]. The latter points out to a non-equivalent ionic adsorption of H^+ and A^{n-} onto apatite: in acidic solutions, adsorption of protons always exceeds that of A^{n-}, while in basic solutions the situation is opposite. Therefore, in acidic solutions there is at least one extra-adsorbed proton per several neighboring unit-cells of apatite when compared to the number of adsorbed anions. The fact of a non-equivalent ionic adsorption might be explained by differences in sizes and mobility between H^+ and A^{n-} [163]. Great difference in the surface sites available is another reason for the non-equivalent ionic adsorption. Namely, the concentration of the lattice ions being on the surface of HA were found to be equal to 2.57, 3.02 and 4.50 ions/nm^2 for OH^-, PO_4^{3-} and Ca^{2+}, respectively [72]. Furthermore, according to drawings of the apatite structure, the most part of its surface (up to 80 – 90%) is covered by oxygen ions of orthophosphate groups [164].

Ions of the crystal lattice of apatites are known to have different crystallographic positions: so-called X-ions (F^-, OH^-, Cl^- and substitutes for them) are located in Ca(2) channels parallel to *c*-axis [1-4, 147, 148]. This position results in a relatively high mobility of X-ions: for example, they might be mutually substituted rather easily [91, 92, 165-167] with formation of intermediate compounds, like fluorhydroxyapatites (FHA) or hydroxyfluorapatites (HFA) [17, 86, 168]. Based on these findings, dissolution of apatite in acids was suggested to start with detachment of X-ions (more likely, as HX) from the surface [105-107] followed by their diffusion away from the crystal steps and further into the bulk solution [144, 160]. Recent computer simulations revealed some differences in the dissolution kinetics between F^- and OH^- anions in apatites: surface fluoride is dissolved not as readily as hydroxide is [148]. Initial leaving of X-ions was also detected in the incongruent dissolution model [17, 71, 80, 86, 91, 92]. Moreover, numerical values for the surface tension found for dissolution of HA appeared to be clearly lower than those found for HA growth (the polynuclear model). The latter data were explained by partial protonation of orthophosphate groups and missing of some hydroxide ions on the crystal surface [63, pp. 299 – 300].

After detachment of X-ions from the surface, there are calcium, orthophosphate, H^+ (as HPO_4 and/or H_2PO_4 groups) and A^{n-} ions left on it (hydration effect is omitted for simplicity). From now on, any differences between FA and HA disappear. In any cases, adsorption of protons onto calcium is impossible due to charge repulsion. Therefore, an electrical double layer with a positive charge on apatite cannot be continuous at the atomic (ionic) scale. There should be local uncharged places in it, say, at the positions of calcium cations. According to the chemical model, an interaction between calcium cations and anions of acid adsorbed from a solution is believed to happen in such places [105-107]. This interaction results in breaking of surface ≡O–Ca bonds ("≡" stands for the surface) and detachment of some calcium from the kink sites followed by their diffusion away from crystal steps and further into the bulk solution (the diffusion-controlled model [144, 160]). Detachment of calcium might occur as calcium-acid complexes (the ion exchange model). Recent computer simulations revealed that a local excess charge of +3 and +4 must be created to cause exothermic Ca^{2+} displacement from Ca(1) and Ca(2) cites, respectively [140]. If so, under otherwise equal conditions, detachment of Ca^{2+} ions from Ca(1) cites of the apatite surface should happen faster and/or easily than that from Ca(2) cites.

Detachment of calcium ions results in formation of dissolution nuclei. These nuclei are defined as collections of vacant sites for Ca^{2+}, PO_4^{3-} and X^- ions on the crystal surface of apatite [59, p. 30]. According to the polynuclear model, removal even of one ion might result in further dissolution because critical nuclei (x*) were calculated to consist of 1 – 26 ions for dissolution of HA and 0.3 – 34 ions for dissolution of FA. (According to the authors, $x^* < 1$ means that there is no nucleation barrier to be overcome [64, p. 309]. Since one unit cell of apatite contains 18 ions, $x^* > 18$ means that more than one unit cell should be dissolved for a nucleus to form). The numeric values for x* were found to depend on solution pH and solution undersaturation [63, 64]. Furthermore, due to charge repulsion, adsorbed protons (as HPO_4 and/or H_2PO_4 groups) might catalyze detachment of calcium ions from the kink sites (the hydrogen catalytic model).

After being detached, calcium cations (possibly, as calcium-acid complexes (the ion exchange model)) diffuse along the surface away from the dissolution steps before entering the solution [144]. For dissolution of octacalcium phosphate (OCP: $Ca_8(HPO_4)_2(PO_4)_4{\cdot}5H_2O$) in slightly acidic (pH = 5.66) solutions numeric values for the mean surface diffusion distance of the lattice ions detached from steps were calculated. These values were found to depend on the edge free energy and be within $(17 \pm 4)\alpha - (41 \pm 10)\alpha$, where $\alpha \approx 3.7$ Å is the size of a growth unit (or mean ionic diameter) [160]. Numeric values of mean ionic diameter for apatites were also calculated. They appeared to be less than that for OCP: 3.09 Å for HA and 3.07 Å for FA [65]. Taking into the consideration great similarities found between the chemical composition and crystal structures of OCP and HA [169], one might assume that the numerical values of surface diffusion distance for HA should be close to the values for OCP.

Ionic detachment of some calcium results in removing of a local positive charge from apatite, which is immediately compensated by adsorption of other protons from the acidic solution. This conclusion is based on the surface charge measurements: no information on influence of dissolution time to values of the surface charge on apatite has been found in literature [73-75, 81, 89-91, 157-161]. So, one may assume, that if a chemical composition of the solution is kept constant the surface charge on apatite will also remain constant during dissolution. Replacement of calcium by protons results in formation of mono-atomic layer acidic calcium orthophosphates on the crystal surface. Chemical phenomena of these transformations are described by the chemical model [105-107] and the kinetics (it strongly depends on the experimental conditions chosen) – by the surface controlled model.

After being desorbed from the surface, calcium cations (presumably, coupled with anions A^{n-}) diffuse through the Nernst layer into the bulk solution (the diffusion controlled model) [144, 160]. However, according to the calcium-rich layer formation model, a part of the already dissolved calcium ions might be adsorbed back onto apatite. Adsorption of calcium was suggested to happen at the very beginning of dissolution only and resulted in formation of a perm-selective layer of calcium cations with or without counter ions (A^{n-} and/or acidic orthophosphate) [19, 51-55, 71]. In this case, there should be a competition between protons and calcium cations for the exposed ions of oxygen on the surface to be adsorbed onto. Experimental data on a shift of the point of surface zero charge into more basic solution pH in presence of calcium in a solution [161] are in favor of such competition.

An idea on formation of some layer on the surface of apatite was raised in several dissolution models. For example, there is either a supposition (the polynuclear, diffusion and kinetically controlled models) or a description (the ion exchange, hydrogen catalytic and chemical models) of a very thin virtual layer of acidic calcium orthophosphates (more exactly presence of HPO_4^{2-} ions) on the surface of apatite. Besides, according to the calcium-rich layer formation model, dissolution of apatite happens by ionic diffusion through a layer containing calcium with or without counter ions [19, 51-55, 71]. In part, this idea can be found within the ion exchange model where a monolayer of calcium citrate [130-132] and/or calcium-acid complexes [135, 136] are assumed to form onto apatite during dissolution. However, no other model requires such suggestion. According to the rest of them, formation of any calcium containing compounds (different from acidic orthophosphates) on apatite is not discussed at all.

According to the chemical model, the initial stages of apatite dissolution consist of calcium detachment from the surface and incorporation of protons instead. Orthophosphate groups are assumed to keep their positions without any relocation [105-107]. Due to the fact, that calcium occupies definite lattice positions, whereas protons are bound to oxygen ions of orthophosphate groups, removing of each calcium results in decreasing of attraction forces between the nearest (to this calcium) orthophosphate group and rest part of the crystals. When all (or almost all) neighboring cations of calcium have been removed, orthophosphate groups (as $H_2PO_4^-$, $CaH_2PO_4^+$ or H_3PO_4 – it is not clear yet) also detach from the surface. Then they diffuse along the surface away from the dissolution steps before entering the solution as described above for calcium [144, 160].

In the case of crystal faces with perfectly smooth surface (dissolution steps are absent), detachment of one or several ions results in formation of dissolution nuclei, which the polynuclear model is based on [58-66]. According to this model, all sites in nuclei edges are the kink positions [65] and, after appearance, the nuclei grow and spread over the surface with a definite lateral rate [59], giving rise to formation of dissolution steps. If dissolution steps already existed on the crystal faces of apatite, detachment of one calcium or one orthophosphate would result in dissolution step movement jump-wise over a distance equal to the sizes of these ions (approximately for 1 and 3 Å, respectively).

The latter is also valid for dissolution of crystal edges and corners. Since the classical paper by Kaischew [170], it is generally considered, that ions are less strongly bound along edges of a crystal than in the middle of faces and still less strongly at the corners [171]. No information about the ionic arrangements on edges and corners of apatite crystals has been found in literature. Therefore, two boundary conditions are possible: the edges and corners mainly consist of ions of either calcium or orthophosphate. However, the edges parallel to the *c*-axis, in principle, may consist of X-ions (F^-, OH^-, Cl^-). The last case is the simplest one. According to the chemical model [105-107], dissolution of apatite starts with protonation and detachment of X-ions, followed by removing of calcium (one should remind here that "walls" of the channels where X-ions are located consist of Ca(2) ions [101]) and afterwards by protonation and removing of orthophosphate. In the case, when the edges and corners of apatite mainly consist of calcium, again, according to the chemical model, detachment of weakly bounded calcium precedes protonation and removing of orthophosphate. Finally, if the edges and corners mainly consist of orthophosphate, the dissolution will start by chemical interaction of protons with these weakly bounded orthophosphate groups. The latter decreases electrostatic attraction forces between orthophosphate anions and neighboring calcium cations (attraction forces between Ca^{2+} and PO_4^{3-} are evidently stronger when compared with those between Ca^{2+} and HPO_4^{2-}). These forces become still weaker after adsorption of the second proton onto the given orthophosphate anion, causing its transformation to $H_2PO_4^-$ followed by detachment.

The Influence of Dislocations and Surface Defects

Before now, no data on crystal defects have been used. For this reason, the above description is valid for dissolution of the perfect single crystals without defects and with the molecularly smooth surface. However, like other solids, crystals of apatite always contain both surface irregularities (e.g., steps, missing ions and dislocation outlets) and structural defects inside the bulk of crystals (e.g., dislocations and inclusions). The schematic depiction of a crystal surface is shown in Figure 5. This model is known as the "Kossel model" [172]. In principle, any surface irregularities may act as dissolution nuclei because sometimes even one missing ion might become the critical nuclei (the polynuclear model) [63, 64].

Each dislocation outlet on apatite was found to be a hollow core with radius within 8.3 – 20 Å for the screw dislocations and even 18 – 43 Å for the edge ones [114]. Dimensions of the hexagonal unit-cell of apatite are: $a \approx 9.43$ Å, $c \approx 6.88$ Å. If the numeric values for the dislocation outlets were calculated correctly, their dimensions (diameter of the hollow cores) would be 1.8 – 4.2 times *a* or 2.4 – 5.8 times *c* (for the screw dislocations) and even 3.8 – 9.1 times *a* or 5.2 – 12.5 times *c* (for the edge dislocations). Thus, each dislocation outlet on the crystal surface looks like a large dissolution nucleus equal to detachment of 3 – 19 unit cells, (if this is taken to be the product *ac*) for the screw dislocations and even of 15 – 90 unit cells for the edge ones.

The process of etch pits formation and growth on apatites was described in details elsewhere [43-50, 95, 101, 111-120]. In the case of single crystals of pure HA, an acid attack was found to start formation of one etch pit in the basal plane which was the preferred dissolution spot or active site. Immediately afterwards the acid formed a longitudinal hole parallel to the *c*-axis [115]. On the contrary, dissolution of natural FA was found to happen with formation of a great number of etch pits (about 5×10^5 cm^{-2}) on the surface (Figure 6) [118-120]. The latter just points out to the great differences in amount of crystal defects and impurities between the chemically pure [43-50, 95, 101, 111-117, 123-127] and natural [118-120] apatites.

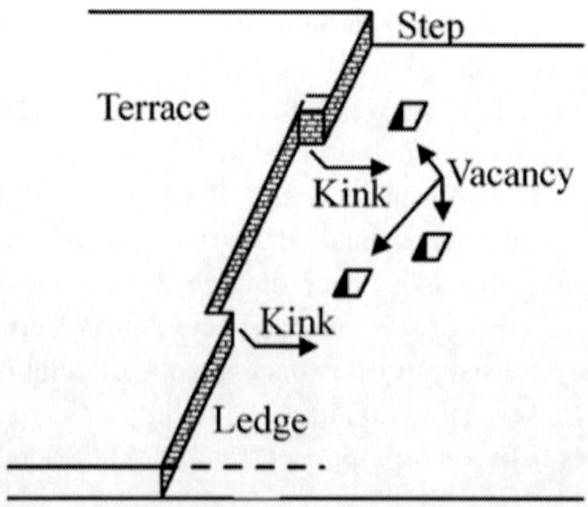

Figure 5. Surface of a crystal according to Kossel [172]. The various types of imperfections are shown. Reprinted from Ref. [35] with permission.

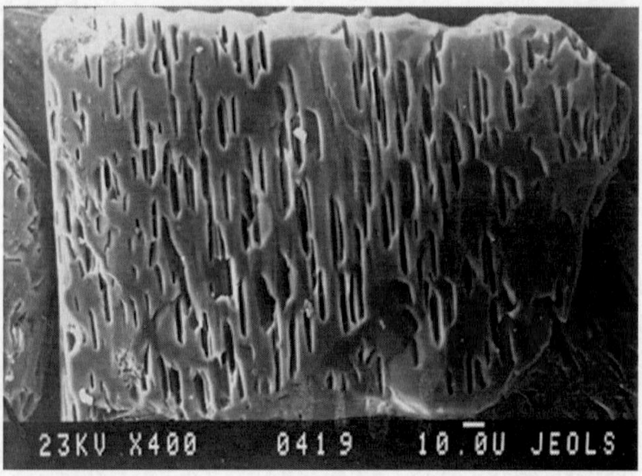

Figure 6. An example of etched crystal surface of natural FA with typical hexagonal etch pits. Bar 10 μm.

In any case, after being formed, the pits were found to grow in all directions (length, width and depth – see Figure 3) with definite rates until they either covered all crystal faces of apatite (dissolution of natural crystals [118-120]) or made longitudinal holes parallel to the *c*-axis (dissolution of the single crystals of pure HA [111-116]). However, according to "a new dissolution model incorporating particle size considerations", the dissolution process becomes spontaneous only when the pits are reached the critical sizes [43-50]. The unit-step movement of etch pit growth was suggested to consist of removing of a single ion, followed by surface diffusion away from the place of ionic detachment, desorption from the surface and diffusion into the bulk solution. All these processes evidently occur according to the basic findings of the ion exchange, chemical, calcium-rich layer formation, hydrogen catalytic and diffusion controlled dissolution models.

To conclude this part, except the dissolution rates increasing and surface etching, no other influences of dislocations on the dissolution mechanism were found.

Conclusion

Thus, based on eight dissolution models drawn from the previously published papers, an attempt to create the general description of apatite dissolution process at the atomic (ionic) level has been made. This dissolution mechanism is based on current level of knowledge and will be updated, corrected and modified simultaneously with appearance of new experimental data. Now some missing and unclear points to be investigated in future can be outlined:

- chemical transformations happening with all ions during diffusion through the Nernst layer;
- chemical composition of the apatite surface during dissolution;
- processes of ionic sorption and detachment;
- processes of dissolution nuclei formation and growth;
- initial stages of etch pits formation at the dislocation outlets;
- specific influence of anions to the ionic sorption, surface diffusion and detachment.

The above topics should be investigated in order to have better understanding of all the processes involved. For example, atomic force microscopy and phase shift interferometry might be useful tools in investigations on dissolution nuclei and etch pits formation [69, 70]. Micron-sized (better nano-sized) electrodes might be helpful in studying ionic transformations during diffusion through the Nernst layer [145]. Modern techniques of the surface state analysis might be helpful in understanding the surface structure and its chemical composition.

References

[1] McConnell, D. Apatite: its crystal chemistry, mineralogy, utilization, and geologic and biologic occurrences. Applied Mineralogy, Vol. 5. Springer-Verlag: Vienna and New York, USA, 1973; 111 pp.

[2] Aoki, H. Science and medical applications of hydroxyapatite. JAAS: Tokyo, Japan, 1991; 245 pp.

[3] Elliott, J.C. Structure and chemistry of the apatites and other calcium orthophosphates. Studies in inorganic chemistry, Vol. 18; Elsevier: Amsterdam, Netherlands, 1994; 389 pp.

[4] LeGeros, R.Z. Calcium phosphates in oral biology and medicine. Monographs in Oral Science. Vol. 15. Karger, Basel, 1991, 201 pp.

[5] Dorozhkin, S.V. Calcium orthophosphate-based bioceramics. *Materials* 2013, 6, 3840-3942.

[6] Dorozhkin, S.V. Calcium orthophosphates: occurrence, properties, biomineralization, pathological calcification and biomimetic applications. *Biomatter* 2011, 1, 121-164.
[7] Dorozhkin, S.V. Biocomposites and hybrid biomaterials based on calcium orthophosphates. *Biomatter* 2011, 1, 3-56.
[8] March, J. March's advanced organic chemistry: reactions, mechanisms, and structure. 5th Ed. Wiley, New York, 2001; 2083 pp.
[9] Nancollas, G.H. Physicochemistry of demineralization and remineralization. *J. Dent. Res.* 1974, 53, 297-302.
[10] Wu, M.S., Higuchi, W.I., Fox, J.L., Friedman, M. Kinetics and mechanism of hydroxyapatite crystal dissolution in weak acid buffers using the rotating disk method. *J. Dent. Res.* 1976, 55, 496-505.
[11] Smith, A.N., Posner, A.M., Quirk, J.P. A model describing the kinetics of dissolution of hydroxyapatite. *J. Colloid Interf. Sci.* 1977, 62, 475-494.
[12] Okazaki, M., Moriwaki, Y., Aoba, T., Doi, Y., Takahashi, J. Dissolution rate behavior of fluoridated apatite pellets. *J. Dent. Res.* 1981, 60, 1907-1911.
[13] Crommelin, D.J., Higuchi, W.I., Fox, J.L. Dissolution rate behavior of hydroxyapatite-fluorapatite mixtures. *Caries Res.* 1983, 17, 289-296.
[14] Nelson, D.G.A., Featherstone, J.D.B., Duncan, J.F., Cutress, T.W. Effect of carbonate and fluoride on the dissolution behaviour of synthetic apatites. *Caries Res.* 1983, 17, 200-211.
[15] Higuchi, W.I., Cesar, E.Y., Cho, P.W., Fox, J.L. Powder suspension method for critically re-examining the two-site model for hydroxyapatite dissolution kinetics. *J. Pharm. Sci.* 1984, 73, 146-153.
[16] Budz, J.A., LoRe, M., Nancollas, G.H. Hydroxyapatite and carbonated apatite as models for the dissolution behavior of human dental enamel. *Adv. Dent. Res.* 1987, 1, 314-321.
[17] Wong, L., Cutress, T.W., Duncan, J.F. The influence of incorporated and adsorbed fluoride on the dissolution of powdered and pelletized hydroxyapatite in fluoridated and non-fluoridated acid buffers. *J. Dent. Res.* 1987, 66, 1735-1741.
[18] Budz, J.A., Nancollas, G.H. The mechanism of dissolution of hydroxyapatite and carbonated apatite in acidic solutions. *J. Cryst. Growth* 1988, 91, 490-496.
[19] Thomann, J.M., Voegel, J.C., Gramain, P. Kinetics of dissolution of calcium hydroxyapatite powder. III: pH and sample conditioning effects. *Calcif. Tiss. Int.* 1990, 46, 121-129.
[20] Chin, K.O.A., Nancollas, G.H. Dissolution of fluorapatite. A constant-composition kinetics study. *Langmuir* 1991, 7, 2175-2179.
[21] Margolis, H.C., Moreno, E.C. Kinetics of hydroxyapatite dissolution in acetic, lactic, and phosphoric acid solutions. *Calcif. Tiss. Int.* 1992, 50, 137-143.
[22] Paschalis, E.P., Wikiel, K., Nancollas, G.H. Dual constant composition kinetics characterization of apatitic surfaces. *J. Biomed. Mater. Res.* 1994, 28, 1411-1418.
[23] Calmanovici, C.E., Gilot, B., Laguerie, C. Mechanism and kinetics for the dissolution of apatitic materials in acid solutions. *Brazilian J. Chem. Eng.* 1997, 14, 95-102.
[24] Mohammadi, Z., Ziaei-Moayyed, A.A., Sheikh-Mehdi, M.A. *In vitro* dissolution of plasma-sprayed hydroxyapatite coatings with different characteristics: experimental study and modeling. *Biomed. Mater.* 2008, 3, 015006 (7 pages).

[25] van der Sluis, S., Meszaros, Y., Marchee, W.G.J., Wesselingh, H.A., van Rosmalen, G.M. The digestion of phosphate ore in phosphoric acid. *Ind. Eng. Chem. Res.* 1987, 26, 2501-2505.

[26] Dobrydnev, S.V., Pochitalkina, S.A., Bogach, V.V., Beskov, V.S. Ionometric study of the kinetics of acid decomposition of fluorapatite. *Russ. J. Appl. Chem.* 2001, 74, 1627-1630.

[27] Dobrydnev, S.V., Beskov, V.S., Bogach, V.V., Pochitalkina, I.A. Ionometric study of the acid decomposition of phosphate minerals. *Theoret. Found. Chem. Eng.* 2001, 35, 292-297.

[28] Dobrydnev, S.V., Bogach, V.V., Beskov, V.S. Analysis of the mass transfer in acid decomposition of phosphorus-containing raw material. *Theoret. Found. Chem. Eng.* 2003, 37, 601-605.

[29] Dobrydnev, S.V., Bogach, V.V., Kol'tsova, E.M., Beskov, V.S. Effect of the charge of solid particles on the activation energy of acid decomposition of phosphorus-containing raw material. *Theoret. Found. Chem. Eng.* 2005, 39, 415-418.

[30] Brahim, K., Antar, K., Khattech, I., Jemal, M. Effect of temperature on the attack of fluorapatite by a phosphoric acid solution. *Sci. Res. Essay* 2008, 3, 035-039.

[31] Zendah, H., Khattech, I., Jemal, M. Thermochemical and kinetic studies of the acid attack of "B" type carbonate fluorapatites at different temperatures (25–55)°C. *Thermochim. Acta* 2013, 565, 46-51.

[32] Fawzi, M.B., Fox, J.L., Dedhiya, M.G., Higuchi, W.I., Hefferren, J.J. A possible second site for hydroxyapatite dissolution in acidic media. *J. Colloid Interf. Sci.* 1978, 67, 304-311.

[33] Fox, J.L., Higuchi, W.I., Fawzi, M.B., Wu, M.S. A new two-site model for hydroxyapatite dissolution in acidic media. *J. Colloid Interf. Sci.* 1978, 67, 312-330.

[34] Griffith, E.N., Katdare, A., Fox, J.L., Higuchi, W.I. Transmission electron microscopic confirmation of the morphological predictions of the two-site model for hydroxyapatite dissolution. *J. Colloid Interf. Sci.* 1978, 67, 331-335.

[35] Ehrlich, H., Koutsoukos, P.G., Demadis, K.D., Pokrovsky, O.S. Principles of demineralization: modern strategies for the isolation of organic frameworks. Part II. Decalcification. *Micron* 2009, 40, 169-193.

[36] Chow, L.C. Response to Dr. E.I.F. Pearce's letter to the Editor. *J. Dent. Res.* 1988, 67, 1058-1059.

[37] Tomson, M.B., Nancollas, G.H. Mineralization kinetics: a constant composition approach. *Science* 1978, 200, 1059-1060.

[38] Tang, R., Henneman, Z.J., Nancollas, G.H. Constant composition kinetics study of carbonated apatite dissolution. *J. Cryst. Growth* 2003, 249, 614-624.

[39] Tang, R., Hass, M., Wu, W., Gulde, S., Nancollas, G.H. Constant composition dissolution of mixed phases. II. Selective dissolution of calcium phosphates. *J. Colloid Interf. Sci.* 2003, 260, 379-384.

[40] Ebrahimpour, A., Zhang, J., Nancollas, G.H. Dual constant composition method and its application to studies of phase transformation and crystallization of mixed phases. *J. Cryst. Growth* 1991, 113, 83-91.

[41] Chow, L.C., Markovic, M., Takagi, S. A dual constant-composition titration system as an *in vitro* resorption model for comparing dissolution rates of calcium phosphate biomaterials. *J. Biomed. Mater. Res. B Appl. Biomater.* 2003, 65B, 245-251.

[42] Guidry, M.W., Mackenzie, F.T. Experimental study of igneous and sedimentary apatite dissolution: control of pH, distance from equilibrium, and temperature on dissolution rates. *Geochim. Cosmochim. Acta* 2003, 67, 2949-2963.
[43] Tang, R., Nancollas, G.H., Orme, C.A. Mechanism of dissolution of sparingly soluble electrolytes. *J. Am. Chem. Soc.* 2001, 123, 5437-5443.
[44] Tang, R., Wang, L., Orme, C.A., Bonstein, T., Bush, P.J., Nancollas, G.H. Dissolution at the nanoscale: self-preservation of biominerals. *Angew. Chem. Int. Ed.* 2004, 43, 2697-2701.
[45] Tang, R., Wang, L., Nancollas, G.H. Size-effects in the dissolution of hydroxyapatite: an understanding of biological demineralization. *J. Mater. Chem.* 2004, 14, 2341-2346.
[46] Tang, R., Orme, C.A., Nancollas, G.H. Dissolution of crystallites: surface energetic control and size effects. *ChemPhysChem* 2004, 5, 688-696.
[47] Wang, L.J., Tang, R., Bonstein, T., Bush, P., Nancollas, G.H. Enamel demineralization in primary and permanent teeth. *J. Dent. Res.* 2006, 85, 359-363.
[48] Wang, L.J., Nancollas, G.H. Calcium orthophosphates: crystallization and dissolution. *Chem. Rev.* 2008, 108, 4628-4669.
[49] Wang, L.J., Nancollas, G.H. Pathways to biomineralization and biodemineralization of calcium phosphates: the thermodynamic and kinetic controls. *Dalton Trans.* 2009, 2665-2672.
[50] Wang, L.J., Nancollas, G.H. Dynamics of biomineralization and biodemineralization. *Met. Ions Life Sci.* 2010, 4, 413-456.
[51] Mafé, S., Manzanares, J.A., Reiss, H., Thomann, J.M., Gramain, P. Model for the dissolution of calcium hydroxyapatite powder. *J. Phys. Chem.* 1992, 96, 861-866.
[52] Gasser, P., Voegel, J.C., Gramain, P. Surface reactions on hydroxyapatite in the presence of fluoride ions. 1. Saturated and congruent conditions. *Coll. Surf. A* 1993, 74, 275-286.
[53] Thomann, J.M., Voegel, J.C., Gramain, P. Quantitative model for the dissolution of calcium hydroxyapatite with a permselective ionic interface. *J. Colloid Interf.* Sci. 1993, 157, 369-374.
[54] Gasser, P., Haikel, Y., Voegel, J.C., Gramain, P. Surface reactions of hydroxyapatite in the presence of fluoride ions. 2. Effects of calcium and phosphate in saturated solutions. *Coll. Surf. A* 1994, 88, 157-168.
[55] Schaad, P., Poumier, F., Voegel, J.C., Gramain, P. Analysis of calcium hydroxyapatite dissolution in non-stoichiometric solutions. *Coll. Surf. A* 1997, 121, 217-228.
[56] Wei, C., Zhu, Y., Yang, F., Li, J., Zhu, Z., Zhu, H. Dissolution and solubility of hydroxylapatite and fluorapatite at 25°C at different pH. *Res. J. Chem. Environ.* 2013, 17, 57-61.
[57] Pan, H.B., Darvell, B.W. Calcium phosphate solubility: the need for re-evaluation. *Cryst. Growth Des.* 2009, 9, 639-645.
[58] Christoffersen, J., Christoffersen, M.R., Kjaergaard, N. The kinetics of dissolution of calcium hydroxyapatite in water at constant pH. *J. Cryst. Growth* 1978, 43, 501-511.
[59] Christoffersen, J. Kinetics of dissolution of calcium hydroxypatite. III. Nucleation-controlled dissolution of a polydisperse sample of crystals. *J. Cryst. Growth* 1980, 49, 29-44.
[60] Christoffersen, J. Dissolution of calcium hydroxyapatite. *Calcif. Tiss. Int.* 1981, 33, 557-560.

[61] Christoffersen, J., Christoffersen, M.R. Kinetics of dissolution of calcium hydroxyapatite. V. The acidity constant for the hydrogen phosphate surface complex. *J. Cryst. Growth* 1982, 57, 21-26.

[62] Christoffersen, M.R., Christoffersen, J. Possible mechanisms for the growth of the biomaterial, calcium hydroxyapatite microcrystals. *J. Cryst. Growth* 1992, 121, 617-630.

[63] Christoffersen, J., Christoffersen, M.R., Johansen, T. Kinetics of growth and dissolution of fluorapatite. *J. Cryst. Growth* 1996, 163, 295-303.

[64] Christoffersen, J., Christoffersen, M.R., Johansen, T. Some new aspects of surface nucleation applied to the growth and dissolution of fluorapatite and hydroxyapatite. *J. Cryst. Growth* 1996, 163, 304-310.

[65] Christoffersen, J., Dohrup, J., Christoffersen, M.R. The importance of formation of hydroxyl ions by dissociation of trapped water molecules for growth of calcium hydroxyapatite crystals. *J. Cryst. Growth* 1998, 186, 275-282.

[66] Christoffersen, M.R., Dohrup, J., Christoffersen, J. Kinetics of growth and dissolution of calcium hydroxyapatite in suspensions with variable calcium to phosphate ratio. *J. Cryst. Growth* 1998, 186, 283-290.

[67] Zhang, H., Li, S., Yan, Y. Dissolution behavior of hydroxyapatite powder in hydrothermal solution. *Ceram. Int.* 2001, 27, 451-454.

[68] Gilmer, G.H., Bennema, P. Simulation of crystal growth with surface diffusion. *J. Appl. Phys.* 1972, 43, 1347-1360.

[69] Onuma, K, Ito, A, Tateishi, T, Kameyama, T. Growth kinetics of hydroxyapatite crystal revealed by atomic force microscopy. J. Cryst. Growth 1995, 154, 118-125.

[70] Onuma, K., Ito, A., Tateishi, T. Investigation of a growth unit of hydroxyapatite crystal from the measurements of step kinetics. *J. Cryst. Growth* 1996, 167, 773-776.

[71] Zhu, Y., Zhang, X., Chen, Y., Xie, Q., Lan, J., Qian, M., He, N. A comparative study on the dissolution and solubility of hydroxylapatite and fluorapatite at 25 °C and 45 °C. *Chem. Geol.* 2009, 268, 89-96.

[72] Kukura, M., Bell, L.C., Posner, A.M., Quirk, J.P. Radioisotope determination of the surface concentrations of calcium and phosphorus on hydroxyapatite in aqueous solution. *J. Phys. Chem.* 1972, 76, 900-904.

[73] Somasundaran, P. Zeta potential of apatite in aqueous solutions and its change during equilibration. *J. Colloid Interf. Sci.* 1968, 27, 659-666.

[74] Bell, L.C., Posner, A.M., Quirk, J.P. Surface charge characteristics of hydroxyapatite and fluorapatite. *Nature* 1972, 239, 515-517.

[75] Bell, L.C., Posner, A.M., Quirk, J.P. The point of zero charge of hydroxyapatite and fluorapatite in aqueous solutions. *J. Colloid Interf. Sci.* 1973, 42, 250-261.

[76] Moreno, E.C., Gregory, T.M., Brown, W.E. Preparation and solubility of hydroxyapatite. *J. Res. Natl. Bur. Stand.* 1968, 72A, 773-782.

[77] Chuong, R. Experimental study of surface and lattice effects on the solubility of hydroxyapatite. *J. Dent. Res.* 1973, 52, 911-914.

[78] Smith, A.N., Posner, A.M., Quirk, J.P. Incongruent dissolution and surface complexes of hydroxyapatite. *J. Colloid Interf. Sci.* 1974, 48, 442-449.

[79] Mika, H., Bell, L.C., Kruger, B.J. The role of surface reactions in the dissolution of stoichiometric hydroxyapatite. *Arch. Oral Biol.* 1976, 21, 697-701.

[80] Moreno, E.C., Kresak, M., Zahradnik, R.T. Physicochemical aspects of fluoride-apatite systems relevant to the study of dental caries. *Caries Res.* 1977, 11, Suppl. 1, 142-171.
[81] Bell, L.C., Mika, H., Kruger, B.J. Synthetic hydroxyapatite-solubility product and stoichiometry of dissolution. Arch. *Oral. Biol. 1978*, 23, 329-336.
[82] Ingram, G.S. Chemical events during tooth dissolution. *J. Dent. Res.* 1999, 69, Spec. Iss., 581-586, discussion 634-636.
[83] Kaufman, H.W., Kleinberg, I. Studies on the incongruent solubility of hydroxyapatite. *Calcif. Tiss. Int.* 1979, 27, 143-151.
[84] Amrah-Bouali, S., Rey, C., Lebugle, A., Bernache, D. Surface modifications of hydroxyapatite ceramics in aqueous media. *Biomaterials* 1994, 15, 269-272.
[85] Nordstrom, E.G., Hara, T., Hero, H. Solubility of hydroxyapatite/mica composites. *Biomed. Mater. Eng.* 1996, 6, 73-78.
[86] Pearce, E.I.F, Guha-Chowdhury, N., Iwami, Y., Cutress, T.W. Stoichiometry of fluoride release from fluorhydroxyapatite during acid dissolution. *Caries Res.* 1995, 29, 130-136.
[87] Shimabayashi, S., Matsumoto, M. Non-stoichiometric dissolution of hydroxyapatite in the presence of simple salts. *Nippon Kagaku Kaishi* 1993, 10, 1118-1122.
[88] Brown, P.W., Martin, R.I. An analysis of hydroxyapatite surface layer formation. *J. Phys. Chem. B* 1999, 103, 1671-1675.
[89] Chaïrat, C., Oelkers, E.H., Schott, J., Lartigue, J.E. Fluorapatite surface composition in aqueous solution deduced from potentiometric, electrokinetic, and solubility measurements, and spectroscopic observations. *Geochim. Cosmochim. Acta* 2007, 71, 5888-5900.
[90] Bengtsson, Å., Shchukarev, A., Persson, P., Sjöberg, S. A solubility and surface complexation study of a non-stoichiometric hydroxyapatite. *Geochim. Cosmochim. Acta* 2009, 73, 257-267.
[91] Bengtsson, Å., Shchukarev, A., Persson, P., Sjöberg, S. Phase transformations, ion-exchange, adsorption, and dissolution processes in aquatic fluorapatite systems. *Langmuir* 2009, 25, 2355-2362.
[92] Bengtsson, Å., Sjöberg, S. Surface complexation and proton-promoted dissolution in aqueous apatite systems. *Pure Appl. Chem.* 2009, 81, 1569-1584.
[93] Bertazzo, S., Zambuzzi, W.F., Campos, D.D.P., Ogeda, T.L., Ferreira, C.V., Bertran, C.A. Hydroxyapatite surface solubility and effect on cell adhesion. *Coll. Surf. B* 2010, 78, 177-184.
[94] Feng, M.H., Ngwenya, B.T., Wang, L., Li, W., Olive, V., Ellam, R.M. Bacterial dissolution of fluorapatite as a possible source of elevated dissolved phosphate in the environment. *Geochim. Cosmochim. Acta* 2011, 75, 5785-5796.
[95] Abe, Y., Okazaki, Y., Hiasa, K., Yasuda, K., Nogami, K., Mizumachi, W., Hirata, I. Bioactive surface modification of hydroxyapatite. *BioMed Res. Int.* 2013, 2013, 626452 (9 pages).
[96] Misra, D.N. Interaction of citric or hydrochloric acid with calcium fluorapatite: precipitation of calcium fluoride. *J. Colloid Interf. Sci.* 1999, 220, 387-391.
[97] Dobrydnev, S.V., Bogach, V.V., Beskov, V.S. Thermodynamic grounds of sulfuric acid dissolution of fluorapatite. *Russ. J. Inorg. Chem.* 2002, 47, 1063-1066.
[98] Dobrydnev, S.V., Bogach, V.V., Beskov, V.S. Thermodynamics of fluoroapatite dissolution in phosphoric acid. *Russ. J. Inorg. Chem.* 2002, 47, 1214-1216.

[99] Dobrydnev, S.V., Bogach, V.V., Beskov, V.S. Thermodynamics of fluoroapatite dissolution in nitric acid. *Russ. J. Inorg. Chem.* 2002, 47, 1217-1220.

[100] Harouiya, N., Chaïrat, C., Köhler, S.J., Gout, R., Oelkers, E.H. The dissolution kinetics and apparent solubility of natural apatite in closed reactors at temperatures from 5 to 50 °C and pH from 1 to 6. *Chem. Geol.* 2007, 244, 554-568.

[101] Chaïrat, C., Schott, J., Oelkers, E.H., Lartigue, J.E., Harouiya, N. Kinetics and mechanism of natural fluorapatite dissolution at 25 °C and pH from 3 to 12. *Geochim. Cosmochim. Acta* 2007, 71, 5901-5912.

[102] Melikhov, I.V., Dorozhkin, S.V., Nikolaev, A.L., Voronina, N.Y. Transition from congruent to incongruent dissolution. *Russ. J. Phys. Chem.* 1992, 66, 1108-1111.

[103] Tanaka, H, Miyajima, K, Nakagaki, M, Shimabayashi, S. Incongruent dissolution of hydroxyapatite in the presence of phosphoserine. *Coll. Polymer Sci.* 1991, 269, 161-165.

[104] Park, C., Fenter, P., Zhang, Z., Cheng, L., Sturchio, N.C. Structure of the fluorapatite (100)-water interface by high-resolution X-ray reflectivity. *Am. Mineral.* 2004, 89, 1647-1654.

[105] Dorozhkin, S.V. Surface reactions of apatite dissolution. *J. Colloid Interf. Sci.* 1997, 191, 489-497.

[106] Dorozhkin, S.V. Acidic dissolution mechanism of natural fluorapatite. II. Nanolevel of investigations. *J. Cryst. Growth* 1997, 182, 133-140.

[107] Dorozhkin, S.V. Inorganic chemistry of the dissolution phenomenon: the dissolution mechanism of calcium apatites at the atomic (ionic) level. *Comments Inorg. Chem.* 1999, 20, 285-299.

[108] Higuchi, W.I., Gray, J.A., Hefferren, J.J., Patel, P.R. Mechanisms of enamel dissolution in acid buffers. *J. Dent. Res.* 1965, 44, 330-341.

[109] Pearce, E.I.F. On the dissolution of hydroxyapatite in acid solutions. *J. Dent. Res.* 1988, 67, 1056-1058.

[110] Eanes, E.D., Meyer, J.L. The maturation of crystalline calcium phosphates in aqueous suspensions at physiologic pH. *Calcif. Tiss. Res.* 1977, 23, 259-269.

[111] Lowell, L.C. Dislocation etch pits in apatite. *Acta Metall.* 1958, 6, 775-778.

[112] Jongebloed, W.L., Molenaar, I., Arends, J. Orientation-dependent etchpit penetration and dissolution of fluoroapatite. *Caries Res.* 1973, 7, 154-165.

[113] Jongebloed, W.L., van den Berg, P.J., Arends, J. The dissolution of single crystals of hydroxyapatite in citric and lactic acids. *Calcif. Tiss. Res.* 1974, 15, 1-9.

[114] Arends, J., Jongebloed, W.L. Dislocations and dissolution in apatites: theoretical considerations. *Caries Res.* 1977, 11, 186-188.

[115] Arends, J., Jongebloed, W.L. Ultrastructural studies of synthetic apatite crystals. *J. Dent. Res.* 1979, 58, Spec. Iss. B, 837-843.

[116] Daculsi, G., Kerebel, B., Kerebel, L.M. Mechanisms of acid dissolution of biological and synthetic apatite crystals at the lattice pattern level. *Caries Res.* 1979, 13, 277-289.

[117] Daculsi, G., LeGeros, R.Z., Mitre, D. Crystal dissolution of biological and ceramic apatites. *Calcif. Tiss. Int.* 1989, 49, 95-103.

[118] Melikhov, I.V., Dorozhkin, S.V., Nikolaev, A.L., Kozlovskaya, E.D., Rudin, V.N. Dislocations and the rate of dissolution of solids. *Russ. J. Phys. Chem.* 1990, 64, 1746-1750.

[119] Dorozhkin, S.V. Chemical etching of natural fluorapatite crystals in acid solutions studied with the scanning electron microscope. *Scanning* 1995, 17, 355-360.

[120] Dorozhkin, S.V. Acidic dissolution mechanism of natural fluorapatite. I. Milli- and microlevels of investigations. J. *Cryst. Growth* 1997, 182, 125-132.

[121] Lasaga, A.C., Luttge, A. Variation of crystal dissolution rate based on a dissolution stepwave model. *Science* 2001, 291, 2400-2404.

[122] Dove, P.M., Han, N., de Yoreo, J.J. Mechanisms of classical crystal growth theory explain quartz and silicate dissolution behavior. *Proc. Natl. Acad. Sci. USA* 2005, 102, 15357-15362.

[123] Kwon, K.Y., Wang, E., Chung, A., Chang, N., Saiz, E., Choe, U.J., Koobatian, M., Lee, S.W. Defect induced asymmetric pit formation on hydroxyapatite. *Langmuir* 2008, 24, 11063-11066.

[124] Kwon, K.Y., Wang, E., Chung, A., Chang, N., Lee, S.W. Effect of salinity on hydroxyapatite dissolution studied by atomic force microscopy. *J. Phys. Chem.* C 2009, 113, 3369-3372.

[125] Kwon, K.Y., Wang, E., Chang, N., Lee, S.W. Characterization of the dominant molecular step orientations on hydroxyapatite (100) surfaces. *Langmuir* 2009, 25, 7205-7208.

[126] Kwon, K.Y., Wang, E., Nofal, M., Lee, S.W. Microscopic study of hydroxyapatite dissolution as affected by fluoride ions. *Langmuir* 2011, 27, 5335-5339.

[127] Dong, S.S., Hwan, K., Jong, K.L. AFM study on the surface dissolution of hydroxyapatite. *Key Eng. Mater*. 2007, 336-338, 1553-1555.

[128] Wang, L.J., Lu, J.W., Xu, F.S., Zhang, F.S. Dynamics of crystallization and dissolution of calcium orthophosphates at the near-molecular level. *Chin. Sci. Bull.* 2011, 56, 713-721.

[129] Dorozhkin, S.V. Process of epitaxial crystal growth for $CaSO_4 \cdot 0.5H_2O$ on a surface of dissolving fluorapatite crystals studied by scanning electron microscopy. *Scanning* 1996, 18, 119-124.

[130] Misra, D.N. Interaction of citric acid with hydroxyapatite: surface exchange of ions and precipitation of calcium citrate. *J. Dent. Res*. 1996, 75, 1418-1425.

[131] López-Macipe, A., Gómez-Morales, J., Rodríguez-Clemente, R. The role of pH in the adsorption of citrate ions on hydroxyapatite. *J. Colloid Interf. Sci*. 1998, 200, 114-120.

[132] Misra, D.N. Adsorption from solutions on synthetic hydroxyapatite: nonaqueous vs. aqueous solvents. *J. Biomed. Mater. Res*. 1999, 48, 848-855.

[133] Filgueiras, M.R.T., Mkhonto, D., de Leeuw, N.H. Computer simulations of the adsorption of citric acid at hydroxyapatite surfaces. *J. Cryst. Growth* 2006, 294, 60-68.

[134] Jiang, W., Pan, H., Cai, Y., Tao, J., Liu, P., Xu, X., Tang, R. Atomic force microscopy reveals hydroxyapatite-citrate interfacial structure at the atomic level. *Langmuir* 2008, 24, 12446-12451.

[135] Yoshida, Y., van Meerbeek, B., Nakayama, Y., Yoshioka, M., Snauwaert, J., Abe, Y., Lambrechts, P., Vanherle, G., Okazaki, M. Adhesion to and decalcification of hydroxyapatite by carboxylic acids. *J. Dent. Res*. 2001, 80, 1565-1569.

[136] Yoshioka, M., Yoshida, Y., Inoue, S., Lambrechts, P., Vanherle, G., Nomura, Y., Okazaki, M., Shintani, H., van Meerbeek, B. Adhesion/decalcification mechanisms of acid interaction with human hard tissues. *J. Biomed. Mater. Res.* 2002, 59, 56-62.

[137] Yoshida, Y., van Meerbeek, B., Nakayama, Y., Snauwaert, J., Hellemans, L., Lambrechts, P., Vannerle, G., Wakasa, K. Evidence of chemical bonding at biomaterial-hard tissue interfaces. *J. Dent. Res*. 2000, 79, 709-714.

[138] Astala, R., Stott, M.J. First-principles study of hydroxyapatite surfaces and water adsorption. *Phys. Rev. B* 2008, 78, 075427 (11 pages).

[139] Posner, A.S. The structure of bone apatite surfaces. *J. Biomed. Mater. Res*. 1985, 19, 241-250.

[140] Hochrein, O., Zahn, D. On the molecular mechanisms of the acid-induced dissociation of hydroxy-apatite in water. *J. Mol. Model*. 2011, 17, 1525-1528.

[141] Jarlbring, M., Gunneriusson, L., Forsling, W. Characterisation of the protolytic properties of synthetic carbonate free fluorapatite. *J. Colloid Interf. Sci*. 2005, 285, 206-211.

[142] Becker, P. Phosphates and phosphoric acid: raw materials technology and economics of the wet process. 2nd Ed. Fertilizer science and technology series. Marcel Dekker: New York, USA, 1989; 760 pp.

[143] Compton, R.G., Harding, M.S., Pluck, M.R., Atherton, J.H., Brennan, C.M. Mechanism of solid/liquid interfacial reactions. The dissolution of benzoic acid in aqueous solution. *J. Phys. Chem*. 1993, 97, 10416-10420.

[144] Zhang, J., Nancollas, G.H. Unexpected pH dependence of dissolution kinetics of dicalcium phosphate dihydrate. *J. Phys. Chem*. 1994, 98, 1689-1694.

[145] de Aza, P.N., Guitian, F., Merlos, A., Lora-Tamayo, E., de Aza, S. Bioceramics – simulated body fluid interfaces: pH and its influence of hydroxyapatite formation. *J. Mater. Sci. Mater. Med*. 1996, 7, 399-402.

[146] Bohner, M., Lemaître, J., Ring, T.A. Kinetics of dissolution of β-tricalcium phosphate. *J. Colloid Interf. Sci*. 1997, 190, 37-48.

[147] Young, R.A. Implications of atomic substitutions and other structural details in apatites. *J. Dent. Res*. 1974, 53, 193-203.

[148] de Leeuw, N.H. Computer simulations of structures and properties of the biomaterial hydroxyapatite. *J. Mater. Chem*. 2010, 20, 5376-5389.

[149] Pareek, A., Torrelles, X., Angermund, K., Rius, J., Magdans, U., Gies, H. Structure of interfacial water on fluorapatite (100) surface. *Langmuir* 2008, 24, 2459-2464.

[150] Pan, H., Tao, J., Wu, T., Tang, R. Molecular simulation of water behaviors on crystal faces of hydroxyapatite. *Front. Chem. Chin*. 2007, 2, 156-163.

[151] Zahn, D., Hochrein, O. Computational study of interfaces between hydroxyapatite and water. *Phys. Chem. Chem. Phys*. 2003, 5, 4004-4007.

[152] Corno, M., Busco, C., Bolis, V., Tosoni, S., Ugliengo, P. Water adsorption on the stoichiometric (001) and (010) surfaces of hydroxyapatite: a periodic B3LYP study. *Langmuir* 2009, 25, 2188-2198.

[153] Pareek, A., Torrelles, X., Rius, J., Magdans, U., Gies, H. Role of water in the surface relaxation of the fluorapatite (100) surface by grazing incidence X-ray diffraction. *Phys. Rev. B* 2007, 75, 035418 (6 pages).

[154] Pan, H., Tao, J., Yu, X., Fu, L., Zhang, J., Zeng, X., Xu, G., Tang, R. Anisotropic demineralization and oriented assembly of hydroxyapatite crystals in enamel: smart structures of biominerals. *J. Phys. Chem*. B 2008, 112, 7162-7165.

[155] Dorozhkin, S.V. Nanodimensional and nanocrystalline calcium orthophosphates. *Int. J. Chem. Mater. Sci.* 2013, 1, 105-174.

[156] Probstein, R.F. Physicochemical hydrodynamics: an introduction. 2nd Ed. Wiley, Hoboken, NJ, USA, 2003, 416 pp.

[157] Wu, L., Forsling, W., Schindler, P.W. Surface complexation of calcium minerals in aqueous solution. 1. Surface protonation at fluorapatite-water interfaces. *J. Colloid Interf. Sci.* 1991, 147, 178-185.

[158] Skartsila, K., Spanos, N. Surface characterization of hydroxyapatite: potentiometric titrations coupled with solubility measurements. *J. Colloid Interf. Sci.* 2007, 308, 405-412.

[159] Vučinić, D.R., Radulović, D.S., Deušić, S.D. Electrokinetic properties of hydroxyapatite under flotation conditions. *J. Colloid Interf. Sci.* 2010, 343, 239-245.

[160] Zhang, J., Nancollas, G.H. Kinetics and mechanisms of octacalcium phosphate dissolution at 37 °C. *J. Phys. Chem.* 1992, 96, 5478-5483.

[161] Doss, S.K. Surface properties of hydroxyapatite: I. The effect of various inorganic ions on the electrophoretic behavior. *J. Dent. Res.* 1976, 55, 1067-1075.

[162] Dobrydnev, S.V., Bogach, V.V., Beskov, V.S. Influence of surfactants on the rate of decomposition of apatites by mineral acids. *Theoret. Found. Chem. Eng.* 2003, 37, 412-415.

[163] Nielsen, A.E. Transport control in crystal growth from solution. *Croat. Chem. Acta* 1980, 53, 255-279.

[164] Okazaki, M., Sato, M., Takahashi, *J. Space-cutting model of hydroxyapatite. Biomaterials* 1995, 16, 45-49.

[165] Iijima, M., Nelson, D.G.A., Pan, Y., Kreinbrink, A.T., Adachi, M., Goto, T., Moriwaki, Y. Fluoride analysis of apatite crystals with a central planar OCP inclusion: Concerning the role of F^- ions on apatite/OCP/apatite structure formation. *Calcif. Tiss. Int.* 1996, 59, 377-384.

[166] Aoba, T. The effect of fluoride on apatite structure and growth. *Crit. Rev. Oral Biol. Med.* 1997, 8, 136-153.

[167] de Leeuw, N.H. Resisting the onset of hydroxyapatite dissolution through the incorporation of fluoride. *J. Phys. Chem. B* 2004, 108, 1809-1811.

[168] Liu, Y., Sethuraman, G., Wu, W., Nancollas, G.H., Grynpas, M. The crystallization of fluorapatite in the presence of hydroxyapatite seeds and of hydroxyapatite in the presence of fluorapatite seeds. *J. Colloid Interf. Sci.* 1997, 186, 102-109.

[169] Dickens, B., Schroeder, L.W. Investigation of epitaxy relationships between $Ca_5(PO_4)_3OH$ and other calcium orthophosphates. *J. Res. Natl. Bur. Stand.* 1980, 85, 347-362.

[170] Kaischew, R. Zur Theorie des Kristallwachstums. *Z. Phys.* 1936, 102, 684-690.

[171] Nangia, S., Garrison, B.J. *Ab initio* study of dissolution and precipitation reactions from the edge, kink, and terrace sites of quartz as a function of pH. *Mol. Phys.* 2009, 107, 831-843.

[172] Kossel, W. Zur Energetik von Oberflächenvorgangen. *Ann. Phys. Leipzig* 1934, 21, 457-480.

In: Apatite
Editors: M. Iafisco and J. M. Delgado-López
ISBN: 978-1-63321-500-9

Chapter 7

HIGH-TEMPERATURE *IN SITU* XRD INVESTIGATIONS OF APATITES. STRUCTURAL INTERPRETATION OF THERMAL DEFORMATIONS

Evgeny N. Bulanov [*] ***and Alexander V. Knyazev***
Lobachevsky State University of Nizhni Novgorod,
Nizhni Novgorod, Russia

ABSTRACT

Thermal expansion is one of the thermophysical characteristics of matter. It provides information about the behaviour of materials in a wide range of temperatures. But the complex analysis of the chemical composition, structure and thermal expansion of a compound allows a conclusion about the causes of thermal deformation to be made, and as a result, predicts the behaviour of a phase and model materials with desired properties. Here we produce the result of such a work for compounds with an apatite structure, as they are one of the most important substances from the industrial point of view.

Keywords: Apatite; structure; high-temperature *in situ* XRD; thermal expansion coefficient; thermal expansion anisotropy; thermal expansion diagram

INTRODUCTION

Apatites are one of the broadest mineral families (Pasero et al., 2010). This fact is made possible by different isomorphic substitutions in the chemical composition of apatites (Masaoka, Kyono, Hatta, & Kimata, 2006; Podsiadlo, 1990a, 1990b; Sghir et al., 2009). In the common case, the general formula of the apatites may be represented as follows: $M_5(AO_4)_3L$, where M – mono-, di-, tris- and tetravalent cations and their combinations; A – atoms, which can form tetrahedra as coordination polyhedra; L – is the position for halogen

[*] Corresponding Author address. Email: bulanoven@gmail.com.

and different groups of atoms such as OH, CO_3 and others. Moreover, in all of the positions various types of defects can be observed.

Recently many publications have been devoted to investigation of apatites due to their exclusive importance for different fields of industry (Figure 1): these compounds are the basis of materials for fertilizers, ionic conductors, matrixes for radioactive wastes and, of course, bioceramic implants (Best, Porter, Thian, & Huang, 2008; Duan et al., 2005; J. Zhang, Liang, Yu, Yuan, & Su, 2009; M. Zhang, Maddrell, Abraitis, & Salje, 2007).

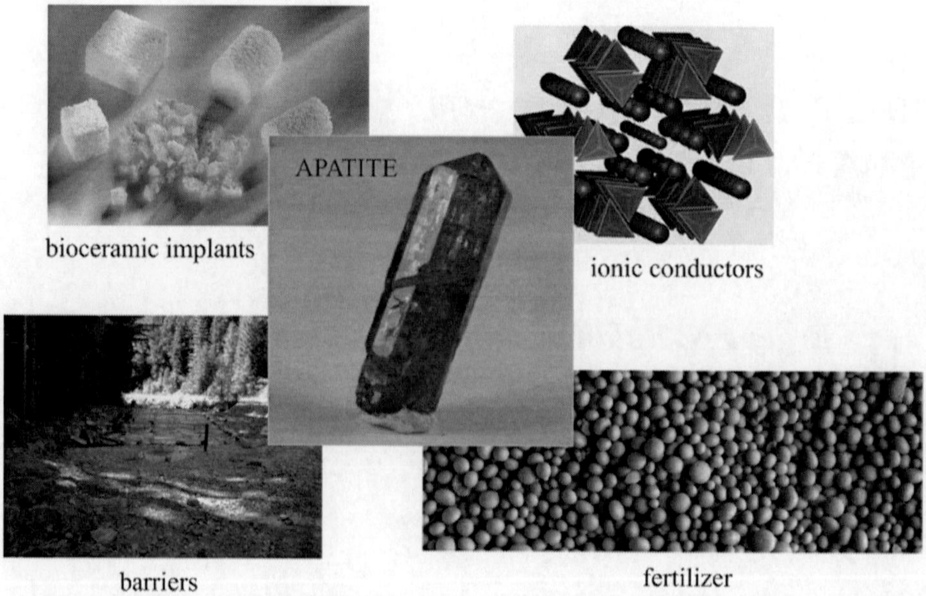

Figure 1. Use of apatites in various industries (Sources: www.crystalvaults.com, www.nl.biomet.be, www.minearnaud.com, (Conca & Wright, 2006; Sansom, Tolchard, Islam, Apperley, & Slater, 2006).

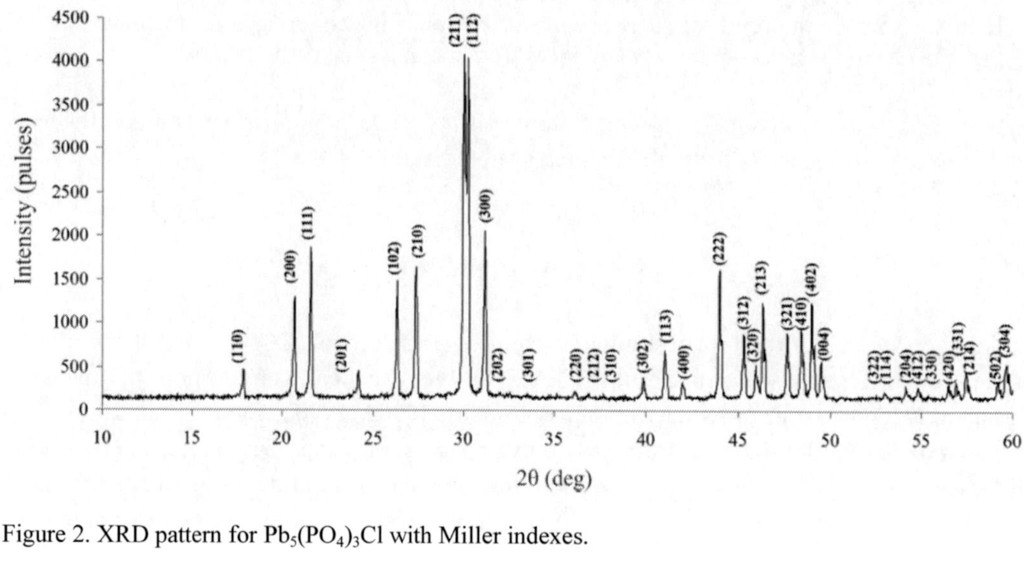

Figure 2. XRD pattern for $Pb_5(PO_4)_3Cl$ with Miller indexes.

As an object of inorganic chemistry, geochemistry and material science, apatites first should be analyzed by XRD methods. These methods allow a number of very important issues to be addressed, such as: whether compound under study belongs to the family of the apatite materials (X-ray phase analysis); what features of the structure does it have (Rietveld or single-crystal refinement); what are the thermal properties of the substance (high- and low-temperature XRD methods?

The first step of XRD investigations is the analysis of the X-ray pattern. On the Figure 2 such picture for the powder of $Pb_5(PO_4)_3Cl$ is presented. Using the model (refined apatite structure) all of the peaks should be attributed to the corresponding *hkl* (Miller's) indexes. Then on the basis of the data matrix "2θ angle – *hkl* indexes" during analytical indexing procedure unit-cell parameters of the compound under study may be calculated.

The next step should be the structure refinement. It may be done on the powder sample using Rietveld refinement procedure or on the single-crystal. The last option is the most difficult due to the complexity of growing single crystals (Masaoka & Kyono, 2006), but it provides the most reliable results. During this procedure, not only the symmetry and parameters of unit-cell may be obtained, but also stoichiometry of the apatite may be also specified. Knowing possible defects in the structure helps to explain some properties of the phase.

Now, when the researcher has the information about the structure and exact crystal-chemical composition of the apatite, he can start to explore its physical properties in particular thermal expansion.

HIGH-TEMPERATURE *IN SITU* XRD INVESTIGATION METHOD

The most objective information about the thermal behavior of a compound in a wide range of temperatures may be acquired using *in situ* methods of investigation. First such method is thermoroentgenography or high-temperature XRD. But in order to get more adequate results this method should be used along with differential scanning calorimetry. These methods allow investigating different types of processes: polymorphic transitions, synthesis and decomposition of solid compounds, crystallization from the amorphous phase, *etc.*

In the investigating of thermal expansion three main principles was formulated (R. S. Bubnova & Filatov, 2008; S. K. Filatov, 1985, 1990; Stanislav K. Filatov, 2008; S. K. Filatov, 2011; S. K. Filatov & Hazen, 1994):

1. The temperature effect is scalar; therefore, the thermal deformation of a crystal is determined by its structure.
2. The anharmonicity of thermal atomic vibrations leads to an increase in the interatomic distances, i.e., to thermal expansion. More anharmonic vibrations with large amplitudes (i.e., with weaker bonds) enhance thermal expansion.
3. An increase in the crystal temperature implies the enhanced thermal motion of atoms and molecules (vibrations, rotations, and jumps) and initiates transformations that tend to increase symmetry with an increase in temperature in the following way: increase in the vibrational symmetry atoms and molecules → structural deformation–

transformation toward higher symmetry → transformation into a more symmetric high-temperature modification.

But during the accumulation of experimental data, as will be shown below, some exceptions have been found.

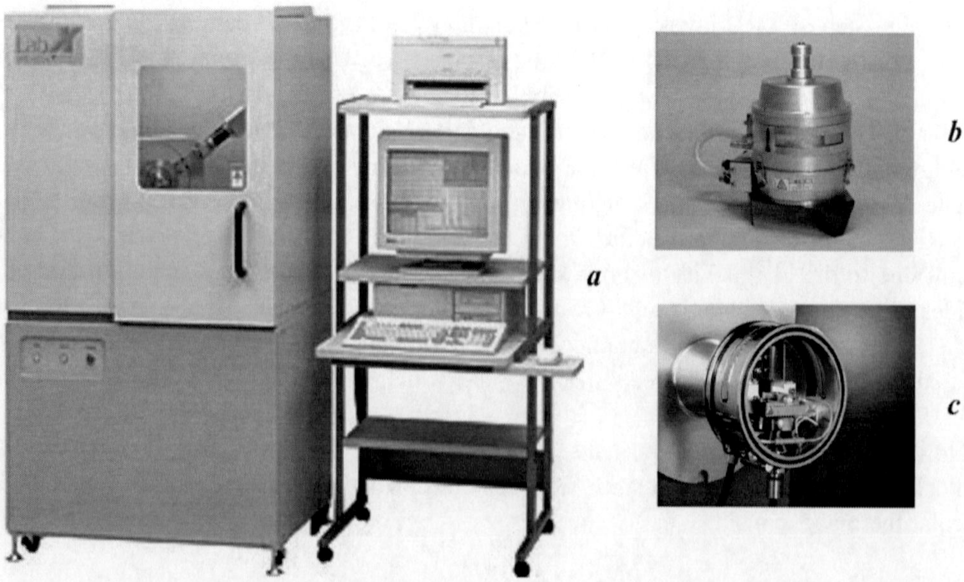

Figure 3. Powder difractometer LabX XRD-6000 (Shumadzu) (***a***), heating attachment HA-1001 (Shimadzu) (***b***) and cooling attachment TTK 450 (Anton Paar) (***c***) (Sources: www.shimadzu.com; www.anton-paar.com).

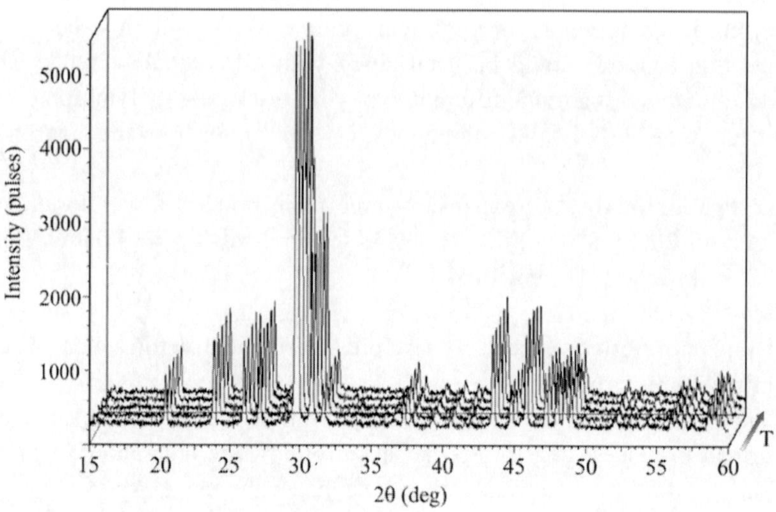

Figure 4. High-temperature XRD pattern for $Sr_5(VO_4)_3F$.

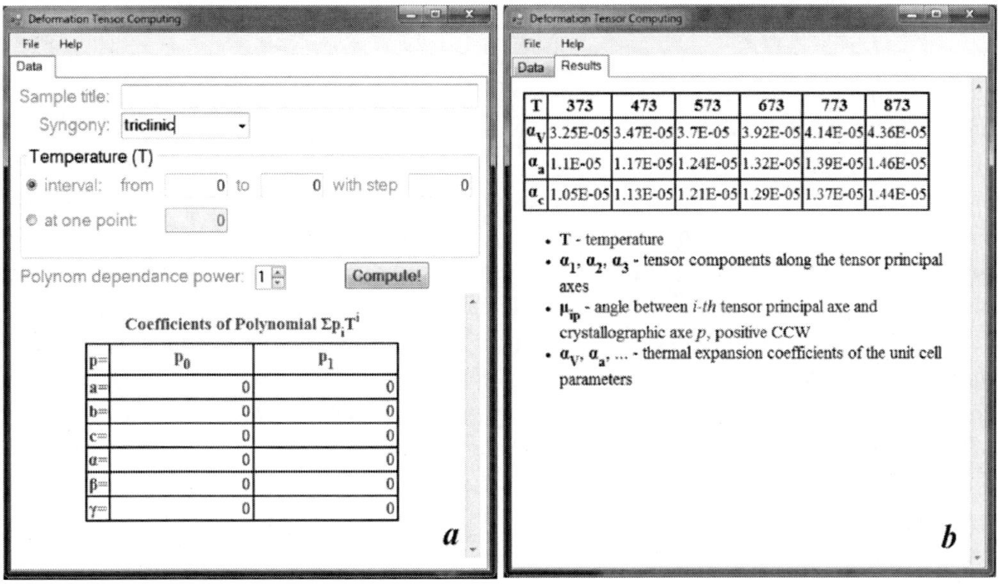

Figure 5. The interface for DTC program.

The following is the experiment as it is. In high-temperature XRD a different kind of difractometer can be used (Fischer, Bardhan, & Geiger, 1983; Miyazaki et al., 2009; Panneerselvam, Krishnan, Nagarajan, & Antony, 2011; Tonegawa et al., 2010). The main condition is the presence of specific heating and cooling attachments. For example, powder difractometer LabX XRD-6100 (Shimadzu) can be equipped with the attachment HA-1001 (Shimadzu) for heating of the samples in the temperature range 298-1173 K or attachment TTK-450 (Anton Paar) for cooling to 153 K (Figure 3). The result of the experiment is a series of XRD patterns of the substance under study which are obtained at selected temperatures (Figure 4). Then using procedure of analytical indexing of powder, XRD patterns for thermal dependencies of unit-cell parameters can be calculated (Kovba & Trunov, 1976; Porai-Koshitz, 1989). Such dependencies (also temperature dependencies of some structural parameters (atomic coordinates, bond lengths, bond angles, *etc.*) may be used) can be approximated by the polynomial, which in general can be written as follows: $l(T) = p_0 + p_1T + p_2T^2 + p_3T^3 + \ldots$ and can be used as a method of describing thermal deformations. But this form of describing thermal deformations is rather cumbersome. That is why thermal expansion coefficients are commonly used. For example for unit-cell parameter l

$$\alpha_l = \varepsilon_l/dT = (dl/l)/dT,$$

where l – volume of the parameter; α_l – thermal expansion of l; ε_l – deformation of l parameter; T – temperature. So, thermal expansion coefficient characterizes temperature dependence of deformation of the parameter (S. K. Filatov, 1990).

There are a few ways to calculate thermal expansion coefficients (Chen et al., 2011; Hovis et al., 2008; *International Tables for Crystallography*, 2003). But the most convenient is using DTC (Deformation Tensor Computing) program, which was developed at St. Petersburg State University by R. I. Belousov and S. K. Filatov (Belousov & Filatov, 2007; R. S. Bubnova, Firsova, & Filatov, 2013). The initial data for thermal expansion calculations are the coefficients of the polynomials approximating the temperature dependence of the linear (and angular if necessary) unit cell parameters, temperature of interest, and unit cell symmetry (Figure 5*a*). After the calculation the user gets a table with linear and volume thermal expansion coefficients at the desired temperatures (Figure 5*b*).

For obtaining more detailed information about high-temperature characteristics of the compound's deformation thermal expansion diagram may be used. These figures constructing at different temperatures demonstrate priority directions of thermal expansion and its anisotropy. Value of thermal expansion coefficient in given direction corresponds to length of radius-vector which is traced from origin of coordinates to edge of figure of expansion (Figure 6). For the construction of 3D version of the thermal expansion diagram DTP (Deformation Tensor Plot) program can be used (Figure 7*a*) (Belousov & Filatov, 2007; R. S. Bubnova, et al., 2013). But in 2013 another algorithm for Maplesoft© was created in Lobachevsky State University (Figure 7*b*). Firstly, the diagrams constructed using the algorithm were published in (A. V. Knyazev, Komshina, Zhidkov, & Plesovskikh, 2013). 2D-section of such diagrams made in different crystallographic planes may be also used to describe thermal expansion.

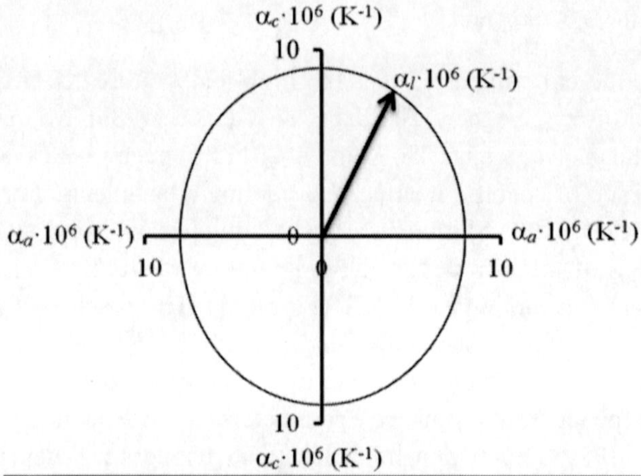

Figure 6. The general view for 2D thermal expansion diagram.

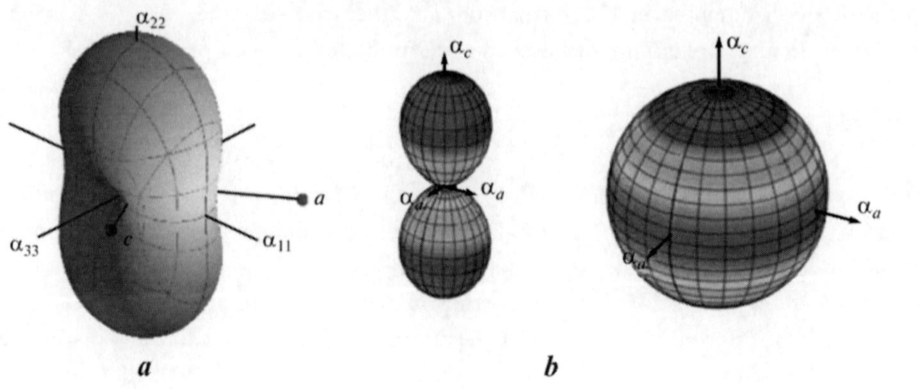

Figure 7. The general view for 3D thermal expansion diagram made in DTP program for $Ba(Si_2O_5)$ (*a*) and Maplesoft© for $RbNaCaTh(Si_8O_{20})$ (*b*).

THERMAL EXPANSION OF DIFFERENT TYPES OF APATITES

The first attempt to make a complex analysis of the apatite thermal expansion was undertaken in the 60s: there was an article about dilatometric finding o thermal expansion characteristics of lead-containing compounds including $Pb_5(SiO_4)(VO_4)_2$, $Pb_5(GeO_4)(VO_4)_2$, and $Pb_5(GeO_4)(PO_4)_2$ (Argyle & Hummel, 1963).

In later papers there was only rare and scattered information on the topic. For example the series of articles with a general title "Thermal Expansion Data" (D. Taylor, 1988) provided a vague description of thermal deformations of a few representatives of apatites with tetra- and pentavalent atoms in the position A. Also some information about unique apatites (with triclinic symmetry of the unit-cell (Whitfield, Le Page, Mercier, & Kim, 2007) or defective silicate oxyapatites (Fukuda, Asaka, & Uchida, 2012; Okudera, Yoshiasa, Masubuchi, Higuchi, & Kikkawa, 2004) can be found in the literature.

In the 2010s, the amount of the thermal expansion data markedly increased due to the work performed on the basis of Lobachevsky State University (Nizhni Novgorod, Russia) and here we present a systematic analysis of data reported in the literature.

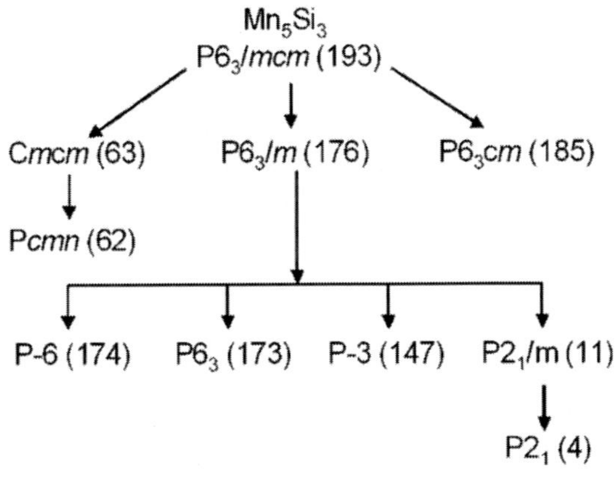

Figure 8. Symmetry relationships for apatite hettotypes that are derived from the Mn_5Si_3 aristotype (from White & Dong, 2003).

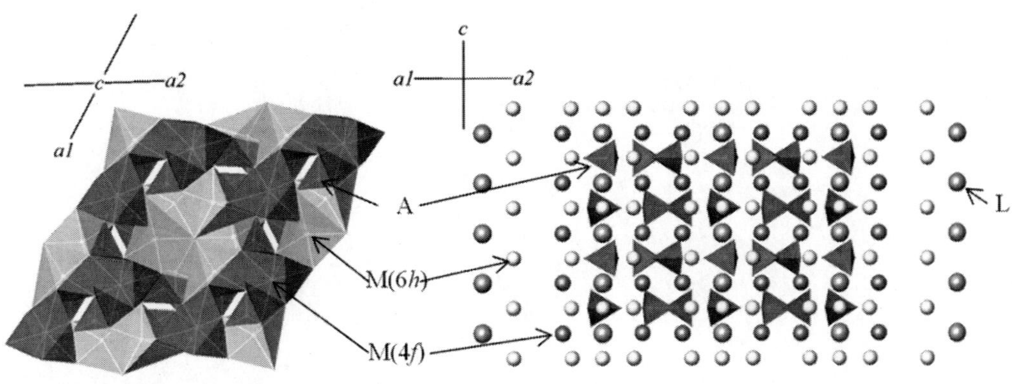

Figure 9. Fragment of the apatite structure (views from the different crystallographic planes).

Thermal deformations as any other physical characteristics of the compound are determined by the structure. Thereby we should consider the apatite structure type in the details.

Most of the apatites crystallize in the space group $P6_3/m$ (J. C. Elliott; McConnell, 1973; Náray-Szabó, 1930; Wyckoff, 1965), but further investigations have shown that these phases can have other symmetries of the unit cell, especially monoclinic (Y. S. Dai, Hughes, & Moore, 1991; J. C. Elliott, Mackie, & Young, 1973) and triclinic (T. Baikie et al., 2007; Whitfield, et al., 2007). Moreover it is possible to unite them into hierarchical tree (Figure 8) (T. White, Ferraris, Kim, & Madhavi, 2005; T. J. White & Dong, 2003). Because of the fact that hexagonal apatites are the most common type, we would like to focus on their structures.

The apatite structure built up from discrete tetrahedra AO_4, which are linked to each other by the cation polyhedra. AO_4 tetrahedra form regular layers perpendicular to crystallographic axis *c* (Figure 9) (Y. Dai & Hughes, 1989; Mathew, Mayer, Dickens, & Schroeder, 1979). That is why the apatites may have properties which are characteristic for both framework and layered structures.

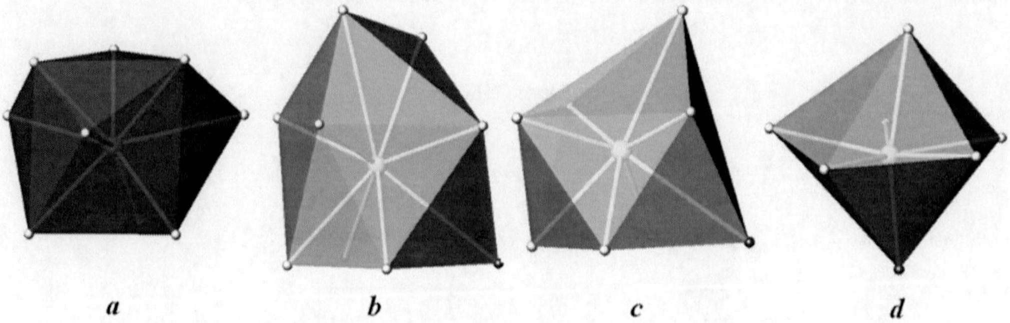

Figure 10. Different types of M-cations polyhedral in the apatite structure.

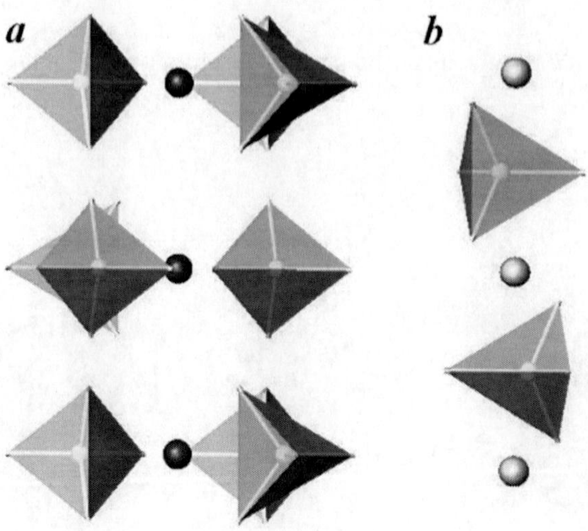

Figure 11. Different crystallographic positions of halogens in apatite structure: *a* – F-atoms (position 2*a*); *b* – Cl (position 2*b*).

One of the key features of the apatite structure is the presence of two crystallographic positions which can be occupied by M cations: 4*f* and 6*h*. Cations in 4*f* position have coordination polyhedron which looks like distorted three-caped trigonal prism with the composition MO_9 (Figure 10*a*). Polyhedron type of M-atoms in 6*h* position depends on the chemical composition of the apatite especially on the type of L-atoms. Analysis of more than 20 refined apatites structures has shown that in the case of fluorine-containing apatites (F-apatites) M-atoms form polyhedron with the coordination number (CN) 7 and can be described as distorted pentagonal bipyramide MO_6F (Figure 10*b*). In Cl-apatites Archimedean antiprisms MO_6Cl_2 can be identified (Figure 10*c*). One-caped Archimedean antiprism MO_7Br_2 forms in bromine-contained apatites (Figure 10*d*). In the case of I-apatites CN of M(6*h*) atoms depends on I-atoms position.

It also should be noted that L atoms (halogens or oxygen atoms in oxyapatites) may occupy different positions in the structure. Particularly F and Cl atoms are often located in 2*a* and 2*b* Wyckoff positions respectively (Figure 11) (Belokoneva, Troneva, Demyanets, Duderov, & Belov, 1982; Hashimoto & Matsumoto, 1998; Herdtweck, 1991; Reinen, Lachwa, & Allmann, 1986; S. B.Hendricks, Jefferson, & Mosley, 1932; Sudarsan.K & Young, 1974). These atoms occupy positions due to their sizes: fluorine atoms (as oxygen atoms of OH-groups too) are located in the aforementioned AO_4 layers ($z = ¼$), whereas Cl and Br which are bigger can be found between such layers ($z = 0$). Iodine as the largest halogen may be located in different sites: from 2*a*, 2*b*, 4*e* in $Cd_5(VO_4)_3I$ (Sudarsan, 1977) to 12*i* in $Pb_5(VO_4)_3I$ (Audubert, Savariault, & Lacout, 1999).

Thermal Expansion of Apatites with A = P, V, Cr, Mn

Thermal expansion of apatites with pentavalent atoms in tetrahedral positions has several features. First of all it is remarkable that thermal dependence of unit cell parameters of such apatites can be approximated primarily by the second-low polynomial. This situation is typical for close-packed structure in comparison to structures with lower density layout of polyhedra in a unit-cell, for example, hollandites (A. V. Knyazev, Maczka, Ladenkov, Bulanov, & Ptak, 2012).

According to the existing classification all compounds are divided into 3 groups depending on the volume of α_l: low-expanding ($0<\alpha_l<2\cdot10^{-6}$ K^{-1}), average-expanding ($2\cdot10^{-6}$ $K^{-1}<\alpha_l<8\cdot10^{-6}$ K^{-1}), and high-expanding compounds ($\alpha_l>8\cdot10^{-6}$ K^{-1}) (S. K. Filatov, 1990). So as it can be seen from the Table all apatites are in high-expanding phases. But the classification was developed for binary and oxide compounds (especially, borates). Therefore it is possible that the application of this classification to apatite group isn't quite legitimate.

Another parameter which should be discussed is the anisotropy parameter α_a/α_c. It shows what is the primary direction of thermal expansion. $\alpha_a/\alpha_c=1$ means that the compound has isotropic thermal deformations. Analysis of published data shows that thermal expansion of s-element-containing phases is more isotropic in comparison with Cd- and Pb-apatites. Among s-element-containing compounds Sr-apatites have the most isotropic expansion. This fact very well correlates with thermal stability of these phases (Alexander V. Knyazev, Chernorukov, & Bulanov, 2012).

Table of thermal expansion coefficients for different types of apatites*

Compound	Parameter	$\alpha_l = \alpha_{l0} + \alpha_{l1}T$ ($\cdot 10^6$ K^{-1})		Temperature interval (K)	Reference
		α_{l0}	α_{l1}		
A = P, V, Cr, Mn					
$Ca_5(PO_4)_3OH$	α_a	10.5	0.0048	298-1173	(Alexander V. Knyazev, et al., 2012)
	α_c	3.0	0.013		
	α_V	24.0	0.023		
	α_a	13.0		298-1173	(D. Taylor, 1988)
	α_c	12.8			
	α_V	38.7			
$Ca_5(PO_4)_3F$	α_a	5.4	0.011	298-1173	(Chernorukov, et al., 2011)
	α_c	5.0	0.011		
	α_V	15.8	0.033		
	α_a	11.0	-	298-1100	(D. Taylor, 1988)
	α_c	11.3	-		
	α_V	33.4	-		
α-$Ca_5(PO_4)_3Cl$	α_a	5.9	0.0084	298-973	(Alexander V. Knyazev, et al., 2012)
	α_c	-0.9	0.022		
	α_V	10.8	0.039		
β-$Ca_5(PO_4)_3Cl$	α_a	-125.9	0.114	1073-1173	
	$\alpha_b=\alpha_c$	112.1	-0.074		
	α_V	-139.8	0.154		
$Ca_5(VO_4)_3F$	α_a	6.9	0.0182	298-1173	New data
	α_c	-7.0	0.0347		
	α_V	6.8	0.0709		
α-$Ca_5(VO_4)_3Cl$	α_a	-0.0202	-0.0072	298-773	(Alexander V. Knyazev, et al., 2012)
	α_c	12.5	0.0353		
	α_V	15.5	0.0184		
β-$Ca_5(VO_4)_3Cl$	α_a	-1.2	0.0036	873-1173	
	$\alpha_b=\alpha_c$	126.5	-0.0943		
	α_V	124.2	-0.0873		
α-$Ca_5(CrO_4)_3F$	α_a	33.1	-0.0233	298-673	New data
	α_c	52.7	-0.0821		
	α_V	119.2	-0.1288		
β-$Ca_5(CrO_4)_3F$	α_a	-96.7	0.1177	773-1173	
	$\alpha_b=\alpha_c$	-1.9	0.0277		
	α_V	-194.8	0.2626		
α-$Ca_5(CrO_4)_3Cl$	α_a	11.0	0.0117	298-873	(Alexander V. Knyazev, et al., 2012)
	α_c	1.9	0.0102		
	α_V	23.8	0.0336		
β-$Ca_5(CrO_4)_3Cl$	α_a	-4.7	-	973-1073	
	$\alpha_b=\alpha_c$	22.7	-		
	α_V	13.2	-		
$Sr_5(PO_4)_3F$	α_a	8.3	0.0072	298-1173	(Chernorukov, et al., 2011)
	α_c	7.6	0.0078		
	α_V	24.24	0.0222		
	α_a	12.4	0.0021	300-1273	(He et al., 2013)
	α_c	11.0	0.0029		
	α_V	35.7	0.0071		

Compound	Parameter	$\alpha_l = \alpha_{l0} + \alpha_{l1}T$ ($\cdot 10^6$ K^{-1})		Temperature interval (K)	Reference
		α_{l0}	α_{l1}		
$Sr_5(PO_4)_3Cl$	α_a	9.8	0.0062	298-1173	(Chernorukov, et al., 2011)
	α_c	6.7	0.0084		
	α_V	26.4	0.0207		
$Sr_5(PO_4)_3Br$	α_a	15.9	-	298-1173	(Alexander V. Knyazev, et al., 2012)
	α_c	15.9	-		
	α_V	47.7	-		
$Sr_5(VO_4)_3F$	α_a	11.0	0.0069	298-1173	(A.V. Knyazev, 2013)
	α_c	3.0	0.0123		
	α_V	25.0	0.0262		
$Sr_5(VO_4)_3Cl$	α_a	9.4	0.0106	298-1173	(Alexander V. Knyazev, et al., 2012)
	α_c	0	0.0229		
	α_V	18.8	0.0441		
$Sr_5(VO_4)_3Br$	α_a	14.0	0.0018	298-1173	(A.V. Knyazev, 2013)
	α_c	4.6	0.017		
	α_V	32.7	0.0205		
$Sr_5(CrO_4)_3F$	α_a	8.8	0.0088	298-1173	(Alexander V. Knyazev, et al., 2012)
	α_c	5.3	0.0094		
	α_V	22.9	0.027		
α-$Sr_5(CrO_4)_3Cl$	α_a	15.3	0.008	298-673	(A.V. Knyazev, 2013)
	α_c	3.7	0.0318		
	α_V	34.4	0.0478		
β-$Sr_5(CrO_4)_3Cl$	α_a	25.1	0.0034	773-1173	
	α_b	71.4	-0.0538		
	α_c	-119.38	0.132		
	α_γ	11.4	-0.0047		
	α_V	-36.7	0.0873		
$Ba_5(PO_4)_3F$	α_a	5.2	0.0109	298-1173	New data
	α_c	4.1	0.0129		
	α_V	14.4	0.0347		
	α_a	13.5	-	300-1123	(He, et al., 2013)
	α_c	18.0	0.0084		
	α_V	45.1	-0.0085		
$Ba_5(PO_4)_3Cl$	α_a	4.1	0.0111	298-1173	(Chernorukov, et al., 2011)
	α_c	4.2	0.0136		
	α_V	12.4	0.0358		
$Ba_5(VO_4)_3F$	α_a	8.3	0.0116	298-1073	New data
	α_c	3.2	0.0118		
	α_V	19.7	0.035		
$Ba_5(VO_4)_3Cl$	α_a	10.0	0.0103	298-1173	(Alexander V. Knyazev, et al., 2012)
	α_c	1.6	0.0163		
	α_V	21.5	0.037		
$Ba_5(MnO_4)_3F$	α_a	3.9	0.0156	298-1073	
	α_c	3.8	0.0114		
	α_V	11.6	0.0425		
$Ba_5(MnO_4)_3Cl$	α_a	5.8	0.0169	298-1073	
	α_c	-1.0	0.0261		
	α_V	10.4	0.0602		

Table. (Continued)

Compound	Parameter	$\alpha_l = \alpha_{l0} + \alpha_{l1}T$ $(\cdot 10^6\ K^{-1})$		Temperature interval (K)	Reference
		α_{l0}	α_{l1}		
$Cd_5(PO_4)_3Cl$	α_a	1.3	0.0006	298-973	(Chernorukov, et al., 2011) recalculated
	α_c	16.8	0.0103		
	α_V	19.5	0.0116		
$Cd_5(VO_4)_3Cl$	α_a	2.4	0.0011	298-973	
	α_c	7.8	0.0198		
	α_V	12.5	0.0219		
α-$Pb_5(PO_4)_3F$	α_a	5.3	0.0167	298-1073	(Alexander V. Knyazev, et al., 2012)
	α_c	-3.0	0.0386		
	α_V	7.7	0.0722		
β-$Pb_5(PO_4)_3F$	α_a	79.8	-	1123-1173	(Chernorukov, et al., 2011)
	α_b	39.0	-		
	α_c	86.9	-		
	α_γ	33.2	-		
	α_V	165	-		
α-$Pb_5(PO_4)_3Cl$	α_a	11.6	0.0053	298-1073	(A.V. Knyazev, 2013)
	α_c	5.4	0.0238		
	α_V	28.6	0.0345		
β-$Pb_5(PO_4)_3Cl$	α_a	11.9	-	1123-1173	(Chernorukov, et al., 2011)
	α_b	24.8	-		
	α_c	50.6	-		
	α_γ	0	-		
	α_V	87.2	-		
$Pb_5(PO_4)_3I$	α_a	$1.49\cdot10^{-4}T^2 - 0.17T + 53.6$		298-1073	(Alexander V. Knyazev, et al., 2012)
	α_c	$1{,}19\cdot10^{-5}T^2 + 0.0\,29T - 5.6$			
	α_V	$3{,}12\cdot10^{-4}T^2 - 0.313T + 102$			
$Pb_5(VO_4)_3F$	α_a	$1{,}20\cdot10^{-4}T^2 - 0.144T + 54.8$		298-973	
	α_c	$1{,}42\cdot10^{-4}T^2 - 0.149T + 52.7$			
	α_V	$3{,}82\cdot10^{-4}T^2 - 0.438T + 162$			
α-$Pb_5(VO_4)_3Cl$	α_a	4.7	0.0163	298-973	
	α_c	12.6	0.0257		
	α_V	22.0	0.0584		
β-$Pb_5(VO_4)_3Cl$	α_a	90.1	-	1073-1173	(Chernorukov, et al., 2011)
	α_b	174.5	-		
	α_c	201.0	-		
	α_γ	18.3	-		
	α_V	444.0	-		
$Pb_5(VO_4)_3I$	α_a	-64.3	0.1888	300-573	(Uno, Shinohara, Kurosaki, & Yamanaka, 2001)
	α_c	-29.7	0.1166		
	α_V	-156.7	0.4907		
A = Si					
$CaY_4(SiO_4)_3O$	α_a	8.2	-	298-1163	(D. Taylor, 1988)
	α_c	7.0	-		
	α_V	23.5	-		
$CaLa_4(SiO_4)_3O$	α_a	10.2	-	298-1253	
	α_c	6.0	-		
	α_V	26.3	-		

Compound	Parameter	$\alpha_l = \alpha_{l0} + \alpha_{l1}T$ ($\cdot 10^6$ K^{-1})		Temperature interval (K)	Reference
		α_{l0}	α_{l1}		
$CaPr_4(SiO_4)_3O$	α_a	8.5	0.0024	298-1173	(A. V. Knyazev, Bulanov, et al., 2013)
	α_c	5.3	0.0036		
	α_V	22.3	0.0085		
$CaSm_4(SiO_4)_3O$	α_a	3.8	0.0078	298-1173	
	α_c	2.7	0.0056		
	α_V	10.3	0.0212		
$CaEu_4(SiO_4)_3O$	α_a	0.1	0.0124	298-1173	
	α_c	3.6	0.0031		
	α_V	3.9	0.028		
$SrPr_4(SiO_4)_3O$	α_a	4.2	0.0082	298-1173	New data
	α_c	8.4	-0.002		
	α_V	16.9	0.0143		
$SrSm_4(SiO_4)_3O$	α_a	8.5	0.002	298-1173	
	α_c	1.5	0.0085		
	α_V	18.4	0.0125		
$SrEu_4(SiO_4)_3O$	α_a	7.4	0.002	298-1173	
	α_c	0.9	0.0085		
	α_V	15.7	0.0125		
$Ca_3Th_2(SiO_4)_3O$	α_a	6.0	0.0053	298-1173	(A.V. Knyazev, 2013)
	α_c	3.1	0.0071		
	α_V	14.9	0.0178		
A = S					
α-$Na_3Ca_2(SO_4)_3F$	α_a	7.2	0.0259	298-873	(A.V. Knyazev, 2013)
	α_c	4.3	0.0348		
	α_V	18.7	0.0863		
β-$Na_3Ca_2(SO_4)_3F$**	α_a	42.8	-	923-973	
	α_b	47.9	-		
	α_c	67.2	-		
	α_V	97.3	-		
α-$Na_3Cd_2(SO_4)_3Cl$	α_a	0.1	0.0622	298-473	
	α_c	61.2	-0.1069		
	α_V	61.7	0.0167		
β-$Na_3Cd_2(SO_4)_3Cl$	α_a	87.2	-0.084	523-773	
	α_b	105.2	-0.1228		
	α_c	-243.4	0.4317		
	α_γ	3.34	0.0094		
	α_V	-54.7	0.2133		
$Na_3Pb_2(SO_4)_3Cl$	α_a	15.6	0.0201	298-873	
	α_c	-11.1	0.0841		
	α_V	20.2	0.124		

* The table shows only the data that allow calculation of the changing of thermal expansion coefficients.

** β-$Na_3Ca_2(SO_4)_3F$ doesn't have an apatite structure (it is similar to $K_3Ca_2(SO_4)_3F$).

It may be explained by absence of priority expansion direction in those compounds which promotes their thermal stability as opposed to other apatites.

Another conclusion is that for phosphates and vanadates crystallographic axis *c* is the primary direction of the thermal deformations whereas for Mn and Cr-containig apatites it is the *a* axis. It can be explained by the different ratio of bonds energy between layers formed by AO_4 tetrahedrons and within them. The basic types of thermal expansion diagrams of apatites are shown on the Figure 12.

As mentioned above, apatites can crystallize in monoclinic symmetry with the space group $P2_1/b$. In (Chernorukov, Knyazev, & Bulanov, 2010) it was shown for the first time that there is a polymorphic transition in some apatites which is accompanied by the lowering of unit-cell symmetry from hexagonal to monoclinic. Such transitions were observed in several apatites: $Pb_5(PO_4)_3F$, $Pb_5(PO_4)_3Cl$, $Pb_5(VO_4)_3Cl$ (Chernorukov, Knyazev, & Bulanov, 2011), $Sr_5(CrO_4)_3Cl$ and $Na_3Ca_2(SO_4)_3Cl$ (A.V. Knyazev, 2013). Further studies have shown that there is a group of apatites ($Ca_5(PO_4)_3Cl$ (A.V. Knyazev, 2013), $Ca_5(VO_4)_3Cl$, $Ca_5(CrO_4)_3Cl$, (Alexander V. Knyazev, et al., 2012) $Ca_5(CrO_4)_3F$) which also has transitions, but the high-temperature modification may be called pseudohexagonal. In this case crystallographic axis *a* of hexagonal modification almost matches with axis *b* of monoclinic modification, $b_2 = 2a_1$, $\gamma \approx 120°$ (Figure 13).

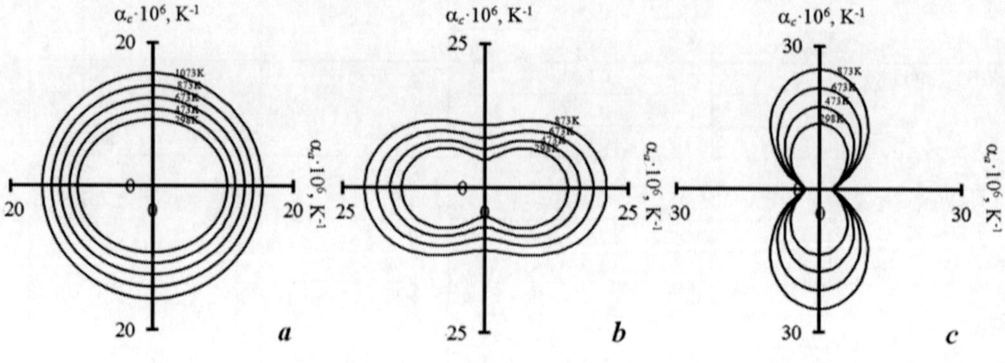

Figure 12. Basic types of apatite thermal expansion diagrams: ***a*** – isotropic expansion of $Sr_5(PO_4)_3Cl$, ***b*** – anisotropic expansion (priority direction *a*) of $Ca_5(CrO_4)_3Cl$, ***c*** – anisotropic expansion (priority direction *c*) of $Cd_5(VO_4)_3Cl$.

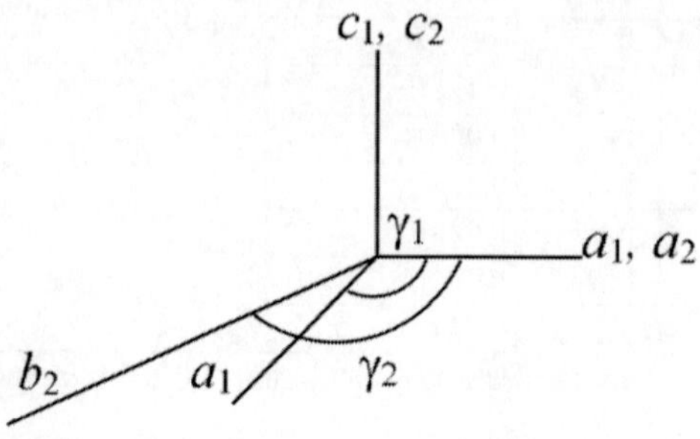

Figure 13. Change from a hexagonal (subscript 1) to a monoclinic (subscript 2) unit cell upon the polymorphic transformation of the apatites.

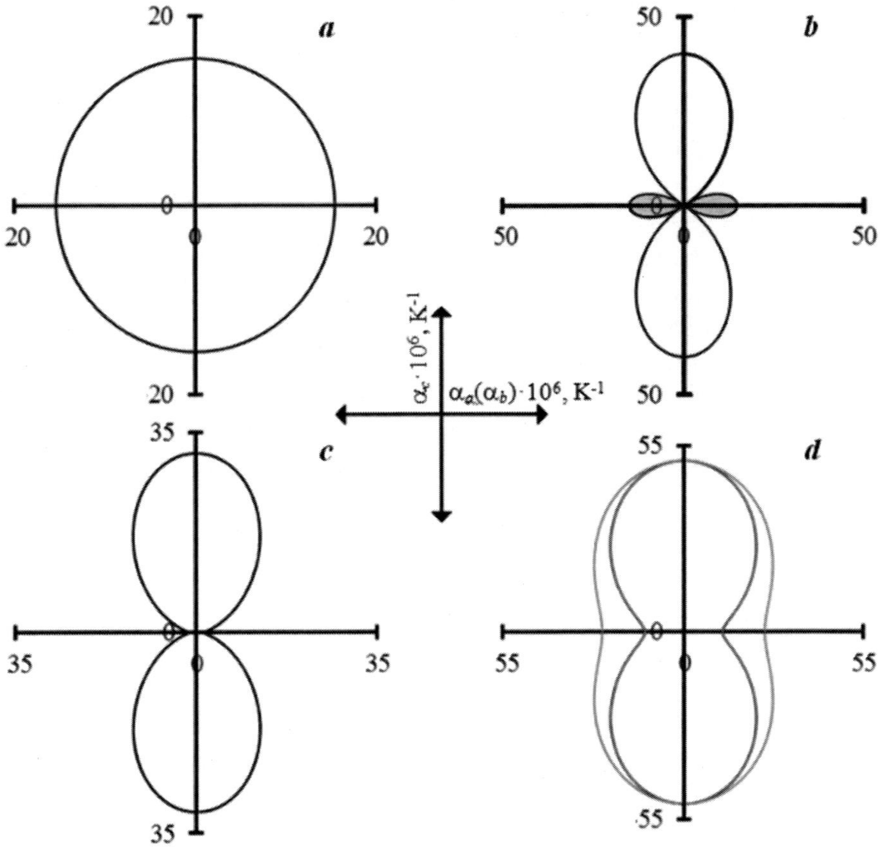

Figure 14. Changing of thermal expansion diagrams during the polymorphic transitions for $Ca_5(PO_4)_3Cl$ (*a*-*b*) and $Pb_5(PO_4)_3Cl$ (*c*-*d*).

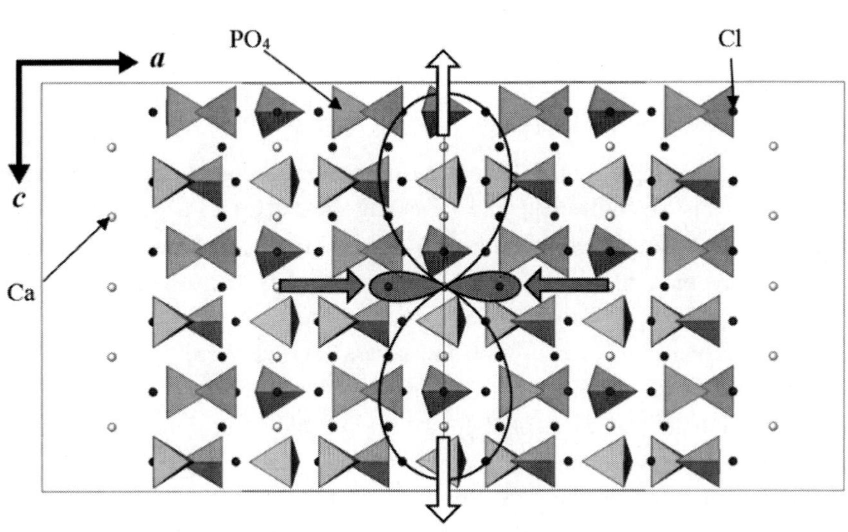

Figure 15. Thermal expansion diagram of $Ca_5(PO_4)_3Cl$ (pseudohexagonal) at 973 K with directions of expansion and structure fragment.

For high-temperature monoclinic modifications larger linear and volume thermal expansion coefficients are typical, whereas for pseudohexagonal modifications of Ca-apatites linear thermal expansion coefficients are similar to hexogonal modification, but they have a considerably greater rate of growth with increasing temperature. It is also notable that abrupt increase of thermal expansion anisotropy was observed for high-temperature modifications of Ca-apatites, whereas for Pb-containing phase's maintenance or in some cases decrease of that characteristic was observed (Figure 14).

In some apatites ($Ca_5(PO_4)_3Cl$, $Ca_5(VO_4)_3Cl$, $Ca_5(CrO_4)_3Cl$ (hexagonal modifications) and $Sr_5(CrO_4)_3Cl$, $Na_3Cd_2(SO_4)_3Cl$ (monoclinic modifications)) negative volumes of thermal expansion coefficients were observed. This effect is rather rare but the probability of its occurrence increases up to 30% at the transition from the cubic to triclinic symmetry (R.S. Bubnova & Filatov, 2008). That remarkable situation can be explained by the structure deformations: In this case crystal structure stretches along direction of greater expansion and shrinks along another direction (Figure 15). It leads to compensation of negative expansion as volume thermal expansion coefficient continues to grow.

Thermal Expansion of Silicate Oxyapatites

Apatites with A=Si are usually called oxyapatites because of the fact that the "halogen-position" contains L oxygen atoms. Recently in the publications, much attention is paid to these compounds due to their importance as potential material for producing SOFC (solid oxide fuel cells) (Nakayama, Kageyama, Aono, & Sadaoka, 1995; Nakayama & Sakamoto, 1998).

Although oxyapatite group with the general formula $M^{II}M^{III}{}_4(SiO_4)_3O$ (M^{II} – divalent cations; M^{III} – lanthanides) has much more representatives as the group of "classic" apatites with A = pentavalent atoms, there is a little information about their properties including thermal expansion. So here we can analyze high-temperature properties of oxyapatites with M^{II} = Ca, Sr and M^{III} = Pr, Sm, Eu.

It was shown using differential scanning calorimetry that in wide temperature range 298-1673 K there is no effects on DTA curve (A. V. Knyazev, Bulanov, Korshunov, & Krasheninnikova, 2013).

During calculation of thermal deformation of such compounds it was discovered that volume thermal expansion coefficient and its change correlate with the size of lanthanide-atom: in particular α_V (average) decreasing with decreasing of lanthanide size.

Anisotropy of the thermal expansion for compounds under consideration is similar to Mn and Cr-contained apatites: crystallographic axis ***a*** is the main direction of the expansion, which means that expansion occurs within the layers formed by SiO_4 tetrahedra. Moreover anisotropy factor α_a/α_c for Ca-oxyapatites is smaller than for Sr-containing phases, which leads to the conclusion about the essential contribution of the size of the divalent cation in the distribution of bond strength in the oxyapatite structure.

Th-containing apatite deserves a special consideration (A. V. Knyazev, 2013). $Ca_3Th_2(SiO_4)_3O$ is a unique compound because of a number of reasons. First of all, it is one of a small number of known silicates of thorium (A. V. Knyazev, Komshina, et al., 2013; Li, Krivovichev, & Burns, 2000; Richard & Perrault, 1972; M. Taylor & Ewing, 1978). Secondly, this apatite is the one known actinide-containing apatite, which can be synthesized as an individual compound (Engel, 1978). And finally, calcium-thorium apatite could be considered as a potential form for binding of radioactive thorium containing wastes.

The Rietveld refinement of the structure shows that Th-apatite has the crystallochemical formula $[Ca_{1.92}Th_{0.08}]^{4f}[Ca_{1.40}{}^{Th}{}_{1.60}]^{6h}(SiO_4)_3O_{1-x}$. This finding is consistent with (Tom Baikie et al., 2010) where it was determined that the chemical species with a greater charge tended to distribute to the tunnel position. This effect was attributed to higher coordination of oxygen in the *6h* site, which helped to distribute the charge more effectively. The summation of the cation charges yields a value of +50.74, which may suggest that there are oxygen vacancies in the channel and/or the presence of protons to achieve an overall charge balance (Pramana et al., 2009). Another remarkable fact is that oxygen atom, which is not included in the tetrahedron SiO_4, occupies position *2a*. The similar situation could be observed in F-containing apatites (Comodi, Liu, Zanazzi, & Montagnoli, 2001) because of the proximity of the radii of the F (1.3 Å) and O (1.35 Å) atoms ("Database of Ionic Radii," 1998).

Thermal deformation of $Ca_3Th_2(SiO_4)_3O$ is also similar to aforementioned oxyapatites. The compound under the study is characterized by not so large thermal expansion coefficient as it was observed in the case of apatites with A = P, V, Cr, Mn (Chernorukov, et al., 2011; Alexander V. Knyazev, et al., 2012).

In $Ca_3Th_2(SiO_4)_3O$ the prior direction of thermal deformations at low temperatures is located along the crystallographic axis *a*. And expansion becomes more isotropic with the increase of the temperature. It can be explained by the fact that in these compounds distortion of the polyhedrons, defined as the difference between the shortest and longest bonds, is not as big enough as in some other apatites (S. B.Hendricks, et al., 1932; Schneider, 1967). So at the high temperatures that difference decreases due to thermal vibrations of atoms.

Anisotropy parameter α_a/α_c shows that with the increasing of the temperature thermal vibrations of atoms along main crystallographic axis become equal and it leads to isotropic expansion. As we reported earlier such situation was observed in the most of the apatites with some reservations: in some apatites anisotropic parameter tends to 1 but then the deviation begins to rise again ($Sr_5(VO_4)_3Cl$) and sometimes anisotropic parameter only rises ($Ba_5(MnO_4)_3F$).

Thermal Expansion of Sulfate Apatites

Sulfates with the apatite structure have a general formula $Na_3M^{II}_2(SO_4)_3L$ (M^{II} = divalent cations; L = F, Cl) and occupy a special place among considered compounds. First of all it should be noted that such phases could be used as sensors and ion exchangers (Laghzizil, Bouhaouss, Barboux, Morineau, & Livage, 1993; Nagai & Nishino, 1988; Pasero, et al., 2010; Schneider, 1967; Tazzoli, 1983). Secondly there are only a few representatives in the group of sulfate-apatites (Klement, 1939; Perret & Bouillet, 1975; T. White, et al., 2005). Moreover the compound with a following composition $K_3Ca_2(SO_4)_3F$ in spite of similar stoichiometry, it doesn't have an apatite structure (it crystallizes in Pna21 space group) and was found in the nature as a mineral of volcanic origin (Fayos, Watkin, & Perezmendez, 1987; Piotrowski, Kahlenberg, & Fischer, 2002).

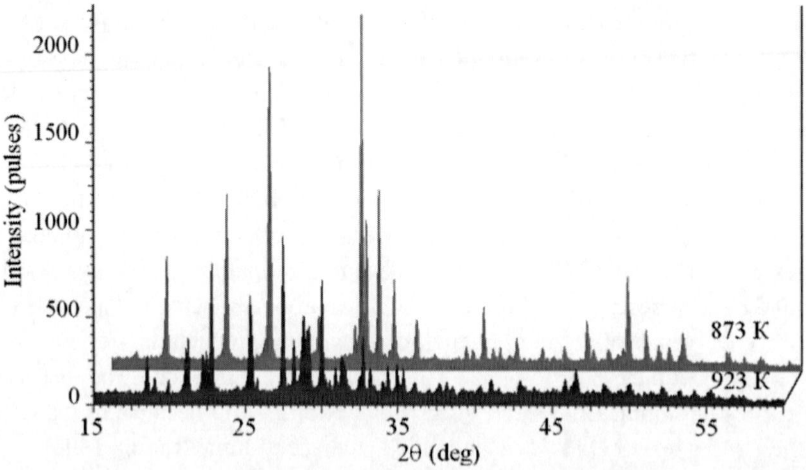

Figure 16. Powder XRD-pattern for $Na_3Ca_2(SO_4)_3F$ above (black) and below (grey) the temperature of polymorphic transition.

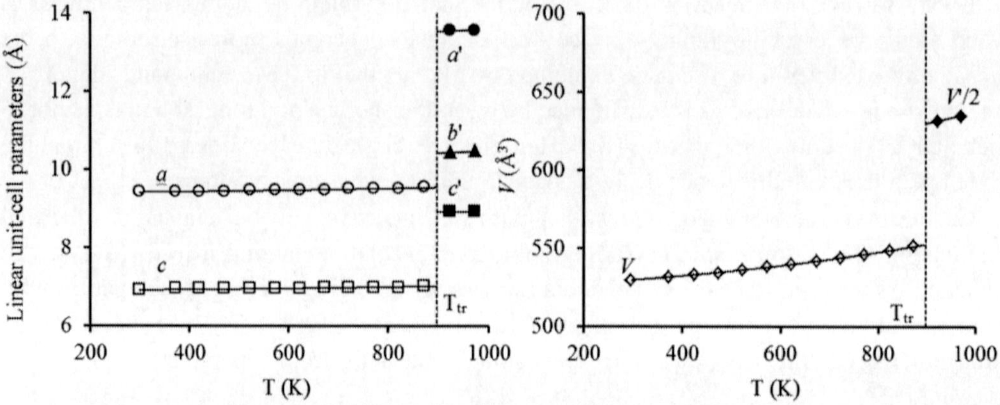

Figure 17. Thermal dependencies for unit-cell parameters of $Na_3Ca_2(SO_4)_3F$.

Here we adduce the data about thermal expansion of three sulfates with the apatite structure $Na_3Ca_2(SO_4)_3F$, $Na_3Cd_2(SO_4)_3Cl$ and $Na_3Pb_2(SO_4)_3Cl$ (A. V. Knyazev, 2013). During the investigation of aforementioned compounds using differential scanning calorimetry it was found that $Na_3Ca_2(SO_4)_3F$ and $Na_3Cd_2(SO_4)_3Cl$ have not only melting effects on DTA-curve, but also some reversible endothermic effects at the lower temperatures (897 K and 494 K respectively).

As it was mentioned above, apatites could undergo some structural distortions accompanied by changing symmetry from hexagonal ($P6_3/m$) to monoclinic ($P2_1/b$). During that polymorphic transformation, the XRD pattern remains practically unchanged as it was shown on $Na_3Cd_2(SO_4)_3Cl$.

On the contrary in the case of $Na_3Ca_2(SO_4)_3F$ XRD pattern changes significantly (Figure 16). Furthermore, endothermic effect on this compound has comparable energy with melting process. Therefore, we concluded that there have been significant structural changes. Similarity of XRD patterns for high-temperature modification of $Na_3Ca_2(SO_4)_3F$ and for

$K_3Ca_2(SO_4)_3F$ also confirms this fact. It was proved by considerable changes of linear and volume parameters of the unit cell too (Figure 17). Thus structural changes in $Na_3Ca_2(SO_4)_3F$ were accompanied not only by lowering of symmetry (to orthorhombic $Pna2_1$) but also by disordering of SO_4^{2-} group's position (Piotrowski, et al., 2002), and it can be called as a transition of reconstructive type (Urusov, 1987). That is why in the $Na_3Ca_2(SO_4)_3F$-$K_3Ca_2(SO_4)_3F$ system morthotropic transitions can be expected.

Regarding the peculiarities of sulfate-apatites' thermal expansion, it is obvious that main direction of expansion is crystallographic axis c.

In the case of $Na_3Ca_2(SO_4)_3F$ increasing of thermal expansion was observed to be anisotropic due to lowering of symmetry from hexagonal (α_a/α_c = 0.86) to orthorhombic (α_a/α_c = 0.64 и α_b/α_c = 0.71). This fact may be explained by distortion of sulfate tetrahedrons (Fayos, et al., 1987). Thermal expansion diagram as it can be found on the Figure 18 changes essentially.

Hexagonal modification of $Na_3Cd_2(SO_4)_3Cl$ will expand isotropically at the 361 K, because in the wide temperature range increasing of α_a is accompanied by decreasing of α_c. Due to proximity of γ-angle of monoclinic modification to 120° α_a and α_b are very close to each other.

As for $Na_3Pb_2(SO_4)_3Cl$ it was shown that it has thermal expansion coefficients similar to other Pb-contaning apatites.

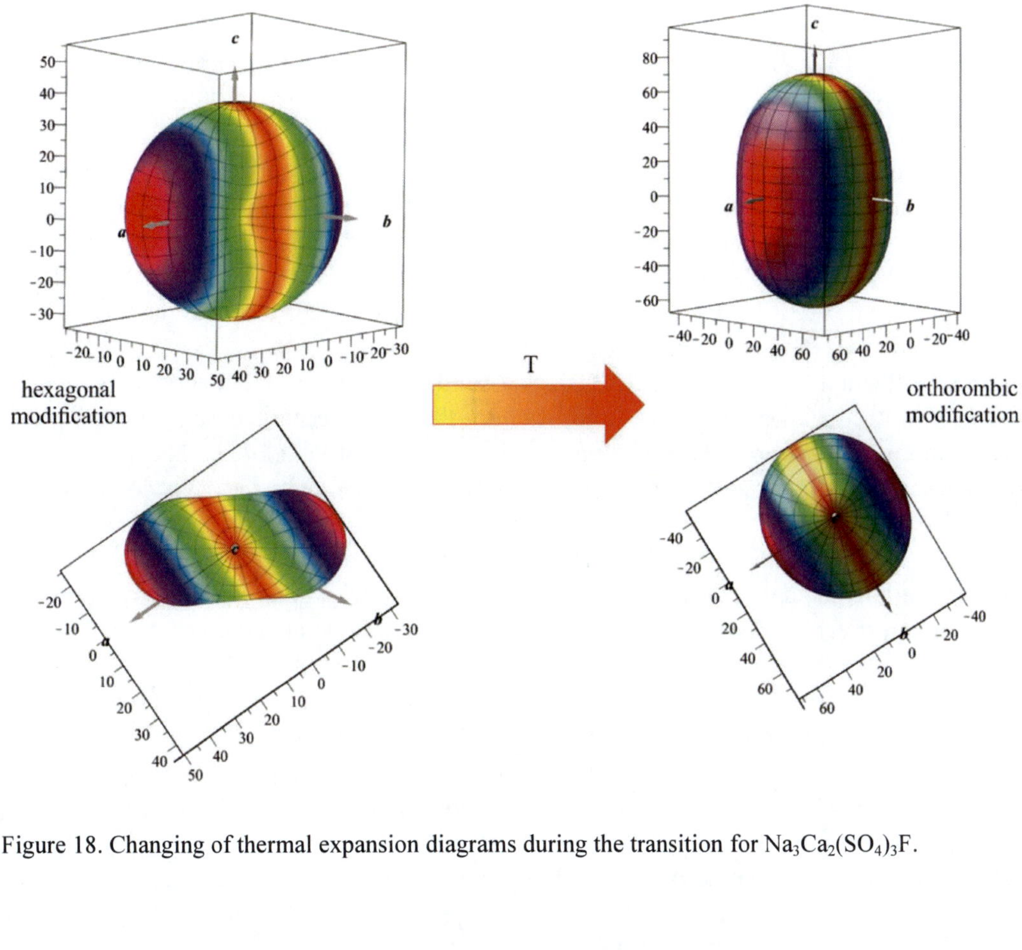

Figure 18. Changing of thermal expansion diagrams during the transition for $Na_3Ca_2(SO_4)_3F$.

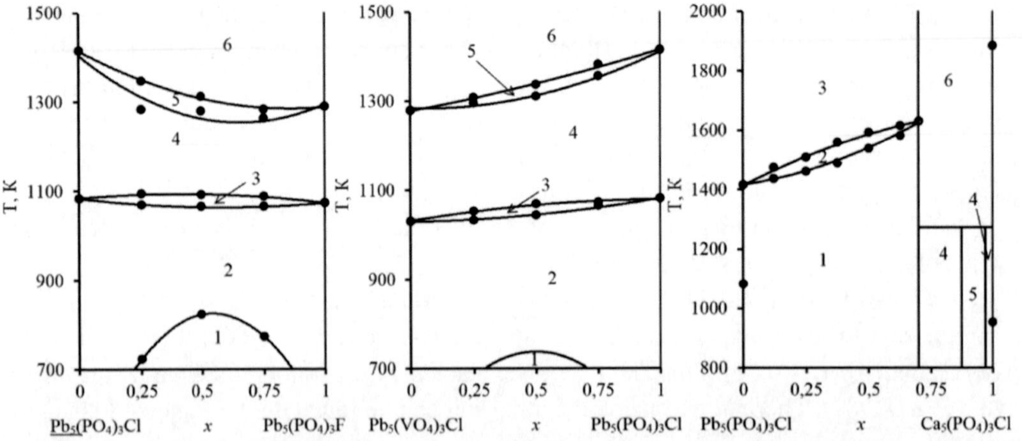

Figure 19. Phase diagrams for binary apatite systems. For $Pb_5(PO_4)_3F$-$Pb_5(PO_4)_3Cl$ and $Pb_5(PO_4)_3Cl$-$Pb_5(VO_4)_3Cl$: 1) mixture of solid solutions, (2) solid solution (the hexagonal phase), (3) mixture of two phases of solid solutions, (4) solid solution (the monoclinic phase), (5) solid solution (monoclinic phase) + liquid, and (6) liquid. For $Ca_5(PO_4)_3Cl$-$Pb_5(PO_4)_3Cl$: (1) solid solutions of the $Pb_{10}(PO_4)_6Cl_2$ – $Ca_7Pb_3(PO_4)_6Cl_2$ system, (2) solid solutions of the $Pb_{10}(PO_4)_6Cl_2$ – $Ca_7Pb_3(PO_4)_6Cl_2$ system + liquid, (3) liquid, (4) solid solutions of the $Ca_7Pb_3(PO_4)_6Cl_2$ – $Ca_{10}(PO_4)_6Cl_2$ system, (5) mixture of solid solutions of the $Ca_7Pb_3(PO_4)_6Cl_2$ – $Ca_{10}(PO_4)_6Cl_2$ system, (6) decomposition products of compounds solutions of the $Ca_7Pb_3(PO_4)_6Cl_2$ – $Ca_{10}(PO_4)_6Cl_2$ system.

High-Temperature *In Situ* X-ray Diffraction As a Method of Investigating of Isomorphism

Apatites have high isomorphic capacity, which causes necessity of investigation not only individual compounds, but various binary systems. Earlier we investigated phase diagrams of three binary systems: $Pb_5(PO_4)_3F$ – $Pb_5(PO_4)_3Cl$ (A. V. Knyazev, Chernorukov, & Bulanov, 2011a), $Pb_5(PO_4)_3Cl$ – $Pb_5(VO_4)_3Cl$ (Chernorukov, et al., 2010), and $Ca_5(PO_4)_3Cl$ – $Pb_5(PO_4)_3Cl$ (A. V. Knyazev, Chernorukov, & Bulanov, 2011b) on the basis of thermodynamic modeling and experimental data (Figure 19).

An investigation of the thermal expansion of solid solutions of aforementioned systems was made in the temperature range from 298 to 1173 K. Because of the existence of polymorphic transitions on some compounds and their different nature (Chernorukov, et al., 2011; Alexander V. Knyazev, et al., 2012), we will consider only the behavior of low-temperature modifications. A visual representation of the obtained data is presented in Figures 20*a-c*. The volume thermal expansion coefficient is an integral quantity, which is characterizing thermal deformation of the compound. That is why we consider this parameter as the function of the composition of the solid solution and the temperature (A.V. Knyazev, 2013).

In the case of the $Pb_5(PO_4)_3F_xCl_{1-x}$ system it can be seen that in the range of $0 < x < 0.5$ volume thermal expansion coefficient doesn't depend on the composition for all temperatures (Figure 20*a*). As follows from the structural data, F and Cl atoms located in different crystallographic positions in the end members: F-atoms occupy 2*a* positions in the layers formed by phosphorus tetrahedron, whereas Cl-atoms are located between the layers (2*b* position)) (Figure 11). So in the range of compositions $0.5 < x < 1$ when the larger Cl-atoms intrude into F-positions of the structure we can observe an increased thermal sensitivity of the structure, which manifests itself in a deviation of studied dependency from the horizontal. Thus, position 2*b* of apatite structure is more tolerant to the content of the larger atoms. This fact can be also confirmed in a number structures, where that position is occupied by Br-atoms (Alberius-Henning, Mattsson, & Lidin, 2000).

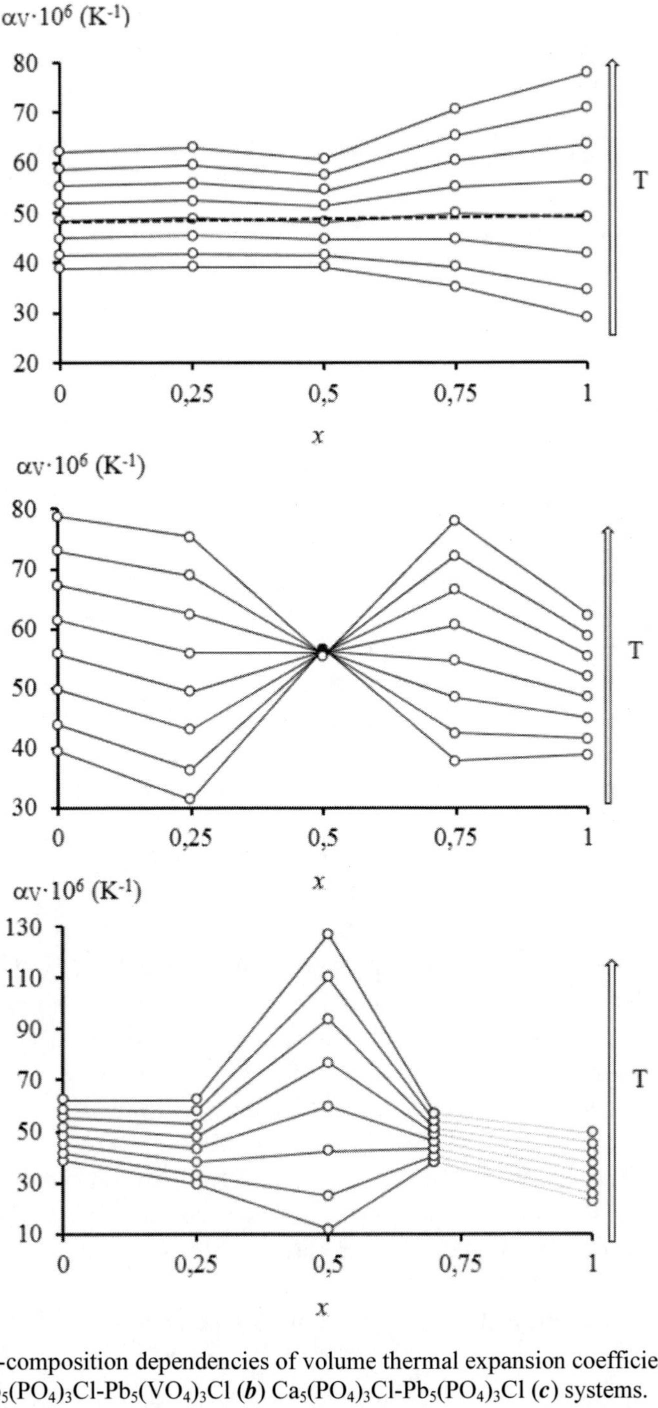

Figure 20. Thermal-composition dependencies of volume thermal expansion coefficient for $Pb_5(PO_4)_3F$-$Pb_5(PO_4)_3Cl$ (***a***), $Pb_5(PO_4)_3Cl$-$Pb_5(VO_4)_3Cl$ (***b***) $Ca_5(PO_4)_3Cl$-$Pb_5(PO_4)_3Cl$ (***c***) systems.

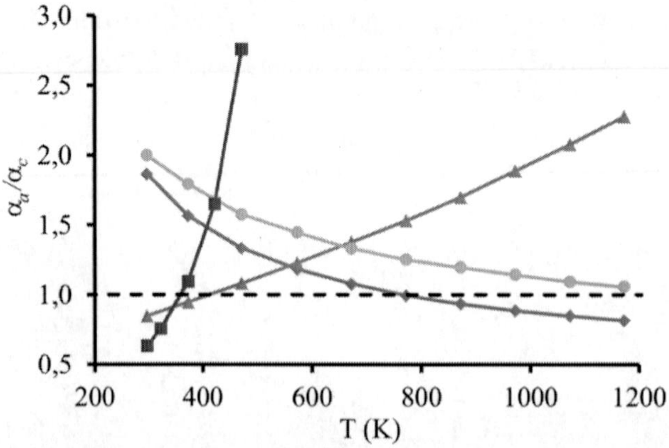

Figure 21. Thermal dependencies of anisotropy parameter α_a/α_c for various apatites: $Na_3Cd_2(SO_4)_3Cl$ (squares); $SrPr_4(SiO_4)_3O$ (triangles); $Sr_5(VO_4)_3Cl$ (rhombs); $Ba_5(VO_4)_3Cl$ (circles).

It also should be noted that at the 573 K volume thermal expansion coefficient in the $Pb_5(PO_4)_3F$ – $Pb_5(PO_4)_3Cl$ system doesn't depend on the composition of the solid solution. In this case increasing the value of linear thermal expansion coefficient α_a (with increasing x) is associated with reducing of the value of linear thermal expansion coefficients α_c. It can be explained by the assumption, that at this temperature F- and Cl-atoms in the apatite structure have equal ellipsoids of thermal vibrations.

In the case of $Pb_5(PO_4)_3Cl$ – $Pb_5(VO_4)_3Cl$ the following situation was observed (Figure 20*b*). Volume thermal expansion coefficient practically doesn't depend on the composition of the solid solution at 673 K. This fact may be proved with the same positions as in the previous case: by the equality of thermal vibrations of the substituted atoms.

Another fact that makes this system interesting is the independence of the volume thermal expansion coefficient from the temperature in the solid solution with equimolar ratio of substituted atoms (P and V). We assume that this phenomenon is connected with the formation of superstructure in the solid solution due to the ordering of the P- and V-atoms in the same crystallographic position 6*h* (Brenner, Engel, & Wondrats.H, 1970).

$(Ca_xPb_{1-x})_5(PO_4)_3Cl$ system is characterized by more complicated character of the solid solution formation in comparison with the other two systems due to the number of substitutable atoms per unit cell: solid solutions can be synthesized only in the range of composition $0 < x < 0.7$ (Chernorukov, et al., 2010; A. V. Knyazev, et al., 2011a, 2011b). Moreover, the substitution takes place in different crystallographic sites 4*f* and 6*h*. That is why in our opinion in the case of equimolar ratio of substituted atoms (Ca and Pb) the strong temperature dependence of volume thermal expansion coefficient was observed (Figure 20*c*). In the previous case, substitution takes place in the same position and P- and V-atoms can be grouped in threes to form superstructure. In $Ca_5(PO_4)_3Cl$ – $Pb_5(PO_4)_3Cl$ system cations in equimolar ratio can't be distributed equally between 4*f* and 6*h* positions because of the fact that there are 4 4*f* positions and 6 6*h* positions in apatite structure. Thus, we have a statistical substitution at each position, which further destabilizes the structure, making it more temperature sensitive.

Also the information about thermal expansion of solid solutions in the $Pb_5(P_xV_{1-x}O_4)_3F$ can be found in the literature (He et al., 2012). For all solid solutions in those systems linear changing of unit-cell parameters was observed, but there are no explicit laws allowing to make any conclusions about structure of such phases.

Conclusion

In conclusion, we present some generalizations about the high-temperature behavior of the apatites.

It was mentioned above that all of the apatites have substantial enough thermal expansion coefficients. This situation leads to the fact that apatites shouldn't be used in the fields where high expansion may cause disruption of a technological process. Moreover, propensity to high anisotropy of thermal expansion may also lead to defect formation even with a small temperature change. On the other hand, apatites may be the chemical basis of sealing materials (e.g. for reactors).

As for thermal expansion anisotropy it should be noted that in all of the apatites described in the literature or investigated by us a tendency to isotropic expansion with increasing temperature was observed. It means that no matter how big or small the anisotropy parameter α_a/α_c at the low temperature it will tend to 1 while the temperature increases (Figure 21).

And the last remarkable fact relates directly to research method. As it was shown in the last chapter high-temperature XRD method is a more sensitive method for the investigation of isomorphism than the accepted calculation of dependencies using Vegard and Retgers rules, in view of the fact that this method fixes the physical properties of the compound.

The reported study was supported by RFBR, research projects No. 13-03-00152 A; 14-03-31234 мол-а.

References

Alberius-Henning, P., Mattsson, C., & Lidin, S. (2000). Crystal structure of pentastrontium tris(phosphate) bromide, $Sr_5(PO_4)_3Br$ and of pentabarium tris(phosphate) bromide $Ba_5(PO_4)_3Br$, two bromoapatites. *Zeitschrift Fur Kristallographie-New Crystal Structures, 215*(3), 345-346.

Argyle, J. F., & Hummel, F. A. (1963). Dilatometric and X-Ray Data for Lead Compounds, II. *Journal of The American Ceramic Society, 46*(1), 10-14.

Audubert, F., Savariault, J. M., & Lacout, J. L. (1999). Pentalead tris(vanadate) iodide, a defect vanadinite-type compound. *Acta Crystallographica Section C-Crystal Structure Communications, 55*, 271-273.

Baikie, T., Mercier, P. H. J., Elcombe, M. M., Kim, J. Y., Le Page, Y., Mitchell, L. D., et al. (2007). Triclinic apatites. *Acta Crystallographica Section B-Structural Science, 63*, 251-256.

Baikie, T., Pramana, S. S., Ferraris, C., Huang, Y., Kendrick, E., Knight, K. S., et al. (2010). Polysomatic apatites. *Acta Crystallographica Section B-Structural Science, 66*, 1-16.

Belokoneva, E. L., Troneva, E. A., Demyanets, L. N., Duderov, N. G., & Belov, N. V. (1982). The crystal-structure of synthetic $Pb_5(PO_4)_3F$ fluor-pyromorphite. *Kristallografiya, 27*(4), 793-794.

Belousov, R. I., & Filatov, S. K. (2007). Algorithm for calculating the thermal expansion tensor and constructing the thermal expansion diagram for crystals. *Glass Physics and Chemistry, 33*(3), 271-275.

Best, S. M., Porter, A. E., Thian, E. S., & Huang, J. (2008). Bioceramics: Past, present and for the future. *Journal of the European Ceramic Society, 28*(7), 1319-1327.

Brenner, P., Engel, G., & Wondrats. H. (1970). Lead apatite with a superstructure. *Zeitschrift Fur Kristallographie Kristallgeometrie Kristallphysik Kristallchemie, 131*(3), 206-&.

Bubnova, R. S., & Filatov, S. K. (2008). *High-temperature crystal chemistry of borates and borosilicates*. St. Petersburg: "Science".

Bubnova, R. S., Firsova, V. A., & Filatov, S. K. (2013). Software for determining the thermal expansion tensor and the graphic representation of its characteristic surface (theta to tensor-TTT). *Glass Physics and Chemistry, 39*(3), 347-350.

Chen, X., Zhou, H. D., Kiswandhi, A., Miotkowski, I., Chen, Y. P., Sharma, P. A., et al. (2011). Thermal expansion coefficients of Bi_2Se_3 and Sb_2Te_3 crystals from 10 K to 270 K. *Applied Physics Letters, 99*(26).

Chernorukov, N. G., Knyazev, A. V., & Bulanov, E. N. (2010). Isomorphism and Phase Diagram of the $Pb_5(PO_4)_3Cl$-$Pb_5(VO_4)_3Cl$ System. *Russian Journal of Inorganic Chemistry, 55*(9), 1463-1470.

Chernorukov, N. G., Knyazev, A. V., & Bulanov, E. N. (2011). Phase Transitions and Thermal Expansion of Apatite-Structured Compounds. *Inorganic Materials, 47*(2), 172-177.

Comodi, P., Liu, Y., Zanazzi, P. F., & Montagnoli, M. (2001). Structural and vibrational behaviour of fluorapatite with pressure. Part I: in situ single-crystal X-ray diffraction investigation. *Physics and Chemistry of Minerals, 28*(4), 219-224.

Conca, J. L., & Wright, J. (2006). An apatite II permeable reactive barrier to remediate groundwater containing Zn, Pb and Cd. *Applied Geochemistry, 21*(12), 2187-2200.

Dai, Y., & Hughes, J. M. (1989). Crystal-structure refinements of vanadinite and pyromorphite. *Canadian Mineralogist CAMIA., 27*, 189-192.

Dai, Y. S., Hughes, J. M., & Moore, P. B. (1991). The crystal-structures of mimetite and clinomimetite $Pb_5(AsO_4)_3Cl$. *Canadian Mineralogist, 29*, 369-376.

Database of Ionic Radii. (1998).

Duan, C. J., Wu, X. Y., Liu, W., Chen, H. H., Yang, X. X., & Zhao, J. T. (2005). X-ray excited luminescent properties of apatitic compounds $Ba_5(PO_4)_3X$ (X : OH^-, Cl^-, Br^-); structure and hydroxyl ion conductivity of barium hydroxylapatite. *Journal of Alloys and Compounds, 396*(1-2), 86-91.

Elliott, J. C. *Structure and chemistry of the apatites and other calcium orthophosphates*.

Elliott, J. C., Mackie, P. E., & Young, R. A. (1973). Monoclinic hydroxyapatite. *Science, 180*(4090), 1055-1057.

Engel, G. (1978). Fluor-berylates with apatite structure and their relation to sulfates and silicates. *Materials Research Bulletin, 13*(1), 43-48.

Fayos, J., Watkin, D. J., & Perezmendez, M. (1987). Crystal-structure of the apatite-like compound $K_3Ca_2(SO_4)_3F$. *American Mineralogist, 72*(1-2), 209-212.

Filatov, S. K. (1985). Crystal-chemical principles and laws of thermal deformations. *Reports of USSR Academy of Science, 280*(2), 369-372.

Filatov, S. K. (1990). *High-temperature crystal chemistry. Theory, methods and research results*. Leningrad: "Subsoils".

Filatov, S. K. (2008). Negative linear thermal expansion of oblique-angle (monoclinic and triclinic) crystals as a common case. *Physica Status Solidi B-Basic Solid State Physics, 245*(11), 2490-2496.

Filatov, S. K. (2011). General concept of increasing crystal symmetry with an increase in temperature. *Crystallography Reports, 56*(6), 953-961.

Filatov, S. K., & Hazen, R. M. (1994). Composition, structure and properties of mineral matter. In A. S. Marphunin (Ed.), (pp. 76-90): Springer-Verlag.

Fischer, G. R., Bardhan, P., & Geiger, J. E. (1983). The lattice thermal-expansion of hydroxyapatite. *Journal of Materials Science Letters, 2*(10), 577-578.

Fukuda, K., Asaka, T., & Uchida, T. (2012). Thermal expansion of lanthanum silicate oxyapatite ($La_{9.33+2x}(SiO_4)_6O_{2+3x}$), lanthanum oxyorthosilicate (La_2SiO_5) and lanthanum sorosilicate ($La_2Si_2O_7$). *Journal of Solid State Chemistry, 194*, 157-161.

Hashimoto, H., & Matsumoto, T. (1998). Structure refinements of two natural pyromorphites, $Pb_5(PO_4)_3Cl$, and crystal chemistry of chlorapatite group, $M_5(PO_4)_3Cl$. *Zeitschrift Fur Kristallographie, 213*(11), 585-590.

He, Q., Liu, X., Hu, X., Deng, L., Chen, Z., Li, B., et al. (2012). Solid solutions between lead fluorapatite and lead fluorvanadate apatite: compressibility determined by using a diamond-anvil cell coupled with synchrotron X-ray diffraction. *Physics and Chemistry of Minerals, 39*(3), 219-226.

He, Q., Liu, X., Li, B., Deng, L., Chen, Z., Liu, X., et al. (2013). Expansivity and compressibility of strontium fluorapatite and barium fluorapatite determined by in situ X-ray diffraction at high-T/P conditions: significance of the M-site cations. *Physics and Chemistry of Minerals, 40*(4), 349-360.

Hendricks, S. B., Jefferson, S. B., & Mosley, V. M. (1932). The crystal structures of some natural and synthetic apatite-like substances. *Zeitschrift fuer Kristallographie, Kristallgeometrie, Kristallphysik, Kristallchemie, 81*, 352-369.

Herdtweck, E. (1991). Structure of decastrontium hexachromate(V) difluoride. *Acta Crystallographica Section C-Crystal Structure Communications, 47*, 1711-1712.

Hovis, G. L., Morabito, J. R., Spooner, A., Mott, A., Person, E. L., Henderson, C. M. B., et al. (2008). A simple predictive model for the thermal expansion of $AlSi_3$ feldspars. *American Mineralogist, 93*(10), 1568-1573.

International Tables for Crystallography. (2003). (Vol. D. Physical properties of crystal). Dordrecht/Boston/London: Kluwer Academic Publishers.

Klement, R. (1939). Natrium-Calcium-sulfatapatit $Na_6Ca_4(SO_4)_6F_2$. *Naturwissenschaften*, 568-570.

Knyazev, A. V. (2013). *Annual report by the RFBR grant № 13-03-00152*: Lobachevsky State University of Nizhni Novgorod.

Knyazev, A. V., Bulanov, E. N., Korshunov, A. O., & Krasheninnikova, O. V. (2013). Synthesis and thermal expansion of some lanthanide-containing apatites. *Inorganic Materials, 49*(11), 1133-1137.

Knyazev, A. V., Chernorukov, N. G., & Bulanov, E. N. (2011a). Isomorphism and phase diagram of $Pb_5(PO_4)_3F$-$Pb_5(PO_4)_3Cl$ system. *Thermochimica Acta, 513*(1-2), 112-118.

Knyazev, A. V., Chernorukov, N. G., & Bulanov, E. N. (2011b). Phase diagram of apatite system $Ca_{10}(PO_4)_6Cl_2$-$Pb_{10}(PO_4)_6Cl_2$. *Thermochimica Acta, 526*(1-2), 72-77.

Knyazev, A. V., Chernorukov, N. G., & Bulanov, E. N. (2012). Apatite-structured compounds: Synthesis and high-temperature investigation. *Materials Chemistry and Physics, 132*(2-3), 773-781.

Knyazev, A. V., Komshina, M. E., Zhidkov, A. V., & Plesovskikh, A. S. (2013). Crystal structure and thermal expansion of $RbNaCaTh(Si_8O_{20})$. *Russian Journal of Inorganic Chemistry, 58*(10), 1172-1176.

Knyazev, A. V., Maczka, M., Ladenkov, I. V., Bulanov, E. N., & Ptak, M. (2012). Crystal structure, spectroscopy, and thermal expansion of compounds in M_2O-Al_2O_3-TiO_2 system. *Journal of Solid State Chemistry, 196*, 110-118.

Kovba, L. M., & Trunov, V. K. (1976). *X-ray analysis*. Moscow: Moscow State University.

Laghzizil, A., Bouhaouss, A., Barboux, P., Morineau, R., & Livage, J. (1993). Mixed ionic conductivities in sodium fluoroapatites. *Solid State Ionics, 67*(1-2), 137-143.

Li, Y. P., Krivovichev, S. V., & Burns, P. C. (2000). The crystal structure of thornasite, $Na_{12}Th_3\ Si_8O_{19}\cdot 4H_2O$: A novel interrupted silicate framework. *American Mineralogist, 85*(10), 1521-1525.

Masaoka, M., & Kyono, A. (2006). Single crystal growth of lead vanado-chlorapatite $Pb_5(VO_4)_3Cl$ using CsCl flux method. *Materials Letters, 60*(29-30), 3922-3926.

Masaoka, M., Kyono, A., Hatta, T., & Kimata, M. (2006). Single crystal growth of $Pb_5(P_xAs_{1-x}O_4)_3Cl$ solid solution with apatite type structure. *Journal of Crystal Growth, 292*(1), 129-135.

Mathew, M., Mayer, I., Dickens, B., & Schroeder, L. W. (1979). Substitution in barium-fluoride apatite - crystal-structures of $Ba_{10}(PO_4)_6F_2$, $Ba_6La_2Na_2(PO_4)_6F_2$ and $Ba_4Nd_3Na_2(PO_4)_6F_2$. *Journal of Solid State Chemistry, 28*(1), 79-95.

McConnell, D. (1973). *Apatite: Its Crystal Chemistry, Mineralogy, Utilization, and Geologic and Biologic Occurrences.* Wein: Springer-Verlag.

Miyazaki, H., Ushiroda, I., Itomura, D., Hirashita, T., Adachi, N., & Ota, T. (2009). Thermal expansion of hydroxyapatite between-100 degrees C and 50 degrees C. *Materials Science & Engineering C-Biomimetic and Supramolecular Systems, 29*(4), 1463-1466.

Nagai, M., & Nishino, T. (1988). Surface conduction of porous hydroxyapatite ceramics at elevated-temperatures. *Solid State Ionics, 28*, 1456-1461.

Nakayama, S., Kageyama, T., Aono, H., & Sadaoka, Y. (1995). Ionic-conductivity of lanthanoid silicates, $Ln_{10}(SiO_4)_6O_3$ (Ln=La, Nd, Sm, Gd, Dy, Y, Ho, Er and Yb). *Journal of Materials Chemistry, 5*(11), 1801-1805.

Nakayama, S., & Sakamoto, M. (1998). Electrical properties of new type high oxide ionic conductor $RE_{10}Si_6O_{27}$ (RE = La, Pr, Nd, Sm, Gd, Dy). *Journal of the European Ceramic Society, 18*(10), 1413-1418.

Náray-Szabó, S. (1930). The Structure of Apatite $(CaF)\cdot Ca_4(PO_4)_3$. *Zeitschrift für Kristallographie, 75*, 387-398.

Okudera, H., Yoshiasa, A., Masubuchi, Y., Higuchi, M., & Kikkawa, S. (2004). Temperature dependence of structural parameters in oxide-ion-conducting $Nd_{9.33}(SiO_4)_6O_2$: single crystal X-ray studies from 295 to 900 K. *Journal of Solid State Chemistry, 177*(12), 4451-4458.

Panneerselvam, G., Krishnan, R. V., Nagarajan, K., & Antony, M. P. (2011). Heat capacity and thermal expansion of samarium titanate. *Materials Letters, 65*(12), 1778-1780.

Pasero, M., Kampf, A. R., Ferraris, C., Pekov, I. V., Rakovan, J., & White, T. J. (2010). Nomenclature of the apatite supergroup minerals. [Article]. *European Journal of Mineralogy, 22*(2), 163-179.

Perret, R., & Bouillet, A. M. (1975). Les apatites-sulfates $Na_3Cd_2(SO_4)_3Cl$ et $Na_3Pb_2(SO_4)_3Cl$. *Bull. Soc. Fr. Mineral. Cryst., 98*, 254-255.

Piotrowski, A., Kahlenberg, V., & Fischer, R. X. (2002). The solid solution series of the sulfate apatite system $Na_{6.45}Ca_{3.55}(SO_4)_6(F_xCl_{1-x})_{1.55}$. *Journal of Solid State Chemistry, 163*(2), 398-405.

Podsiadlo, H. (1990a). Polymorphic transition in the binary-system lead chlorapatite $Pb_{10}(PO_4)_6Cl_2$ - calcium fluorapatite $Ca_{10}(PO_4)_6F_2$. *Journal of Thermal Analysis, 36*(3), 1205-1211.

Podsiadlo, H. (1990b). Polymorphic transition in the binary-system lead fluorapatite $Pb_{10}(PO_4)_6F_2$ - calcium fluorapatite $Ca_{10}(PO_4)_6F_2$.. *Journal of Thermal Analysis, 36*(2), 569-575.

Porai-Koshitz, M. A. (1989). *Basic of structural analisys of chemical compounds*. Moscow: High School.

Pramana, S. S., White, T. J., Schreyer, M. K., Ferraris, C., Slater, P. R., Orera, A., et al. (2009). Pseudomorphic 2A -> 2M -> 2H phase transitions in lanthanum strontium germanate electrolyte apatites. *Dalton Transactions*(39), 8280-8291.

Reinen, D., Lachwa, H., & Allmann, R. (1986). Color and constitution for Mn(V) in tetrahedral oxygen coordination. 1. An electron-paramagnetic-res and ligand-field spectroscopic investigation of Mn(V) in apatite phases and the structure of $Ba_5(MnO_4)_3Cl$. *Zeitschrift Fur Anorganische Und Allgemeine Chemie, 542*(11), 71-88.

Richard, P., & Perrault, G. (1972). Crystal-structure of ekanite from St. Hilaire, Quebec. *Acta Crystallographica Section B-Structural Crystallography and Crystal Chemistry, B 28*(JUL15), 1994-&.

Sansom, J. E. H., Tolchard, J. R., Islam, M. S., Apperley, D., & Slater, P. R. (2006). Solid state Si-29 NMR studies of apatite-type oxide ion conductors. *Journal of Materials Chemistry, 16*(15), 1410-1413.

Schneider, W. (1967). Caracolit, das $Na_3Pb_2(SO_4)_3Cl$ mit Apatitstruktur. *Neues Jahrbuch fuer Mineralogie.*, 284-289.

Sghir, B., Hlil, E. K., Laghzizil, A., Boujrhal, F. Z., El Moursli, R. C., & Fruchart, D. (2009). Structure electronic and ionic conductivity study versus Ca content in $Ca_{10-x}Sr_x(PO_4)_6F_2$ apatites. *Materials Research Bulletin, 44*(7), 1592-1595.

Sudarsan, K. (1977). The structures of some cadmium 'apatites' $Cd_5(MO_4)_3X$. I. Determination of the structures of $Cd_5(VO_4)_3I$, $Cd_5(PO_4)_3Br$, $Cd_5(AsO_4)_3Br$ and $Cd_5(VO_4)_3Br$. In R. A. Young (Ed.) (Vol. 33, pp. 3136-3142). Acta Crysrallographica B.

Sudarsan.K, & Young, R. A. (1974). Structure refinement and random error analysis for strontium chlorapatite $Sr_5(PO_4)_3Cl$. *Acta Crystallographica Section B-Structural Science, B 30*(JUN15), 1381-1386.

Taylor, D. (1988). Thermal-expansion data. 13. Complex oxides with chain, ring and layer structures and the apatites. *British Ceramic Transactions and Journal, 87*(3), 88-95.

Taylor, M., & Ewing, R. C. (1978). Crystal-structures of $ThSiO_4$ polymorphs - huttonite and thorite. *Acta Crystallographica Section B-Structural Science, 34*(APR), 1074-1079.

Tazzoli, V. (1983). The crystal-structure of cesanite, $Ca_{1+x}Na_{4-x}(SO_4)_3(OH)_x \cdot (1-x)H_2O$, a sulfate isotypic to apatite. *Mineralogical Magazine, 47*(342), 59-63.

Tonegawa, T., Ikoma, T., Suetsugu, Y., Igawa, N., Matsushita, Y., Yoshioka, T., et al. (2010). Thermal expansion of type A carbonate apatite. *Materials Science and Engineering B-Advanced Functional Solid-State Materials, 173*(1-3), 171-175.

Uno, M., Shinohara, M., Kurosaki, K., & Yamanaka, S. (2001). Some properties of a lead vanado-iodoapatite $Pb_{10}(VO_4)_6I2$. *Journal of Nuclear Materials, 294*(1-2), 119-122.

Urusov, V. S. (1987). *Theoretical Crystal Chemistry*. Moscow: "Moscow State University".

White, T., Ferraris, C., Kim, J., & Madhavi, S. (2005). Apatite - An adaptive framework structure. In G. Ferraris & S. Merlino (Eds.), *Micro- and Mesoporous Mineral Phases* (Vol. 57, pp. 307-401).

White, T. J., & Dong, Z. L. (2003). Structural derivation and crystal chemistry of apatites. *Acta Crystallographica Section B-Structural Science, 59*, 1-16.

Whitfield, P. S., Le Page, Y., Mercier, P. H. J., & Kim, J. Y. (2007). Least-squares thermal expansion tensor of vanadate and arsenate triclinic apatites derived from laboratory X-ray powder diffraction cell data. *Journal of Applied Crystallography, 40*, 1019-1026.

Wyckoff, R. W. G. (1965). *Crystal Structures. Vol. 3. Inorganic Compounds $R_x(MX_4)_y$, $R_x(M_nX_p)_y$, Hydrates and Ammoniates.* New York: John Wiley and Sons.

Zhang, J., Liang, H., Yu, R., Yuan, H., & Su, Q. (2009). Luminescence of Ce3+-activated chalcogenide apatites $Ca_{10}(PO_4)_6Y$ (Y = S, Se). *Materials Chemistry and Physics, 114*(1), 242-246.

Zhang, M., Maddrell, E. R., Abraitis, P. K., & Salje, E. K. H. (2007). Impact of leach on lead vanado-iodoapatite $Pb_5(VO_4)_3I$: An infrared and Raman spectroscopic study. *Materials Science and Engineering B-Solid State Materials for Advanced Technology, 137*(1-3), 149-155.

In: Apatite
Editors: M. Iafisco and J. M. Delgado-López
ISBN: 978-1-63321-500-9

Chapter 8

NANOCRYSTALLINE APATITES FUNCTIONALIZED WITH MONOCLONAL ANTIBODIES FOR TARGETED CANCER THERAPIES

Maria Prat*[*1], Francesca Oltolina[1], Luca Gregoletto[1], José Manuel Delgado-López[2] and Jaime Gómez-Morales[2]
[1]Dipartimento di Scienze della Salute,
Università del Piemonte Orientale "A. Avogadro", Novara, Italy
[2]Laboratorio de Estudios Cristalográficos, IACT (CSIC-UGR),
Armilla, Granada, Spain

ABSTRACT

Nano-sized drug carriers conjugated with moieties specifically targeting tumor cells are a promising tool in oncology, improving anticancer drug pharmacodynamic and pharmacokinetics, therapeutic index and reducing adverse side effects. Due to their high biocompatibility, physical properties, low cost and easy synthesis, nanocrystalline apatite appears to be an appealing biomaterial for drug delivery, assuring passive targeting effect and active targeting functionalization. Growth factor tyrosine kinase receptors, which may be over-expressed at the surface of tumor cells of several cancers, represent interesting targets for this kind of delivery strategy. In this chapter we show the potential of ternary nanoparticles prepared with bioinspired carbonated apatite nanocrystals as nanocarriers, and coupled with monoclonal antibodies directed against the tyrosine kinase receptor for Hepatocyte Growth Factor (Met) and the chemotherapeutic drug doxorubicin, as a multifunctional platform for such anticancer therapies. Functionalized nanocarriers were found to be immunocompetent, being able to react with the Met receptor solubilized from cancer cells over-expressing it. Confocal microscopy showed selective binding on Met-positive cancer cells and their subsequent internalization. Moreover doxorubicin was released and localized in the nucleus. Ternary nanoparticles displayed a cytotoxic effect against cancer cells. Our final perspective is to obtain smart targeted drug delivery systems and these results indicate that the specific delivery of chemotherapeutics with ternary nanoparticles targeting HGF receptor-positive tumors

* Corresponding Author address: mprat@med.unipmn.it.

may be a promising strategy to enhance anti-cancer drug efficacy with reduced side effects.

Keywords: Nanocrystalline apatites, targeted cancer therapy, monoclonal antibodies, doxorubicin, nanocarriers

INTRODUCTION

Nanotechnology applied to medicine is a wide research field appealing for scientists and for investors because of its great variety of new possibilities ranging from biological discovery to clinical practice [1, 2]. In particular, nano-scaled carriers have revolutionized the drug delivery field, allowing the production of multifunctional diagnostic and therapeutic agents able to overcome biological barriers and reach selectively targeted cells, tissues and organs, while minimizing the exposure of healthy tissues to the probes and drugs [3]. It is beyond dispute that cancer remains one of the major causes of morbidity and mortality in the industrialized countries and its incidence is increasing worldwide even if there have been undeniable advances in the development of therapies in recent years. At present the main techniques used to treat tumours are surgical treatment, radiotherapy and chemotherapy, and other treatments, namely biological therapies and hyperthermia, are under active development. The main problem associated with most cancer therapies is their lack of specificity.

Indeed, most pharmacological approaches to cancer therapy are based on chemotherapeutic substances, which generally exhibit high cytotoxic activities, but poor specificity for the intended biological target. This results because of a systemic distribution of the cytotoxic agents leading to undesired interaction of the antitumor drug with healthy tissues [4, 5]. Moreover, the non-selective action of anticancer drugs can result in substantial toxicity to many normal rapidly dividing cells (e.g. bone marrow cells, hair follicles and epithelial cells in general) [6, 7]. There are thus urgent requirements for new treatments, which should be both highly effective and have low toxic side effects.

Drug delivery systems are expected to reduce the toxic side effects of the anticancer drugs, by providing a physical barrier between the drug and normal tissues, favouring their delivery to the tumor site, thus reducing the administered drug dose and sparing their interaction with normal tissues [8]. Moreover, the ideal drug delivery system should maintain the drug within a desired therapeutic range after a single dose, which might be released in a temporally controlled way [9, 10] besides being released in a spatially controlled context.

In this scenario nanotechnology has provided new tools to develop new strategies for drug delivery in general and cancer therapy in particular. Nanotechnology is a multidisciplinary topic, based on concepts and approaches established in physics, colloidal and polymeric chemistry, pharmaceutics, biomaterials, molecular biology, biophysics, developed from an original hypothesis advanced by Richard Feynman in 1959 [11]. On this basis nano-structured materials acquire new properties not present in the original microstructured corresponding materials. Indeed, when compared to traditional drug delivery systems, nano-based drug delivery systems (size $\leq$ 100-150 nm) have greater potential in a variety of areas, such as multiple targeting functionalization, *in vivo* imaging, combined drug delivery, extended circulation time, and systemic control release. This greatly increased potential and efficiency depend only on the nanosized dimensions of the drug carriers, for this reason called nanoparticles (NPs).

Half-Life and Passive Targeting of Nanoparticles

The size allows the prolonged circulation of NPs in the blood stream escaping the capture from cells of the reticuloendothelial system (RES), a cell family responsible for the innate defense comprising circulating monocytes, bone marrow progenitors and tissue macrophages, which is deputed to the first clearance activity in mammalian organism [12].

The fact that NPs can easily circulate in the blood stream, together with the peculiar structural features of capillaries in the tumor sites, facilitates their selective accumulation within the tumour microenvironment via the so-called enhanced permeability and retention (EPR) effect (Figure 1). This phenomenon, also called passive targeting, is the consequence of the highly disorganized vessel architecture, with leaky endothelium and compromised lymphatic drainage in these sites [13–15].

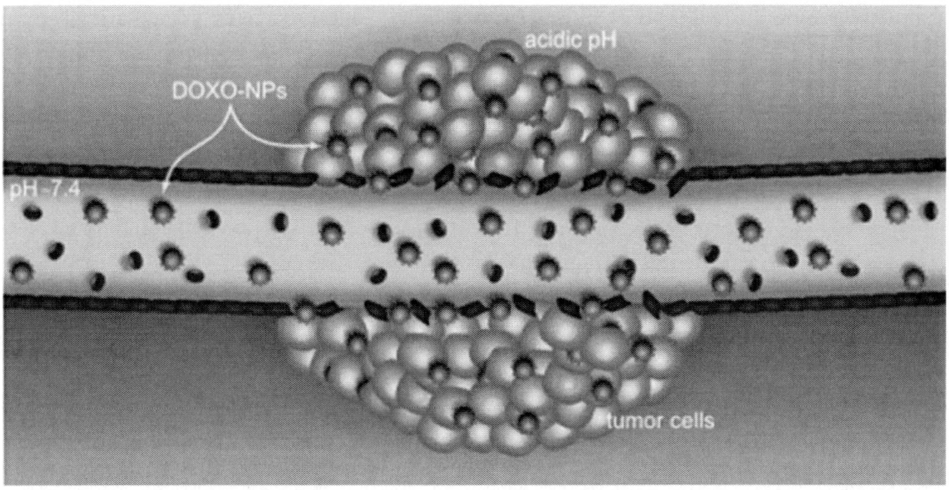

Figure 1. Schematic representation of the use of nanoparticles in passive targeting due to the enhanced permeability and retention (EPR) effect in the tumor region. The extracellular tumor environment and the intracellular lysosome exhibit acidic pH facilitating the selective release of the drug within tumor cells. Reprinted with permission from reference [96]. Copyright (2013) American Chemical Society.

In addition to their size, the surface characteristics of nanoparticles are also an important factor determining their life span and fate during circulation relating to their capture by macrophages. Indeed, unprotected NPs circulating in the blood stream can be recognized as invading agents by plasma proteins, which are then adsorbed on their surfaces and facilitate their endocytosis by the RES (opsonization), resulting in their removal from blood circulation and accumulation in organs with high phagocytic activity, such as liver and spleen [16, 17].

Nanoparticles should ideally have a hydrophilic surface to escape macrophage capture [18]. This can be achieved in different ways: coating the nanoparticles' surface with a hydrophilic polymer, such as polyethylene-glycol (PEG), which protects them from opsonization by repelling plasma proteins. Alternatively, nanoparticles can be obtained from block amphiphilic copolymers with hydrophilic moieties exposed at the surface [19–22]. More recently another strategy was suggested by Tasciotti and coworkers, the so-called biomimetic camouflage of drug-carriers with cellular membranes isolated from freshly harvested leukocytes. Indeed these engineered vectors are able to avoid opsonisation, to delay uptake by the mononuclear phagocyte system, to bind preferentially inflamed endothelium, and to facilitate chemotherapeutics transport across the endothelium while eluding the lysosomal pathway [23]. Particles endowed with these properties have a longer half-life in the bloodstream, i.e. they have low clearance, and a greater chance to reach the targeted tumor tissues.

Nanocrystaline Apatites as Multifunctional Nanocarriers

Nanocrystalline apatite (Ap) possesses ideal features of biocompatibility, because of its structural and chemical similarity with the mineral component of mammalian bones and dentin [24, 25]. In fact, its first and wider applications are in the field of bone repair, due to its well-known properties of osteoinductivity and bioactivity [26, 27]. Being highly similar to a natural body component, it displays high degradability properties, which makes it preferable to other more toxic materials used for nanomedicine such as silica, quantum dots, carbon nanotubes, or metallic magnetic particles. Under this aspect Ap appear to be one of the safest nanomaterial evaluated so far [28].

Ap nanoparticles are easy to prepare, and can be obtained with different tunable morphologies, ranging from spheres to needles, wires and rods on the basis of the conditions used [29]. In the case of carriers projected for targeted drug delivery through vessels the needle-shape seems to perform better, because of its higher propensity to adhere [30] and to be internalized within cells [31, 32]. They can be prepared at relatively low costs and display excellent storage abilities, being resistant to microbial degradation. Another advantage of Ap NPs is their stability in biological fluids, such as plasma, because of their insolubility there; at the same time they dissolve at acidic pH, a condition found in the lysosomal cell compartment and in the tumor microenvironment [33]. The above pH-dependent dissolution is shown in Figure 2, where the percentage of Ca^{2+} ions released from Ap nanoparticles as a function of time at physiological pH 7.4 and at acidic pH 5.0, is reported.

These particular features make this kind of nanoparticles particularly suitable for controlled drug release and moreover their dissolution should avoid undesirable nanoparticle accumulation in cells and tissues, a drawback often encountered with other inorganic and metallic nanosystems [34].

Finally, as already discussed above in general for nanoparticles, also in this case high surface to volume ratios favour their use as multifunctional platforms. They can be derivatized through isothermal adsorption, since they carry both $>Ca^{\delta+}$ and $>PO_x^{\delta-}$ surface groups able to bind many different types of molecules [35–38]. As an example the $>PO_4^{\delta-}$ negative groups can bind efficiently the $-NH_3^+$ groups of the doxorubicin [39]. Historically calcium phosphates found broad application as carrier for nucleic acids in cell transfection experiments, at the beginning for DNA molecules for the expression of exogenous genes. Ap NPs were then used as vehicles for gene therapy [40] and for siRNA molecules designed specifically to silence genes [41], either to prove their role in physiological cell activities [42], or to interfere with them in pathological contexts, such as cancers [43–45]. Similarly

DNAzymes, synthetic single stranded nucleic acid that bind and degrade specifically targeted mRNA could inhibit nasopharyngeal cancer in a tumor xenograft model [46]. Ap NPs are also being used as vehicles for antibiotics [47], gold moieties [48], cancer chemotherapeutics [49, 50], magnetic molecules for cancer hyperthermia therapy [51, 52] and doped with fluorescent lanthanides for *in vivo* imaging applications [53, 54].

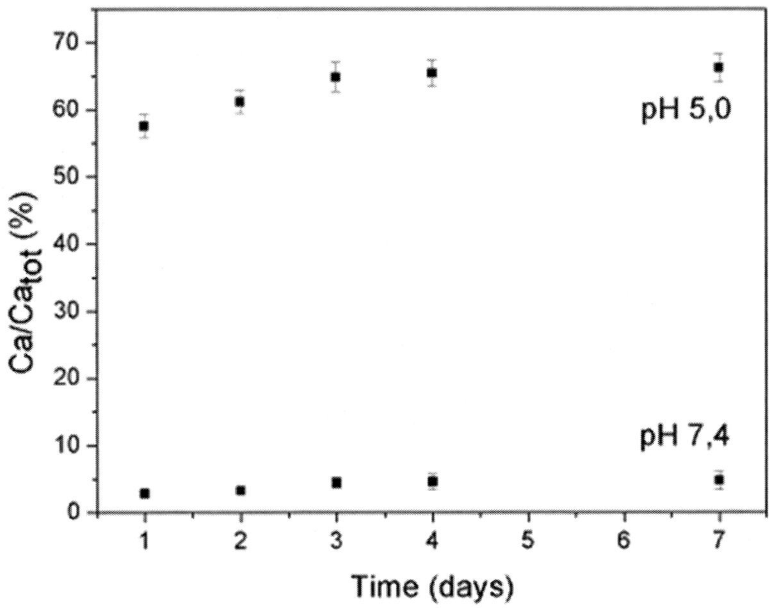

Figure 2. Percentage of calcium ions released from Ap nanoparticles as a function of time at pH 7.4 and 5.0.

Nanoparticles as Multifunctional Platforms (Active Targeting)

The high surface/volume ratio of NPs, consequence of their nanometric size, allows a highly efficient functionalization, i.e. coupling with high amounts of different types of molecules. NPs can thus be viewed as multifunctional platforms. They can be more effective drug carriers, also improving the circulation of drugs otherwise insoluble in the hydrophilic plasma, protecting them from degradation and furthermore avoiding direct interaction with healthy tissues if the drug is carried inside the NPs. Functionalization can also be exploited for the active targeting. In this case NPs are coupled with moieties specifically recognizing tumor cells or their microenvironment, turning closer to reality the concept of "magic bullet" proposed in 1908 by Paul Ehrlich and William Coley [55]. Both passive and active targeting drug nanocarriers should display improved anti-cancer activity (therapeutic index), by increasing the concentration of drug in the primary and in the metastatic tumor, while significantly reducing nonspecific toxicity towards normal cells and thus the severe toxic side effects.

By definition, *tumor specific markers* should be molecules expressed only on tumor cells or sites, but not on their normal counterparts. However, very few, if any, of absolute specificity have been identified. Nonetheless, *tumor associated markers* can be valuable alternatives. These are molecules or signatures expressed preferentially or in significantly higher amounts on tumor cells or in the tumor microenvironment. The same increased permeability of vessels, which indeed can be associated not only to tumor vessels, but also to sites of inflammation, can be exploited to facilitate the extravasion of drug-loaded nanoparticles, which will not be able to cross the endothelial layer lining normal blood vessels [14]. This is a physical trait, but new vessels originating in the tumor sites can be targeted also on the basis of more subtle chemical differences, such as the presence of the $\alpha_V\beta_3$ integrin, which is a recognized useful target for imaging, chemotherapy and anti-angiogenic therapy [56]. Tumor associated vessels have been targeted also for other molecules, such as CXC Chemokine Receptor 4 [57], Neuropilin [58], Ang 1 [59, 60], VEGF receptor [61, 62]. The fact that in most cases tumor vasculature expresses the same markers independently of the cancer type makes this tissue a good target for cancer therapies.

The finding that in some cases tumor associated vessels and tumor cells overexpress the same molecular markers has open to the so-called dual-targeting drug delivery, in which the same carrier targets both endothelial and tumor cells [63]. Indeed cancer-derived endothelial-like cells have been found in several malignancies [64].

Cell surface receptors are another category of tumor-associated markers for growth factors or scatter/migration factors. In fact, their overexpression may be responsible for the uncontrolled cell proliferation accompanying the tumorigenic process. Examples are the Folate Receptor [65, 66], the Human Epidermal Growth Factor Receptor family [67, 68], the Transferrin Receptor [69], the Met/Hepatocyte Growth Factor Receptor [70], the Insulin Receptor [71], the Platelet Derived Growth Factor Receptor [72] and the Cytokines CXCR family [73, 74].

While being functional for the growing tumors and their metastatic progression, these markers may also be targets for multifunctional nanocarriers engineered with specific probes and loaded with chemotherapeutics or moieties for imaging. Growth factor receptors, which are exposed at the cell surface, can be recognized and bound by the cognate probes and can mediate the internalization of the payload [75]. The tumor-associated markers can be recognized and targeted by different types of molecules, which thus act as specific probes and possibly also as therapeutic agents. Monoclonal antibodies (mAbs) are the most used reagents in this context, since by definition they offer endless specificities [76]. Indeed, the use of monoclonal antibodies has revolutionized both cancer therapy and cancer imaging because a whole series of mAbs have been used to target tumor markers, namely growth factor receptors and integrins [77–79]. Full size antibodies are relatively large molecules composed of two heavy chains and two light chains with different functional domains. For this reason, antibody molecules consisting only of the Fab domain (the antigen binding Fragment) obtained by proteolytic digestion of the full size mAb or even of only the two regions of the two polypeptidic chains involved in the variable antigenic binding site obtained by molecular biology engineering have been used [80]. These are the so-called single chain fragment variable (scFv), which penetrate better in the solid tumors due to their reduced size [81]. Under this aspect nanobodies, naturally produced in animals of the Camelidae family and consisting only of the heavy chains are under active study, since they should still reduce the size of the targeting probe [82]. Receptor ligands have been also used; this is exemplified by folate [83, 84].

The use of tumor homing or tumor penetrating peptides (TPP) is another possibility to target tumors and their microenvironment as well as to deliver drugs there [85]. The prototypic peptide of this class is iRGD (CRGDKGPDC), which contains the integrin-binding RGD motif mediating tumor-homing through binding to α_V integrins, present both on tumor cells, but especially on tumor endothelial cells. This strategy is enhanced if a second sequence motif, R/KXXR/K, containing a basic residue is exposed at the C-terminus of the peptide (C-end Rule or CendR). Upon iRGD binding, a proteolytic processing expose the cryptic CendR motif, which can binds to neuropilin-1, activating an endocytic delivery pathway through tumor tissue. Based on this principle other homing or tumor penetrating peptides have been identified by phage library screening in live mice [86].

Another attractive class of tumor targeting molecules is aptamers, a class of oligonucleotides that can specifically bind to different targets with high affinity. These are generated by molecular evolution and selected for their high affinity and specific binding ability to a predefined target [87]. For example they have been used both as chemotherapeutic drugs and as siRNA delivery systems in tumor treatment [88, 89], either directly linked to the drug, thus acting as a carrier or as a ligand, or in aptamer-nanoparticles systems, acting only as ligands [90]. We propose the use of drug-loaded antibody targeted nanoapatites as platforms for targeted delivery strategies against cancer. The preparation and biological responses of nanocrystalline apatites coupled with monoclonal antibodies (mAb) directed against the Met receptor and with the chemotherapeutic drug Doxorubicin (DOXO) are described below.

METHODS

Preparation of Biomimetic Nanocrystalline Apatites

Numerous methods for the preparation of nanocrystalline apatite can be found in the literature [25]. For the results described here, the nanocrystalline apatite were crystallized either in the presence of acetate following the method described by Liou et al. [91] or in the presence of citrate [92]. Briefly, by using the first method, the nanocrystals are precipitated from a solution containing 0.35 M $Ca(CH_3COO)_2$ by the slow addition (1 drop per second) of an aqueous solution of 0.21 M H_3PO_4, keeping the pH at a constant value of 10 by the addition of a NH_4OH solution. The reaction mixture is stirred for 24 h at room temperature, then left stand for 2 hours to allow deposition of the inorganic phase. The latter is isolated by centrifugation, repeatedly washed with water, and freeze-dried at −60 °C under vacuum (3 mbar) overnight. Plate-shaped biomimetic apatite (Ap) nanocrystals with a bulk Ca/P ratio of 1.62, slightly lower than that of stoichiometric hydroxyapatite (1.67), and a limited amount of carbonate groups (about 2% wt), derived from the CO_2 dissolved in the preparation media, have been obtained by using this method [93]. The average length of the nanocrystals, measured by Transmission Electron Microscopy (TEM) (Figure 3A) and Dynamic Light scattering (DLS), is around 35-45 nm. Detailed studies of the FT-IR bands of carbonate ν_3 (1350–1600 cm^{-1}) suggested a B-type carbonate substitution (i.e., carbonate replacing phosphate) in the apatite structure. The nanocrystals exhibited a high specific surface area (SSA), between 150–170 $m^2 g^{-1}$.

The second method, based on thermal decomplexing of Ca-citrate-phosphate solutions, consists basically in mixing equal volumes of two solutions at 4 °C: (i) 0.1 M $CaCl_2$ + 0.4 M

Na_3Cit and (ii) 0.12 M Na_2HPO_4 + x mM Na_2CO_3 (x = 0 or 100). The pH is adjusted to 8.5 with HCl and the mixture is then introduced in a 250 mL round-bottom flask, sealed with a glass stopper and immersed in a water bath at 80ºC. The heating time is approximately 96 h. After the precipitation, the particles are repeatedly washed with ultrapure water (MilliQ, Millipore) by centrifugation, and then dried at 37 ºC.

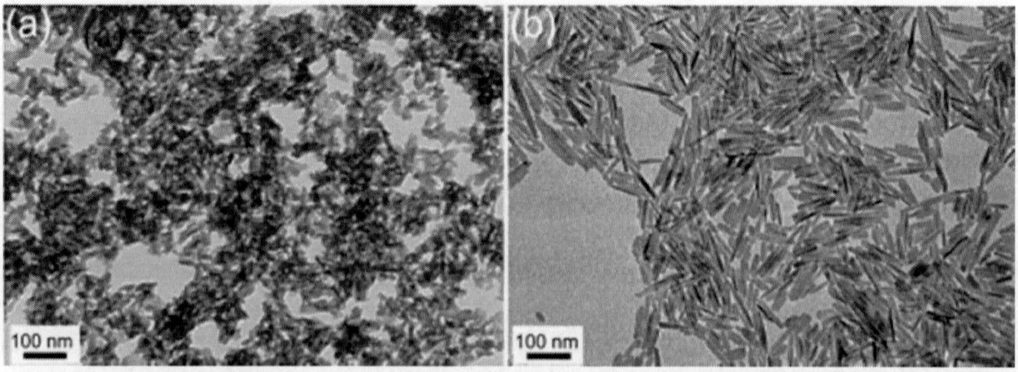

Figure 3. TEM micrographs of Ap nanoparticles prepared by the two different protocols, i.e. in the presence of acetate (a) or citrate (b).

Then the nanocrystals are maturated in the reaction mixture at 80 ºC during 96 hours. After the precipitation, the particles are repeatedly washed with ultrapure water by centrifugation, and then freeze-dried overnight [92]. This method provides platy citrate-coated apatite nanoparticles (cit-Ap). The amount of carbonate incorporated in the crystal lattice can be tailored by the initial sodium carbonate concentration [92]. The morphology and chemical composition, including carbonate content and the presence of adsorbed citrate, are both similar to those of bone nanocrystalline apatites. Their size can be also controlled by both the initial carbonate content and the maturation time. Indeed, the average length as measured by TEM ranges between 45 and 120 nm depending on these latter experimental conditions [92]. Figure 3B represents a TEM image of cit-Ap synthesized in the presence of sodium carbonate. The presence of sodium carbonate in the initial solution produced smaller and more isometric nanoparticles in comparison with those obtained in the absence of sodium carbonate, irrespective of the maturation time [92]. Moreover, the mechanism of formation of this platy-shaped morphology, breaking the typical hexagonal symmetry of apatite, during amorphous-to-crystal transformation in the presence of citrate ions has been recently revealed by using advanced X-ray total scattering techniques [94].

Functionalization of Nanocrystalline Apatites with DOXO and mAb (Binary) or both Molecules (Ternary Nanoparticles)

Nanocrystalline apatites can be easily coupled with DOXO and/or mAb by isothermal adsorption at 37°C [93, 95]. The used mAb is an IgG 2a/k monoclonal antibody produced against the ectodomain of the human Met/HGF receptor [78, 96]. The functionalization was carried out by incubating separately a non-saturating amount of reagents, i.e. 2 mg mL^{-1} of mAb in Hepes buffered saline solution (0.01 M Hepes, 0.15 M NaCl) or 1 mg/mL of DOXO in deionized water, with 5 mg of apatite nanoparticles (so-called binary nanoparticles) (Figure 4A).

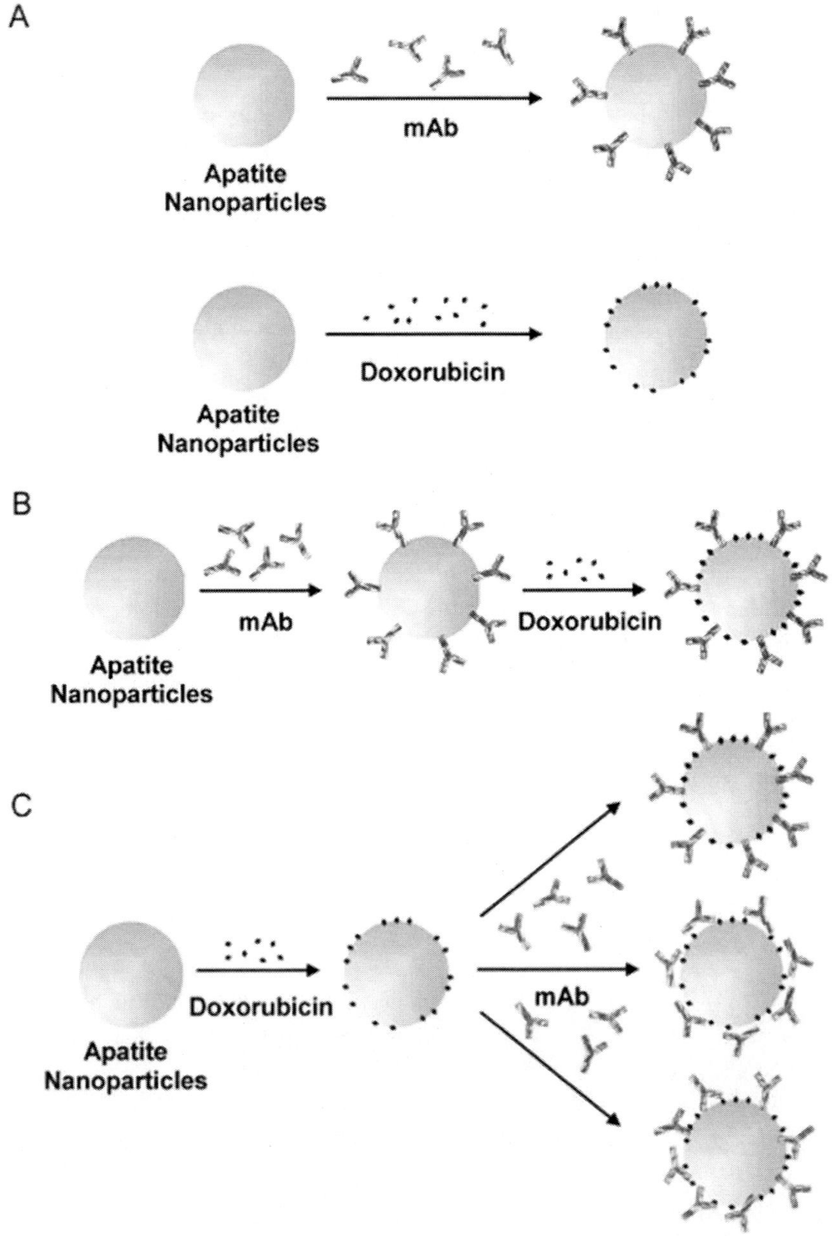

Figure 4. Models of functionalization of Ap nanoparticles with DO-24 mAb or DOXO (binary nanoparticles (A)) and with both reagents (ternary nanoparticles (B-C)) according to the two different methods used in this work. Apatite nanocarriers are represented as spheres for clarity, though they are plate-shaped.

The amounts of adsorbed molecules DOXO and mAb have been calculated from the differences between the concentrations of the corresponding molecule solutions before and after the adsorption on the nanocarriers. The concentration of both molecules is determined by UV-Vis spectroscopy (λ = 495 and 280 nm, ε = 10650 and 64000 $M^{-1}cm^{-1}$, respectively) using standard curves. The solid component is washed twice with ultrapure water, recovered by centrifuging at 10000 rpm (12700 g) for 3 min and then freeze-dried at −60 ° C under vacuum (3 mbar) overnight. In the case of apatites conjugated with both moieties (DOXO and mAb) (the so-called ternary nanoparticles), the same protocols are sequentially applied, each one followed by extensive washings.

Protonated DOXO molecules form dimers in aqueous solutions, being the anti-parallel configuration, with the $-NH_3^+$ groups in opposite directions, the most stable. Based on ζ-potential and dynamic light scattering (DLS) measurements of Ap and Ap functionalized with DOXO in physiological conditions, the most probable configuration of the adsorbed molecules is the one in which the DOXO dimer interacts through its positive $-NH_3^+$ groups with the negative $>PO_x^{\delta-}$ groups of the Ap surface and the resulting binary nano-assemblies display the remaining $-NH_3^+$groups directed toward the surrounding solution (positive ζ-potential of +29 .3 mV) [97]. On the contrary, a ζ-potential of −17.9 mV, measured after coupling Ap with mAb indicates a preferential affinity of the mAb for the positive surface $>Ca^{\delta+}$, therefore mAb must bind to Ap surface through its Fc domain, exposing the Fab fragments toward the surrounding solutions [93].

For ternary nanoparticles different models explain the observed ζ-potential values, depending on the functionalization sequence [93]. When the sequence is Ap-mAb-DOXO (Figure 4B), in the first step, leaded mainly by steric hindrance, mAb adsorbs on Ap through its Fc fragment, displaying the Fab fragment towards the surrounding solution. In the second step, DOXO adsorbs on the remaining free adsorption sites of Ap nanocrystals, displaying positive $-NH_3^+$ residues toward the surrounding solution. When the sequence is Ap-DOXO-mAb (Figure 4C), then different configurations can be responsible of the broad distribution of ζ -potential values found, i.e. mAb can bind directly to Ap through its Fc fragment or to the adsorbed DOXO, either through its Fc or Fab fragments.

pH-Responsive Delivery of mAb and Doxorubicin from Nanocrystalline Apatites

The stability of the adsorbed moieties under physiological conditions was tested by incubating the ternary NPs at 37 °C for different times up to 3 days in a buffer at pH 7.4 (physiological conditions), or in a buffer at pH 5.0, mimicking the acidic environment typical of endosomes/lysosomes. At pH 7.4 no significant release of mAb and DOXO was detected, whereas about the 80% of the bound DOXO, as well as of the bound mAb, were released upon 3 days at pH 5.0 [93]. These values, in combination with data relative to the dissolution of unfunctionalized Ap at the same pHs (Figure 2) demonstrate the integrity of the two types of ternary NPs at pH 7.4, thus suggesting that both can reach the target cells without losing the functional moieties. Once they reach the target, the payload is progressively lost and at the same time the nanocarrier dissolves.

A similar situation is found when the release of DOXO from binary cit-Ap-DOXO nano-assemblies is analyzed in detail (Figure 5). In this case the release of DOXO is found also to be strongly pH dependent, being higher at pH 5.0 than at pH 7.4. Specifically, after 7 days the amount of DOXO released at physiological pH 7.4 was 17% of the total drug amount loaded on cit-Ap. In the acidic media, the drug was released faster, reaching values of about 29 and of the total DOXO loaded [97]. Since the release of DOXO was much higher from Ap crystallized in the presence of acetate (80 % of the bound DOXO) rather than from cit-Ap (29 %), we can foresee that those bearing citrate molecules on its surface shall assure a slowest drug delivery once they have reached the acidic environment.

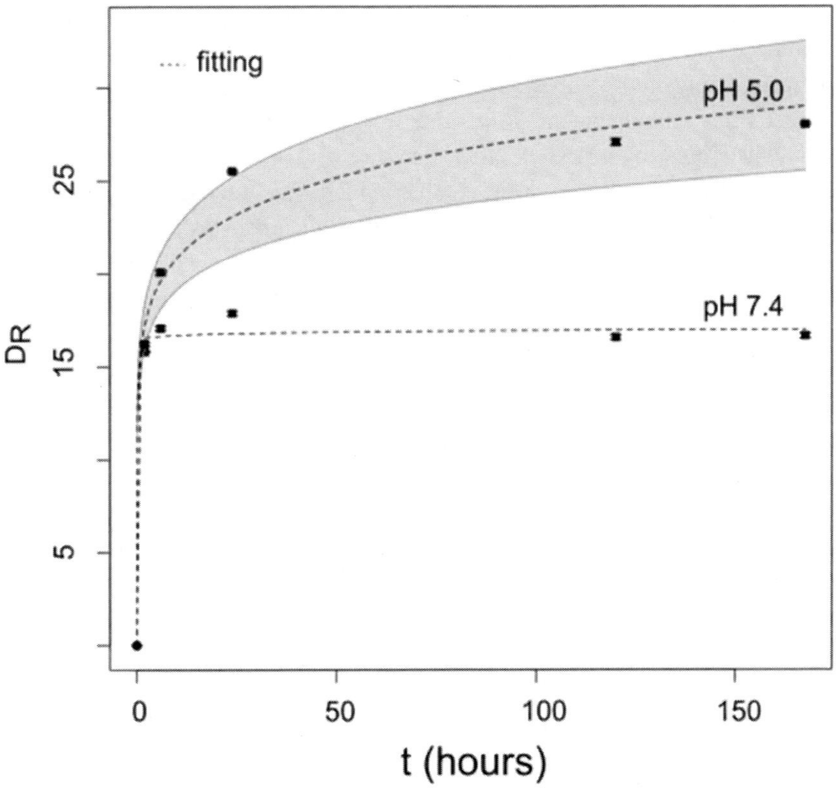

Figure 5. DOXO release from cit-Ap pH 7.4 and pH 5.0. Adapted with permission from reference [96]. Copyright (2013) American Chemical Society.

Targeted Therapies using Nanocrystalline Apatites

Since nanocrystalline apatites are the main mineral component of bone, Ap NPs were used first for bone replacement, augmentation, and/or regeneration, directly even without specific targeting functionalization [26, 98] or after coupling with an osteoinductive growth factor [99]. The fact that the liver is the first organ encountered by microparticulate or nanoparticulate injected in the blood and thus can entrap them, was exploited for delivering there unfunctionalized hydroxyapatite nanoparticles in experimental models of hepatocellular carcinoma therapy [100–102]. The anti-tumor activity, relaying on cell apoptosis, was found strongly dependent on the size of apatite nanoparticles, the ones of medium size (45 nm) being the most effective among the ones tested, which ranged from 26 to 175 nm.

Drug delivery to bone, in view of a possible therapy for osteosarcoma or carcinoma metastasis in bone marrow, could be enhanced by functionalizing cisplatinum loaded Ap with bisphosphonate, due to the high affinity of this moiety for calcium ions and hence for bones

[103]. Moreover, since Ap is a biocompatible material, it has been used as carrier for chemiotherapic agents in colon cancer treatments [104]. In addition, the double functionalization of Ap with Paclitaxel and collagen was shown to be a valid delivery system for metastatic cancer cells [105].

In any case the field of Ap nanoparticles used in cancer therapy is an emerging and promising one, so any innovation is always well accepted in this constantly growing branch.

For example, specific tumor targeting can be achieved by coupling Ap with ligands specifically recognizing tumor markers. Ap were also engineered to obtain tools for simultaneous contrast enhancement for three major molecular imaging techniques, e.g. magnetic resonance imaging (MRI), X-ray imaging and near-infrared (NIR) fluorescence imaging, and specific targeting to tumor cells, which was realized by grafting the cell-membrane receptor ligand folic acid [106]. These multifunctionalized nanoparticles were shown to be fully biocompatible and not toxic *in vitro* on different cell lines up to 500 μg mL^{-1}. This combinatorial strategy could be exploited to produce theranostic tools, in which also the therapeutic function is present.

A monoclonal antibody, called DO-24, directed against the Met/HGF receptor [78] was coupled as targeting moiety on Ap nanoparticles. This antibody was produced upon immunization with human cells overexpressing the Met/HGF receptor and found to recognize an extracellular conformational epitope shared also by the receptor of other species, such as mouse, pig, and rat. The Met/HGF receptor is overexpressed at the surface of many types of cancer cells, and thus can be viewed as a tumor marker. The behaviour of both targeted binary NPs, composed of the Ap and mAb, and of ternary NPs, to which also the Doxorubicin drug was coupled, has been evaluated on cells established from a gastric carcinoma (GTL-16 cells) by confocal microscopy analysis (Figure 6).

Upon incubation at 37°C for 2 hours the binary NPs, which were detected by means of fluorescent antibodies against the mouse mAb anti-Met/HGF receptor, interacted efficiently with GTL-16 cells (Figure 6A), while they did not bind to NIH-3T3 fibroblasts, which do not express the Met/HGF-R (Figure 6B). Upon binding the NPs were internalized in the endosomal-lysosomal compartment, since they co-localized with Lyso-traker, a marker of the lysosomes (Figure 6A). When the interaction of ternary NPs with GTL-16 cells was analysed, a similar localization of the mAb-traceable NPs was observed. Indeed the green fluorescence revealing the mAb associated to the NPs was detected at the cell surface or within the cell in proximity of the cell membrane (Figure 6C). By contrast Doxorubicin was found dispersed in the whole cytoplasm and, moreover, also in the nucleus.

As described before, two types of ternary NPs were prepared, depending on which moiety was first coupled. We found that Ap-mAb-DOXO, in which the mAb was coupled first (Fig 4B), were more efficient than Ap-DOXO-mAb for the different biological properties tested, such as interaction with the surface of target cells, internalization and cytotoxic activity. This higher efficiency can be explained on the basis of more functional exposition of mAb at the nanoparticles surface, as well as of a higher propensity of Doxorucibin to be discharged, since possibly present in a more external position on the nanoparticles.

Upon incubation for 72 hours these NPs exerted cytotoxic activity, by decreasing cell viability to about 30-40% as revealed in a standard tetrazolium assay, displaying a somehow lower activity than free Doxorubicin (Figure 7A). Nanoparticles functionalized only with the antibody or unfunctionalized NPs did not display significant toxicity, reducing cell viability to about only 85-90% (Figure 7B).

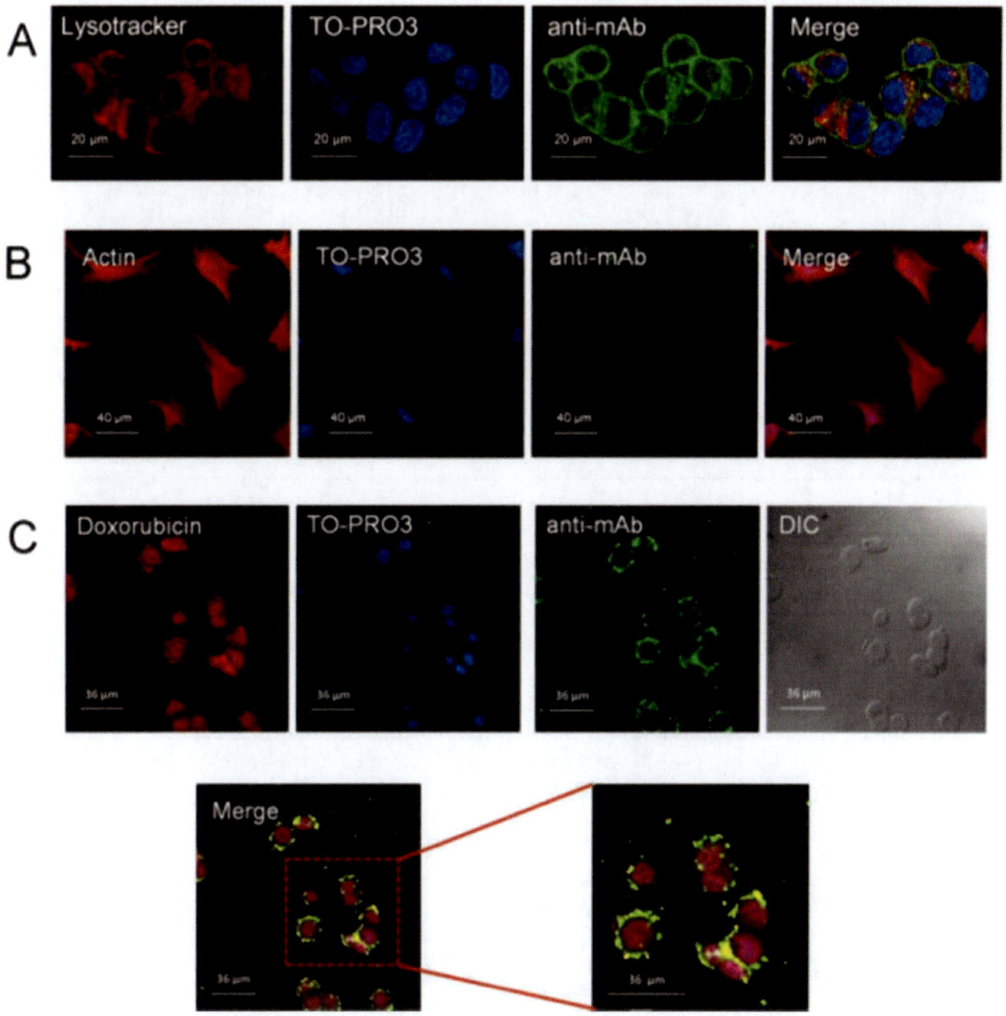

Figure 6. Internalization of the functionalized nanoparticles in GTL-16 and NIH-3T3 cells by confocal microscopy analysis. Met^+ GTL-16 cells were incubated with binary Ap-mAb (A) or ternary Ap-mAb-DOXO (C) nanoparticles at 37 ° C for 3 h, washed, fixed, permeabilized and saturated. Met^- NIH-3T3 cells were incubated with binary Ap-mAb (B) nanoparticles at the same conditions. Nanoparticles are visualized in green (staining with FITC-labeled anti-mouse IgG), nuclei in blue (TO-PRO®-3) and Lysotracker (A), Actin (B) or DOXO (C) in red. Cell morphology was evaluated by Differential Interference Contrast (DIC) microscopy. Reprinted with permission from reference [92]. Copyright (2013) John Wiley & Sons.

These findings demonstrate that Ap NPs represent an efficient multifunctional platform for drug delivery. Indeed they can bind at the cell surface in a specific way, can be internalized within the cell in the endosome/lysosome compartment, where they release Doxorubicin, possibly because of the acidic microenvironment. The drug then diffuses in the cytoplasm and in the nucleus, where it exerts its cytotoxic activity (Figure 8). mAb-related signals were never detected within the nucleus, thus strongly suggesting that also NPs, to which they are bound, were not translocated there and, possibly were recycled to the cell membrane through a receptor-mediated mechanism. Indeed it is acknowledged that nuclear translocation is strictly controlled and meets many requirements [107, 108] and in other drug delivery systems based on NPs, the latter were never detected in this cellular compartment [109].

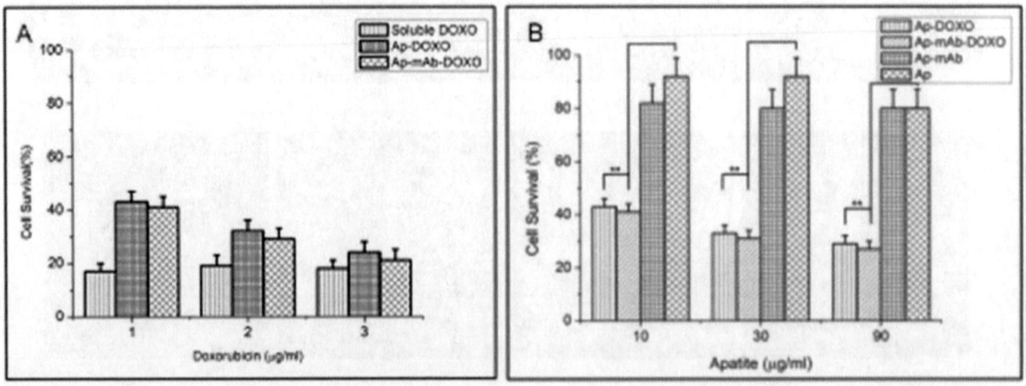

Figure 7. Cytotoxic activity of the different functionalized nanoparticles on GTL-16 cells. Cells were incubated with equimolar amounts of DOXO, either soluble or loaded on nanoparticles (A), as well as with equivalent amounts of Ap nanoparticles (B) for 3 days, before performing the MTT assay. Untreated cells, receiving medium without nanoparticles and DOXO, were taken as reference value (100%) of viable cells to which refer the values of treated cells. Statistical significance: Panel A: ANOVA one-way revealed significant differences ($p < 0.001$). Post-hoc test (Sheffè's test) showed highly significant difference in all two-samples comparisons ($p \leq 0.001$). Panel B: significance is referred to unfunctionalized nanoparticles (** $p \leq 0.001$; unlabelled: no significance). Data are the mean ± standard error of the mean (SEM) of 4 experiments performed in triplicates. Adapted with permission from reference [92]. Copyright (2013) John Wiley & Sons.

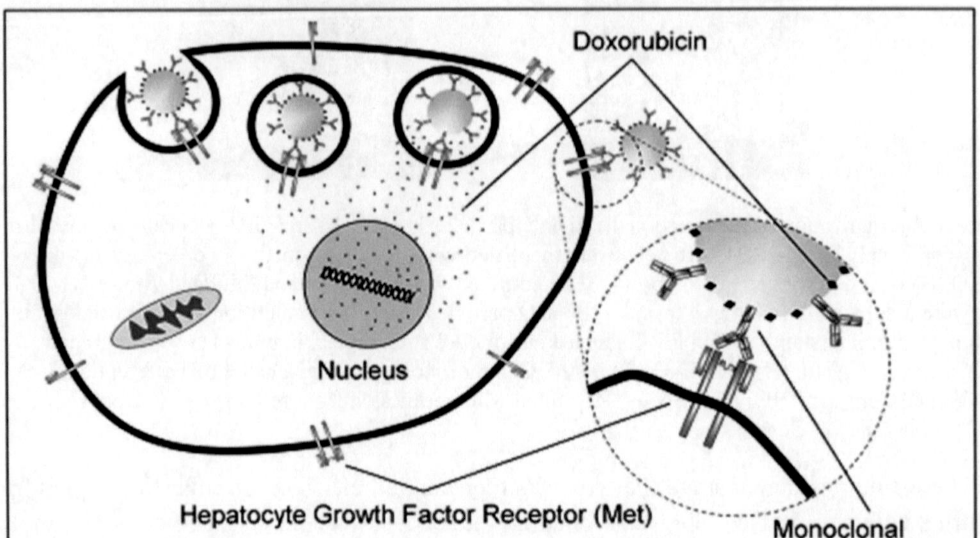

Figure 8. Scheme of functionalized nanoparticles interactions with cells. Ternary apatite nanoparticles specifically bind tumor cells overexpressing the tumor associated marker Met/Hepatocyte Growth Factor receptor and are internalized within the cells. DOXO is then discharged and reaches the nucleus exerting cytotoxicity. This work opens new perspectives in the use of nanocrystalline apatites as new platforms for theranostic applications in nanomedicine.

Conclusion and Future Perspectives

The encouraging data relative to the *in vitro* specific targeting and cytotoxic activity of the functionalized Ap NPs are the required basis to proceed to *in vivo* experiments. Preliminary assays with these NPs, either unfunctionalized or covalently labeled with a luminescent probe, were found to be not toxic to mice of different strains (BALB/c, C57/B6, CD1) when injected intravein up to 500 μg per animal. The animals were checked for up to 2 months after the administration of the NPs, they were fine and they did not lose weight. After cervical dislocation organs were collected and analysed histologically. No alterations were detected relative to organs of untreated healthy animals of the same age.

Labeling allowed us to follow the biodistribution of NPs, which were detectable soon after injection mainly in the liver, and later also in the spleen and sometime later also in kidneys. Within two days no signal was detectable, strongly suggesting that most of the injected NPs were eliminated from the animal body. This result supports the high level of biocompatibility and biodegradability of Ap NPs and highlights their potential as multifunctional platform carriers. It is expected that once injected in the body the targeted Ap NPs will bind to tumor cells, which express a higher level of the recognized Met/HGF receptor. Indeed this receptor is present at low levels also on normal healthy cells. This differential reactivity has already been reported in the case of a mAb against HER2, a membrane orphan receptor whose expression is highly increased in a category of breast cancers [110] and it is now approved by FDA. In this case the mAb itself is toxic to the cells.

Finally, the targeting activity of the functionalized NPs discussed above could be improved by coupling NPs with other functional groups targeting other tumor markers, as already reported in the case of the so called dual-targeting NPs [111] prepared with different carriers.

Acknowledgments

Supported by the Project "Multifunctional drug-loaded antibody-targeted inorganic nanoparticles for biomedical use" (Azioni Integrate Italia-Spagna, IT10M4FC89) and Progetto Ricerca Sanitaria Finalizzata Regione Piemonte *"Nanoparticelle multifunzionali per la terapia dei carcinomi: studi in vitro e in vivo in modelli sperimentali"*– (30258/DB2001). Projects from Spanish MINECO co-funded with FEDER (Projects MAT2011-28543 and Factoría de Cristalización, Consolider-Ingenio 2010) and International Campus of Excellence (CEI-BioTic, University of Granada) are also acknowledged.

References

[1] Nijhara, R., and Balakrishnan, K. 2006. Bringing nanomedicines to market: regulatory challenges, opportunities, and uncertainties. *Nanomedicine 2*, 127–36.

[2] Doane, T. L., and Burda, C. 2012. The unique role of nanoparticles in nanomedicine: imaging, drug delivery and therapy. *Chem. Soc. Rev. 41*, 2885–911.

[3] Bhaskar, S., Tian, F., Stoeger, T., Kreyling, W., de la Fuente, J. M., Grazú, V., Borm, P., Estrada, G., Ntziachristos, V., and Razansky, D. 2010. Multifunctional Nanocarriers for diagnostics, drug delivery and targeted treatment across blood-brain barrier: perspectives on tracking and neuroimaging. *Part. Fibre Toxicol. 7*, 3.

[4] Dobson, J. 2006. Magnetic micro- and nano-particle-based targeting for drug and gene delivery. *Nanomedicine (Lond). 1*, 31–7.

[5] Sun, C., Lee, J. S. H., and Zhang, M. 2008. Magnetic nanoparticles in MR imaging and drug delivery. *Adv. Drug Deliv. Rev. 60*, 1252–65.

[6] Malam, Y., Loizidou, M., and Seifalian, A. M. 2009. Liposomes and nanoparticles: nanosized vehicles for drug delivery in cancer. *Trends Pharmacol. Sci. 30*, 592–9.

[7] Sanna, V., Pala, N., and Sechi, M. 2014. Targeted therapy using nanotechnology: focus on cancer. *Int. J. Nanomedicine 9*, 467–483.

[8] Shahin, M., Soudy, R., Aliabadi, H. M., Kneteman, N., Kaur, K., and Lavasanifar, A. 2013. Engineered breast tumor targeting peptide ligand modified liposomal doxorubicin and the effect of peptide density on anticancer activity. *Biomaterials 34*, 4089–97.

[9] Langer, R. 2003. Where a pill won't reach. *Sci. Am. 288*, 50–7.

[10] Nicolas, J., Mura, S., Brambilla, D., Mackiewicz, N., and Couvreur, P. 2013. Design, functionalization strategies and biomedical applications of targeted biodegradable/biocompatible polymer-based nanocarriers for drug delivery. *Chem. Soc. Rev. 42*, 1147–235.

[11] Feynman, R. 1959. There's plenty of room at the bottom. Available at: http://www.zyvex.com/nanotech/feynman.html.

[12] Hume, D. A., Ross, I. L., Himes, S. R., Sasmono, R. T., Wells, C. A., and Ravasi, T. 2002. The mononuclear phagocyte system revisited. *J. Leukoc. Biol. 72*, 621–7.

[13] Pietronave, S., Iafisco, M., Locarno, D., Rimondini, L., and Maria Prat, M. 2009. Functionalized nanomaterials for diagnosis and therapy of cancer. *J. Appl. Biomater. Biomech. 7*, 77–89.

[14] Maeda, H., Nakamura, H., and Fang, J. 2013. The EPR effect for macromolecular drug delivery to solid tumors: Improvement of tumor uptake, lowering of systemic toxicity, and distinct tumor imaging in vivo. *Adv. Drug Deliv. Rev. 65*, 71–9.

[15] Kunjachan, S., Pola, R., Gremse, F., Theek, B., Ehling, J., Moeckel, D., Hermanns-Sachweh, B., Pechar, M., Ulbrich, K., Hennink, W. E., et al. 2014. Passive versus Active Tumor Targeting Using RGD- and NGR-Modified Polymeric Nanomedicines. *Nano Lett. 14*, 972–81.

[16] Moghimi, S. M., Hunter, A. C., and Murray, J. C. 2001. Long-circulating and target-specific nanoparticles: theory to practice. *Pharmacol. Rev. 53*, 283–318.

[17] Colombo, M., Carregal-Romero, S., Casula, M. F., Gutiérrez, L., Morales, M. P., Böhm, I. B., Heverhagen, J. T., Prosperi, D., and Parak, W. J. 2012. Biological applications of magnetic nanoparticles. *Chem. Soc. Rev. 41*, 4306–34.

[18] Moghimi, S. M., and Szebeni, J. 2003. Stealth liposomes and long circulating nanoparticles: critical issues in pharmacokinetics, opsonization and protein-binding properties. *Prog. Lipid Res. 42*, 463–78.

[19] Adams, M. L., Lavasanifar, A., and Kwon, G. S. 2003. Amphiphilic block copolymers for drug delivery. *J. Pharm. Sci. 92*, 1343–55.

[20] Kobayashi, K., Niikura, K., Takeuchi, C., Sekiguchi, S., Ninomiya, T., Hagiwara, K., Mitomo, H., Ito, Y., Osada, Y., and Ijiro, K. 2014. Enhanced cellular uptake of amphiphilic gold nanoparticles with ester functionality. *Chem. Commun. (Camb). 50*, 1265–7.

[21] Yang, B., Cai, H., Qin, W., Zhang, B., Zhai, C., Jiang, B., and Wu, Y. 2013. Bcl-2-functionalized ultrasmall superparamagnetic iron oxide nanoparticles coated with amphiphilic polymer enhance the labeling efficiency of islets for detection by magnetic resonance imaging. *Int. J. Nanomedicine 8*, 3977–90.

[22] Zhang, Y., Yin, Q., Lu, H., Xia, H., Lin, Y., and Cheng, J. 2013. PEG-Polypeptide Dual Brush Block Copolymers: Synthesis and Application in Nanoparticle Surface PEGylation. *ACS Macro Lett. 2*, 809–813.

[23] Parodi, A., Quattrocchi, N., van de Ven, A. L., Chiappini, C., Evangelopoulos, M., Martinez, J. O., Brown, B. S., Khaled, S. Z., Yazdi, I. K., Enzo, M. V., et al. 2013. Synthetic nanoparticles functionalized with biomimetic leukocyte membranes possess cell-like functions. *Nat. Nanotechnol. 8*, 61–8.

[24] Roveri, N., Palazzo, B., and Iafisco, M. 2008. The role of biomimetism in developing nanostructured inorganic matrices for drug delivery. *Expert Opin. Drug Deliv. 5*, 861–77.

[25] Gómez-Morales, J., Iafisco, M., Delgado-López, J. M., Sarda, S., and Drouet, C. 2013. Progress on the preparation of nanocrystalline apatites and surface characterization: Overview of fundamental and applied aspects. *Prog. Cryst. Growth Charact. Mater. 59*, 1–46.

[26] Bose, S., and Tarafder, S. 2012. Calcium phosphate ceramic systems in growth factor and drug delivery for bone tissue engineering: a review. *Acta Biomater. 8*, 1401–21.

[27] Vallet-Regí, M., and Ruiz-Hernández, E. 2011. Bioceramics: from bone regeneration to cancer nanomedicine. *Adv. Mater. 23*, 5177–218.

[28] Singh, R., Pantarotto, D., Lacerda, L., Pastorin, G., Klumpp, C., Prato, M., Bianco, A., and Kostarelos, K. 2006. Tissue biodistribution and blood clearance rates of intravenously administered carbon nanotube radiotracers. *Proc. Natl. Acad. Sci. U. S. A. 103*, 3357–62.

[29] Uskoković, V., and Uskoković, D. P. 2011. Nanosized hydroxyapatite and other calcium phosphates: chemistry of formation and application as drug and gene delivery agents. *J. Biomed. Mater. Res. B. Appl. Biomater. 96*, 152–91.

[30] Fisher, K. D., and Seymour, L. W. 2010. HPMA copolymers for masking and retargeting of therapeutic viruses. *Adv. Drug Deliv. Rev. 62*, 240–5.

[31] Faraji, A. H., and Wipf, P. 2009. Nanoparticles in cellular drug delivery. *Bioorg. Med. Chem. 17*, 2950–62.

[32] Albanese, A., Tang, P. S., and Chan, W. C. W. 2012. The effect of nanoparticle size, shape, and surface chemistry on biological systems. *Annu. Rev. Biomed. Eng. 14*, 1–16.

[33] Kester, M., Heakal, Y., Sharma, A., Robertson, G. P., Morgan, T. T., Altinoğlu, E. İ., Tabaković, A., Parette, M. R., Rouse, S., Ruiz-Velasco, V., et al. 2008. Calcium phosphate nanocomposite particles for in vitro imaging and encapsulated chemotherapeutic drug delivery to cancer cells. *Nano Lett. 8*, 4373–9.

[34] Lacerda, L., Bianco, A., Prato, M., and Kostarelos, K. 2006. Carbon nanotubes as nanomedicines: from toxicology to pharmacology. *Adv. Drug Deliv. Rev. 58*, 1460–70.

[35] Iafisco, M., Varoni, E., Di Foggia, M., Pietronave, S., Fini, M., Roveri, N., Rimondini, L., and Prat, M. 2012. Conjugation of hydroxyapatite nanocrystals with human immunoglobulin G for nanomedical applications. *Colloids Surf. B. Biointerfaces 90*, 1–7.

[36] Iafisco, M., Di Foggia, M., Bonora, S., Prat, M., and Roveri, N. 2011. Adsorption and spectroscopic characterization of lactoferrin on hydroxyapatite nanocrystals. *Dalton Trans. 40*, 820–7.

[37] Iafisco, M., Sabatino, P., Lesci, I. G., Prat, M., Rimondini, L., and Roveri, N. 2010. Conformational modifications of serum albumins adsorbed on different kinds of biomimetic hydroxyapatite nanocrystals. *Colloids Surf. B. Biointerfaces 81*, 274–84.

[38] Boix, T., Gómez-Morales, J., Torrent-Burgués, J., Monfort, A., Puigdomènech, P., and Rodríguez-Clemente, R. 2005. Adsorption of recombinant human bone morphogenetic protein rhBMP-2m onto hydroxyapatite. *J. Inorg. Biochem. 99*, 1043–50.

[39] Delgado-López, J. M., Iafisco, M., Rodríguez-Ruiz, I., and Gómez-Morales, J. 2013. Bio-inspired citrate-functionalized apatite thin films crystallized on Ti-6Al-4V implants pre-coated with corrosion resistant layers. *J. Inorg. Biochem. 127*, 261–8.

[40] Do, T. N. T., Lee, W.-H., Loo, C.-Y., Zavgorodniy, A. V, and Rohanizadeh, R. 2012. Hydroxyapatite nanoparticles as vectors for gene delivery. *Ther. Deliv. 3*, 623–32.

[41] Asai, T., and Oku, N. 2014. Systemic delivery of small RNA using lipid nanoparticles. *Biol. Pharm. Bull. 37*, 201–5.

[42] Hossain, S., Stanislaus, A., Chua, M. J., Tada, S., Tagawa, Y., Chowdhury, E. H., and Akaike, T. 2010. Carbonate apatite-facilitated intracellularly delivered siRNA for efficient knockdown of functional genes. *J. Control. Release 147*, 101–8.

[43] Liang, Z.-W., Guo, B.-F., Li, Y., Li, X.-J., Li, X., Zhao, L.-J., Gao, L.-F., Yu, H., Zhao, X.-J., Zhang, L., et al. 2011. Plasmid-based Stat3 siRNA delivered by hydroxyapatite nanoparticles suppresses mouse prostate tumour growth in vivo. *Asian J. Androl. 13*, 481–6.

[44] Tan, Y. F., Mundargi, R. C., Chen, M. H. A., Lessig, J., Neu, B., Venkatraman, S. S., and Wong, T. T. 2014. Layer-by-Layer Nanoparticles as an Efficient siRNA Delivery Vehicle for SPARC Silencing. *Small*.

[45] Chu, S.-H., Zhou, Z.-M., Feng, D.-F., and Ma, Y.-B. 2014. Inhibition of human glioma U251 cells growth in vitro and in vivo by hydroxyapatite nanoparticle-assisted delivery of short hairpin RNAs against SATB1. *Mol. Biol. Rep. 41*, 977–86.

[46] Chen, Y., Yang, L., Huang, S., Li, Z., Zhang, L., He, J., Xu, Z., Liu, L., Cao, Y., and Sun, L. 2013. Delivery system for DNAzymes using arginine-modified hydroxyapatite nanoparticles for therapeutic application in a nasopharyngeal carcinoma model. *Int. J. Nanomedicine 8*, 3107–18.

[47] Uskoković, V., and Desai, T. a 2013. Phase composition control of calcium phosphate nanoparticles for tunable drug delivery kinetics and treatment of osteomyelitis. I. Preparation and drug release. *J. Biomed. Mater. Res. A 101*, 1416–26.

[48] Huang, Y.-T., Yamauchi, Y., Lai, C.-W., and Chen, W.-J. 2013. Evaluating the antibacterial property of gold-coated hydroxyapatite: A molecular biological approach. *J. Hazard. Mater.*

[49] Luo, Y., Ling, Y., Guo, W., Pang, J., Liu, W., Fang, Y., Wen, X., Wei, K., and Gao, X. 2010. Docetaxel loaded oleic acid-coated hydroxyapatite nanoparticles enhance the

docetaxel-induced apoptosis through activation of caspase-2 in androgen independent prostate cancer cells. *J. Control. Release 147*, 278–88.

[50] Venkatesan, P., Puvvada, N., Dash, R., Prashanth Kumar, B. N., Sarkar, D., Azab, B., Pathak, A., Kundu, S. C., Fisher, P. B., and Mandal, M. 2011. The potential of celecoxib-loaded hydroxyapatite-chitosan nanocomposite for the treatment of colon cancer. *Biomaterials 32*, 3794–806.

[51] Hou, C.-H., Hou, S.-M., Hsueh, Y.-S., Lin, J., Wu, H.-C., and Lin, F.-H. 2009. The in vivo performance of biomagnetic hydroxyapatite nanoparticles in cancer hyperthermia therapy. *Biomaterials 30*, 3956–60.

[52] Iwasaki, T., Nakatsuka, R., Murase, K., Takata, H., Nakamura, H., and Watano, S. 2013. Simple and Rapid Synthesis of Magnetite/Hydroxyapatite Composites for Hyperthermia Treatments via a Mechanochemical Route. *Int. J. Mol. Sci. 14*, 9365–78.

[53] Hasna, K., Kumar, S. S., Komath, M., Varma, M. R., Jayaraj, M. K., and Kumar, K. R. 2013. Synthesis of chemically pure, luminescent Eu3+ doped HAp nanoparticles: a promising fluorescent probe for in vivo imaging applications. *Phys. Chem. Chem. Phys. 15*, 8106–11.

[54] Escudero, A., Calvo, M. E., Rivera-Fernández, S., de la Fuente, J. M., and Ocaña, M. 2013. Microwave-assisted synthesis of biocompatible europium-doped calcium hydroxyapatite and fluoroapatite luminescent nanospindles functionalized with poly(acrylic acid). *Langmuir 29*, 1985–94.

[55] Strebhardt, K., and Ullrich, A. 2008. Paul Ehrlich's magic bullet concept: 100 years of progress. *Nat. Rev. Cancer 8*, 473–80.

[56] Shuhendler, A. J., Prasad, P., Leung, M., Rauth, A. M., Dacosta, R. S., and Wu, X. Y. 2012. A novel solid lipid nanoparticle formulation for active targeting to tumor α(v) β(3) integrin receptors reveals cyclic RGD as a double-edged sword. *Adv. Healthc. Mater. 1*, 600–8.

[57] Wong, D., Kandagatla, P., Korz, W., and Chinni, S. R. 2014. Targeting CXCR4 with CTCE-9908 inhibits prostate tumor metastasis. *BMC Urol. 14*, 12.

[58] Shin, T.-H., Sung, E.-S., Kim, Y.-J., Kim, K.-S., Kim, S.-H., Kim, S.-K., Lee, Y.-D., and Kim, Y.-S. 2014. Enhancement of the tumor penetration of monoclonal antibody by fusion of a neuropilin-targeting Peptide improves the antitumor efficacy. *Mol. Cancer Ther. 13*, 651–61.

[59] Ding, Y., Li, S., and Nie, G. 2013. Nanotechnological strategies for therapeutic targeting of tumor vasculature. *Nanomedicine (Lond). 8*, 1209–22.

[60] Zhang, Z.-L., Zhang, J.-F., Yuan, Y.-F., He, Y.-M., Liu, Q.-Y., Mao, X.-W., Ai, Y.-B., and Liu, Z.-S. 2014. Suppression of angiogenesis and tumor growth in vitro and in vivo using an anti-angiopoietin-2 single-chain antibody. *Exp. Ther. Med. 7*, 543–552.

[61] Chen, C.-T., and Hung, M.-C. 2013. Beyond anti-VEGF: dual-targeting antiangiogenic and antiproliferative therapy. *Am. J. Transl. Res. 5*, 393–403.

[62] Navis, A. C., Bourgonje, A., Wesseling, P., Wright, A., Hendriks, W., Verrijp, K., Laak, J. A. W. M. Van Der, Heerschap, A., and Leenders, W. P. J. 2013. Effects of Dual Targeting of Tumor Cells and Stroma in Human Glioblastoma Xenografts with a Tyrosine Kinase Inhibitor against c-MET and VEGFR2. *8*.

[63] Hu, Q., Gu, G., Liu, Z., Jiang, M., Kang, T., Miao, D., Tu, Y., Pang, Z., Song, Q., Yao, L., et al. 2013. F3 peptide-functionalized PEG-PLA nanoparticles co-administrated with tLyp-1 peptide for anti-glioma drug delivery. *Biomaterials 34*, 1135–45.

[64] Ricci-Vitiani, L., Pallini, R., Biffoni, M., Todaro, M., Invernici, G., Cenci, T., Maira, G., Parati, E. A., Stassi, G., Larocca, L. M., et al. 2010. Tumour vascularization via endothelial differentiation of glioblastoma stem-like cells. *Nature 468*, 824–8.

[65] Leamon, C. P., and Reddy, J. A. 2004. Folate-targeted chemotherapy. *Adv. Drug Deliv. Rev. 56*, 1127–41.

[66] Yoo, H. S., and Park, T. G. 2004. Folate-receptor-targeted delivery of doxorubicin nano-aggregates stabilized by doxorubicin-PEG-folate conjugate. *J. Control. Release 100*, 247–56.

[67] Kol, A., Terwisscha van Scheltinga, A. G. T., Timmer-Bosscha, H., Lamberts, L. E., Bensch, F., de Vries, E. G. E., and Schröder, C. P. 2014. HER3, serious partner in crime: Therapeutic approaches and potential biomarkers for effect of HER3-targeting. *Pharmacol. Ther.*

[68] Montemurro, F., and Scaltriti, M. 2014. Biomarkers of drugs targeting HER-family signalling in cancer. *J. Pathol. 232*, 219–29.

[69] Tortorella, S., and Karagiannis, T. C. 2014. The Significance of Transferrin Receptors in Oncology: the Development of Functional Nano-Based Drug Delivery Systems. *Curr. Drug Deliv.*

[70] Gelsomino, F., Facchinetti, F., Haspinger, E. R., Garassino, M. C., Trusolino, L., De Braud, F., and Tiseo, M. 2014. Targeting the MET gene for the treatment of non-small-cell lung cancer. *Crit. Rev. Oncol. Hematol. 89*, 284–99.

[71] Singh, P., Alex, J. M., and Bast, F. 2014. Insulin receptor (IR) and insulin-like growth factor receptor 1 (IGF-1R) signaling systems: novel treatment strategies for cancer. *Med. Oncol. 31*, 805.

[72] Ehnman, M., and Östman, A. 2014. Therapeutic targeting of platelet-derived growth factor receptors in solid tumors. *Expert Opin. Investig. Drugs 23*, 211–26.

[73] Liao, Y.-X., Zhou, C.-H., Zeng, H., Zuo, D.-Q., Wang, Z.-Y., Yin, F., Hua, Y.-Q., and Cai, Z.-D. 2013. The role of the CXCL12-CXCR4/CXCR7 axis in the progression and metastasis of bone sarcomas (Review). *Int. J. Mol. Med. 32*, 1239–46.

[74] Cojoc, M., Peitzsch, C., Trautmann, F., Polishchuk, L., Telegeev, G. D., and Dubrovska, A. 2013. Emerging targets in cancer management: role of the CXCL12/CXCR4 axis. *Onco. Targets. Ther. 6*, 1347–1361.

[75] Ayyappan, S., Prabhakar, D., and Sharma, N. 2013. Epidermal growth factor receptor (EGFR)-targeted therapies in esophagogastric cancer. *Anticancer Res. 33*, 4139–55.

[76] Köhler, G., and Milstein, C. 1975. Continuous cultures of fused cells secreting antibody of predefined specificity. *Nature 256*, 495–7.

[77] Corsi, F., Fiandra, L., De Palma, C., Colombo, M., Mazzucchelli, S., Verderio, P., Allevi, R., Tosoni, A., Nebuloni, M., Clementi, E., et al. 2011. HER2 expression in breast cancer cells is downregulated upon active targeting by antibody-engineered multifunctional nanoparticles in mice. *ACS Nano 5*, 6383–93.

[78] Prat, M., Crepaldi, T., Pennacchietti, S., Bussolino, F., and Comoglio, P. M. 1998. Agonistic monoclonal antibodies against the Met receptor dissect the biological responses to HGF. *J. Cell Sci. 111 (Pt 2*, 237–47.

[79] Liu, Y., Yang, Y., and Zhang, C. 2013. A concise review of magnetic resonance molecular imaging of tumor angiogenesis by targeting integrin αvβ3 with magnetic probes. *Int. J. Nanomedicine 8*, 1083–93.

[80] Bird, R. E., Hardman, K. D., Jacobson, J. W., Johnson, S., Kaufman, B. M., Lee, S. M., Lee, T., Pope, S. H., Riordan, G. S., and Whitlow, M. 1988. Single-chain antigen-binding proteins. *Science 242*, 423–6.

[81] Huston, J. S., Levinson, D., Mudgett-Hunter, M., Tai, M. S., Novotný, J., Margolies, M. N., Ridge, R. J., Bruccoleri, R. E., Haber, E., and Crea, R. 1988. Protein engineering of antibody binding sites: recovery of specific activity in an anti-digoxin single-chain Fv analogue produced in Escherichia coli. *Proc. Natl. Acad. Sci. U. S. A. 85*, 5879–83.

[82] Oliveira, S., Heukers, R., Sornkom, J., Kok, R. J., and van Bergen En Henegouwen, P. M. P. 2013. Targeting tumors with nanobodies for cancer imaging and therapy. *J. Control. Release 172*, 607–17.

[83] Müller, C., and Schibli, R. 2013. Prospects in Folate Receptor-Targeted Radionuclide Therapy. *Front. Oncol. 3*, 249.

[84] Serpe, L., Gallicchio, M., Canaparo, R., and Dosio, F. 2014. Targeted treatment of folate receptor-positive platinum-resistant ovarian cancer and companion diagnostics, with specific focus on vintafolide and etarfolatide. *Pharmgenomics. Pers. Med. 7*, 31–42.

[85] Teesalu, T., Sugahara, K. N., and Ruoslahti, E. 2013. Tumor-penetrating peptides. *Front. Oncol. 3*, 216.

[86] Li, Z. J., and Cho, C. H. 2012. Peptides as targeting probes against tumor vasculature for diagnosis and drug delivery. *J. Transl. Med. 10 Suppl 1*, S1.

[87] Ara, M. N., Hyodo, M., Ohga, N., Hida, K., and Harashima, H. 2012. Development of a novel DNA aptamer ligand targeting to primary cultured tumor endothelial cells by a cell-based SELEX method. *PLoS One 7*, e50174.

[88] Hu, M., and Zhang, K. 2013. The application of aptamers in cancer research: an up-to-date review. *Future Oncol. 9*, 369–76.

[89] Pala, K., Serwotka, A., Jeleń, F., Jakimowicz, P., and Otlewski, J. 2014. Tumor-specific hyperthermia with aptamer-tagged superparamagnetic nanoparticles. *Int. J. Nanomedicine 9*, 67–76.

[90] Li, X., Zhao, Q., and Qiu, L. 2013. Smart ligand: aptamer-mediated targeted delivery of chemotherapeutic drugs and siRNA for cancer therapy. *J. Control. Release 171*, 152–62.

[91] Liou, S.-C., Chen, S.-Y., Lee, H.-Y., and Bow, J.-S. 2004. Structural characterization of nano-sized calcium deficient apatite powders. *Biomaterials 25*, 189–96.

[92] Delgado-López, J. M., Iafisco, M., Rodríguez, I., Tampieri, A., Prat, M., and Gómez-Morales, J. 2012. Crystallization of bioinspired citrate-functionalized nanoapatite with tailored carbonate content. *Acta Biomater. 8*, 3491–9.

[93] Iafisco, M., Delgado-Lopez, J. M., Varoni, E. M., Tampieri, A., Rimondini, L., Gomez-Morales, J., and Prat, M. 2013. Cell surface receptor targeted biomimetic apatite nanocrystals for cancer therapy. *Small 9*, 3834–44.

[94] Delgado-López, J. M., Frison, R., Cervellino, A., Gómez-Morales, J., Guagliardi, A., and Masciocchi, N. 2014. Crystal Size, Morphology, and Growth Mechanism in Bio-Inspired Apatite Nanocrystals. *Adv. Funct. Mater. 24*, 1090–1099.

[95] Iafisco, M., Palazzo, B., Martra, G., Margiotta, N., Piccinonna, S., Natile, G., Gandin, V., Marzano, C., and Roveri, N. 2012. Nanocrystalline carbonate-apatites: role of Ca/P

ratio on the upload and release of anticancer platinum bisphosphonates. *Nanoscale 4*, 206–17.

[96] Prat, M., Narsimhan, R. P., Crepaldi, T., Nicotra, M. R., Natali, P. G., and Comoglio, P. M. 1991. The receptor encoded by the human c-MET oncogene is expressed in hepatocytes, epithelial cells and solid tumors. *Int. J. Cancer 49*, 323–8.

[97] Rodríguez-Ruiz, I., Delgado-López, J. M., Durán-Olivencia, M. a, Iafisco, M., Tampieri, A., Colangelo, D., Prat, M., and Gómez-Morales, J. 2013. pH-responsive delivery of doxorubicin from citrate-apatite nanocrystals with tailored carbonate content. *Langmuir 29*, 8213–21.

[98] Zakaria, S. M., Sharif Zein, S. H., Othman, M. R., Yang, F., and Jansen, J. A. 2013. Nanophase hydroxyapatite as a biomaterial in advanced hard tissue engineering: a review. *Tissue Eng. Part B. Rev. 19*, 431–41.

[99] Hannink, G., Geutjes, P. J., Daamen, W. F., and Buma, P. 2013. Evaluation of collagen/heparin coated TCP/HA granules for long-term delivery of BMP-2. *J. Mater. Sci. Mater. Med. 24*, 325–32.

[100] Yuan, Y., Liu, C., Qian, J., Wang, J., and Zhang, Y. 2010. Size-mediated cytotoxicity and apoptosis of hydroxyapatite nanoparticles in human hepatoma HepG2 cells. *Biomaterials 31*, 730–40.

[101] Chakraborty, S., Das, T., Sarma, H. D., Venkatesh, M., and Banerjee, S. 2008. Preparation and preliminary studies on 177Lu-labeled hydroxyapatite particles for possible use in the therapy of liver cancer. *Nucl. Med. Biol. 35*, 589–97.

[102] Das, T., Chakraborty, S., Sarma, H. D., Venkatesh, M., and Banerjee, S. 2009. 166Ho-labeled hydroxyapatite particles: a possible agent for liver cancer therapy. *Cancer Biother. Radiopharm. 24*, 7–14.

[103] Iafisco, M., and Margiotta, N. 2012. Silica xerogels and hydroxyapatite nanocrystals for the local delivery of platinum-bisphosphonate complexes in the treatment of bone tumors: a mini-review. *J. Inorg. Biochem. 117*, 237–47.

[104] Hossain, S., Yamamoto, H., Chowdhury, E. H., Wu, X., Hirose, H., Haque, A., Doki, Y., Mori, M., and Akaike, T. 2013. Fabrication and intracellular delivery of doxorubicin/carbonate apatite nanocomposites: effect on growth retardation of established colon tumor. *PLoS One 8*, e60428.

[105] Watanabe, K., Nishio, Y., Makiura, R., Nakahira, A., and Kojima, C. 2013. Paclitaxel-loaded hydroxyapatite/collagen hybrid gels as drug delivery systems for metastatic cancer cells. *Int. J. Pharm. 446*, 81–6.

[106] Ashokan, A., Menon, D., Nair, S., and Koyakutty, M. 2010. A molecular receptor targeted, hydroxyapatite nanocrystal based multi-modal contrast agent. *Biomaterials 31*, 2606–16.

[107] Rajendran, L., Knölker, H.-J., and Simons, K. 2010. Subcellular targeting strategies for drug design and delivery. *Nat. Rev. Drug Discov. 9*, 29–42.

[108] Turner, J. G., Dawson, J., and Sullivan, D. M. 2012. Nuclear export of proteins and drug resistance in cancer. *Biochem. Pharmacol. 83*, 1021–32.

[109] Shi, M., Ho, K., Keating, A., and Shoichet, M. S. 2009. Doxorubicin-Conjugated Immuno-Nanoparticles for Intracellular Anticancer Drug Delivery. *Adv. Funct. Mater. 19*, 1689–1696.

[110] Witton, C. J., Reeves, J. R., Going, J. J., Cooke, T. G., and Bartlett, J. M. S. 2003. Expression of the HER1-4 family of receptor tyrosine kinases in breast cancer. *J. Pathol. 200*, 290–7.

[111] Kümler, I., Tuxen, M. K., and Nielsen, D. L. 2014. A systematic review of dual targeting in HER2-positive breast cancer. *Cancer Treat. Rev. 40*, 259–70.

In: Apatite
Editors: M. Iafisco and J. M. Delgado-López
ISBN: 978-1-63321-500-9

Chapter 9

BIOCOMPATIBLE APATITE NANOPARTICLES: A VERSATILE TOOL FOR CELLULAR DELIVERY OF GENES, siRNAS, DRUGS AND PROTEINS

Sharif Hossain[1,2], Ezharul Hoque Chowdhury[3], Nihad Adnan[1] and Toshihiro Akaike[1,2*]
[1]Department of Biomolecular Engineering,
Graduate School of Bioscience and Biotechnology,
Tokyo Institute of Technology, Nagatsuta-cho, Midori-ku,
Yokohama, Japan
[2]Biomaterials Centre for Regenerative Medical Engineering,
Foundation for Advancement of International Science, Tsukuba, Japan
[3]Jeffrey Cheah School of Medicine and Health Sciences,
Monash University, Malaysia

ABSTRACT

Genetic manipulation of human cells through delivery of a functional gene or a gene-silencing element is an attractive approach to treat critical diseases very precisely and effectively. Genes encode proteins through messenger RNAs (mRNAs) to coordinate and perform the major functions in our body and an abnormality in terms of gene mutation resulting in over expression or suppression of vital cellular gene(s) could lead to a critical human disease. By virtue of the recombinant DNA technology enabling insertion of desirable genetic sequences into a plasmid vector and production of tailor-made proteins, and the antisense technology paving the way to specific knockdown of a mRNA transcript and thus precisely inhibiting a functional gene with small interfering RNA (siRNA), there is growing enthusiasm to effectively treat those genetic abnormalities while diminishing the off-target effects usually observed with classical therapy. However, direct delivery of the gene- and protein-based therapeutics is generally inefficient and suffers from serious shortcomings, such as, rapid clearance from blood circulation and

* All correspondence should be addressed to; Professor Toshihiro Akaike, Tel: +81-45-924-5146; Fax: +81-45-924-5790, E-mail: takaike@bio.titech.ac.jp, akaike@fais.or.jp

enzymatic degradation. Despite existence of both genetically engineered viral and synthetically designed non-viral nanocarriers for delivery of the therapeutics, an ideal delivery system in terms of safety and efficacy is still lacking. We have developed super effective pH-sensitive carbonate apatite based on some salient and unique features, namely, strong electrostatic affinity toward genes/proteins/drugs, ability of reducing particle growth and generating nano-sized particles and high dissolution rate in acidic endosomal environment, contributing to enhanced stability, improved cellular uptake and controlled release of the therapeutics to the intracellular destination(s). We elucidated how $CaCl_2$ and NaCl, existing as the major electrolytes in the bicarbonate-buffered solution, dose-dependently govern particle size and eventually internalization and expression of particle-associated DNA with a proper understanding of internalization route of nanoparticles as well as the intracellular fate of calcium, a key constituent of the particles. This chapter will thus focus on the development of biocompatible inorganic nanoparticles of carbonate apatite as a powerful tool for efficient intracellular delivery of DNA, siRNA, proteins and anti-cancer drugs in various primary, cancer, embryonic stem (ES) or induced pluripotent stem (iPS) cell lines along with its potential roles in many biomedical and regenerative medicine-related applications.

Keywords: Nanoparticles, drug delivery system, carbonate apatite, gene therapy, endocytosis

INTRODUCTION

Intracellular transfer of classical drugs and emerging bio-active molecules, such as, peptides, proteins, DNAs and RNAs etc. through endocytosis across the cell membrane has attracted increasing attention because of its immense importance in clinical medicine. Particularly, regulation of gene expression through the delivery of exogenous gene(s) or gene silencing element(s) could assist in restoring the regular physiological functions for treatment of genetic or acquired diseases. Following delivery to the cytoplasm, a foreign gene enters the nucleus and is transcribed to the corresponding mRNA, which is subsequently transported to the cytoplasm for translation into a specific protein. On the other hand, a gene-silencing element, such as an antisense oligonucleotide or a small interfering RNA, blocks the transcription of a target mRNA. Nucleic acid delivery has, therefore, been an essential tool to turn on and off the expression of a particular gene in basic research laboratories and is highly promising for the development of new therapeutic concepts, such as gene therapy and DNA vaccination, which are likely to have an impact on clinical medicine and biotechnology in this century [1, 2]. Plasmid DNA containing a gene of interest and siRNA (a short duplex RNA) could have immense impacts as promising therapeutic drugs for treating critical human diseases at the molecular level with high level of efficacy and precision by expressing a desirable protein and silencing expression of a harmful gene, respectively while a therapeutic recombinant protein could confer an immediate effect both at the extracellular and intracellular levels.

The short half-lives due to enzymatic degrdation in blood, the lack of tissue targetability and the incapability to passively diffuse across the plasma membrane and traffic across the harsh intracelluar environment are the major shortcomings for systemic delivery of nucleic acid-and protein-based potential therapeutics, such as recombinant plasmid, small interferring RNA (siRNA) and genetically engineered proteins. Considering the tremendous and potential impact on enhancement of stability and cell membrane permeability of DNA, siRNA and proteins, a substantial level of efforts were made in the past to develop viral as well as non-viral carriers addressing the aforementioned challenges to deliver the therapeutics [3]. The carcinogenicity and immunigenicity issues of the viral vectors pose serious challenges for their clinical applications. Additionally, viral vectors are not suitable for carring long DNA, siRNA, proteins and classical aniticancer drugs. In non-viral approach, DNA or siRNA being anionic in nature is electrostatically associated with a cationic lipid or polymer forming lipoplex or polyplex, respectively, or a cationized inorganic gold, silica or iron oxide particle. However, complextion of proteins with nanoparticles is usually complex requiring a tedious procedure under a harsh chemical environment thus affecting the protein integrity. The main concern of using non-biodegradable particles, specially the inorganic ones, is the adverse effects owing to their long-term interactions with body components [4]. Recently, biodegradable pH-sensitive inorganic nanoparticles of carbonate apatite have been developed for efficient transgene delivery to primary, cancer and embryonic stem cells, by virtue of their high affinity binding with DNA, ability to contact the cell membrane by ionic or ligand-receptor interactions and fast dissolution kinectis in endosomal acidic pH facilitating release of the DNA from the dissolving particles and also from the endosomes [5, 6]. In addition, siRNA and proteins have successfully been delivered into cancer cells with carbonate apatite in apparently less challenging manner than the DNA delivery which possesses an additional barrier, i.e., nuclear translocation. Moreover, given that the primary research goals of nanoparticles in therapeutics delivery are directed to highly specific drug targeting and delivery, reduced toxicity while maintaining therapeutic effects, greater safety and biocompatibility [7] particularly in cancer treatment. Notably, apatite-based efficient and quick intracellular delivery of anticancer drugs into cancer cells has enabled us to establish apatite-doxorubicin composite particles of desirable size for efficient cellular uptake and significant *in vitro* and *in vivo* tumor inhibition. Overall, carbonate-substituted apatite nano-carriers are highly promising candidates as efficient and cell-targetable non-viral vectors for therapeutic delivery in pre-clinical and clinical trials.

1. Nanoparticle-Mediated Gene/Drug Delivery

Gene delivery is the process of introducing foreign DNA into host cells with a variety of methods developed for uptake of the genetic material(s) by a various types of cells and tissues, from bacterial to mammalian [8]. The successful design of a gene delivery system requires complete understanding of the interaction mechanism between the target cell and delivery system as well as the intercellular traffic and targeting mechanism. Cell targeting refers to delivery of the therapeutic agent to a specific compartment or organelle of the cell [9]. Following cellular uptake of the delivery system by endocytosis, intracellular release takes place to initiate DNA transcription and translation, and produce the related protein. A successful gene delivery procedure involves minimizing potential inhibitory inflammatory response while also overcoming certain barriers at each step of the gene delivery procedure, in order to optimize gene activity. Nanoparticles are widely used for gene/drug delivery to target cells or organs. The basic prerequisites for the design of new materials for drug delivery include drug incorporation, cellular interaction and intracellular release, formulation stability and shelf-life, biocompatibility, biodistribution and targeting, and functionality. Moreover, the possible adverse effects of residual material after drug delivery should be considered as well when used solely as a carrier. Therefore, biodegradable nanoparticles with a lifespan as long as therapeutically needed would be optimal. The basic objective of drug incorporation within the nanoparticles is to promote the enhanced delivery to, or uptake by, target cells and/or to reduce the toxicity of the free drug to non-target organs [10]. Depending on the carrier of therapeutics, gene/drug delivery systems can be generally classified into four major groups: viral carriers, organic cationic compounds, recombinant proteins and inorganic nanoparticles. Viral delivery systems that predominantly include adenoviruses, retroviruses, and lentiviruses are modified to be replication-deficient by redesigning with the ability to deliver the genes into the cells. Despite having advantages such as constant expression of therapeutic genes, there are some major limitations that restrict the use of these viral systems, which particularly include immunogenicity, toxicity and lack of optimization in large-scale production [11]. Non-viral delivery systems, on the other hand, were developed as an alternative to viral-based systems and are basically categorized into physical and chemical groups. Microinjection, electroporation, gene gun, ultrasound-mediated methods, and hydrodynamic systems are the most widely used physical methods [12]. Physical methods involve the use of physical force to increase the permeability of the cell membrane allowing the gene to enter the cell. The primary advantage of physical methods is that they are easy to use and reliable. However, they also have the disadvantage of causing tissue damage in some applications. In contrast, chemical methods involve the use of carriers prepared from synthetic or natural compounds for gene delivery into the cell, including synthetic and natural polymers, liposomes, dendrimers, synthetic proteins, and cationic lipids. The biggest advantages of these systems are that they are non-immunogenic and generally have low toxicity [12].

Carriers consisting of cationic compounds mainly include cationic lipids, cationic polysaccharides, polycationic polymers, where the positive charges are resulted from protonation of various amino/imino groups. A particular type of DNA vectors receiving attention recently is recombinant proteins. These proteins mimic the various viral properties by combining diverse peptide segments that are required for efficient gene delivery, into a single molecule through protein engineering. For example, the multifunctional proteins may include polylysine segments, protamine, histones and amphipathic cationic peptides to bind DNA to form thermodynamically stable complexes. They may also contain antibodies or antibody segments for targeting cell surface receptors and some short peptide sequences acting as nuclear localization signals [13]. Although the formation of vector-DNA complexes follows the similar mechanism among the cationic compounds, protein vectors are much more biocompatible. Inorganic nanoparticles as new non-viral carriers have attracted much attention only recently. Among many inorganic materials, carbonate apatite, calcium phosphate, gold, carbon materials, silicon oxide, iron oxide and layered double hydroxide have been well studied.

2. Major Intracellular Barriers for Non-Viral Gene/Drug Delivery

The number of DNA molecules decreases at each step of the journey from outside the cell to inside the nucleus, resulting in low transgene expression. There are three major barriers which are responsible for inefficient DNA delivery via the non-viral route: low uptake across the plasma membrane, lack of endosomal escape resulting inadequate release of DNA molecules with limited stability, and absence of nuclear targeting (Figure 1). These anatomical and cellular barriers limit the overall efficiency of gene transfer by non-viral methods. Briefly, anatomical barriers are epithelial, endothelial cell linings and the extracellular matrix surrounding the cells that prevent direct access of macromolecules to the target cells. Professional phagocytes such as Kupffer cells in the liver and residential macrophages in the spleen are largely responsible for the clearance of the DNA-loaded colloidal particles administered through blood circulation. In addition, various nucleases existing in blood and extracellular matrix can rapidly degrade free and unprotected nucleic acids following systemic administration. Crossing plasma membrane is considered as the most critical limiting step for an efficient DNA transfection.

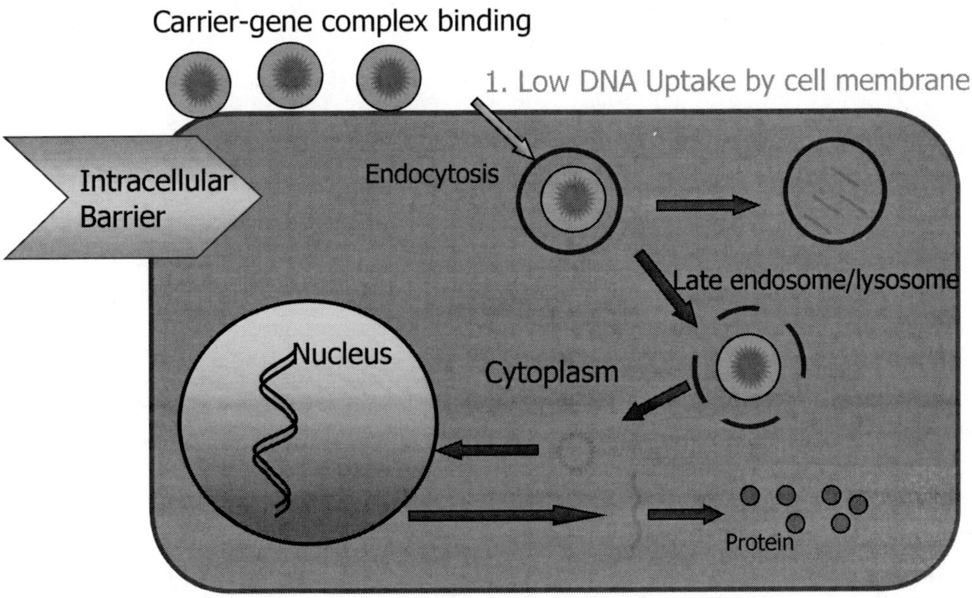

Figure 1. Major barriers to gene delivery by non-viral methods include low uptake across the plasma membrane, lack of endosomal escaping resulting inadequate release of DNA molecules with limited stability, and absence of nuclear targeting.

Nucleic acids typically cannot pass through cell membrane unless their entry is facilitated by creating transient holes by physical meanings, or through various active cell uptake mechanisms such as endocytosis, pinocytosis, or phagocytosis [14]. Upon being taken up via endocytosis, macromolecules captured within the endosomes usually transform into digestive lysosomes unless some escape mechanisms are used to intercept this maturation process. Two escape mechanisms have been explored. The first involves the use of membrane active or fusogenic molecules such as fusion peptides or lipid components with acid-sensitive bond and large hydrophobic portion of the molecules to rupture the endosome membrane [15]. The other mechanism acts on building up osmotic pressure within the endosomal compartment that eventually triggers the swelling or burst of endosomal vesicles [16]. Upon their release from endosomes, DNA molecules in their free form or as complexes must travel through cytoplasmic space filled with viscous protein solution and a network of cytoskeleton matrix toward the nucleus where transcription takes place. Observations made by direct intracellular microinjection of naked DNA prove that the movement by diffusion is slow and inefficient, and the resulting levels of gene expression are very weak [17]. Understanding the major barrier of nonviral gene/drug delivery, the following key parameters should be taken into account during cellular delivery using nanoparticle [4] :i). effective particle size for endocytosis; ii). proper surface charge distribution for cellular interaction with nanoparticles and surface fictionalization for targetibility; iii) biocompatibility and cytotoxicity of nanoparticles [18]; iv). protection capability from enzyme, endosomal escaping and controlled release of therapeutics for consequent efficacy. Apart from cancer application using nanoparticle-mediated gene/drug delivery, another current challenge is to develop drug delivery systems that ensure the safe and effective passage of drugs across the blood brain barrier (BBB). Therefore, nanoparticles able to ferry drugs across the BBB to CNS tumors have been reported and some of the necessary characteristics of such tools have been defined [19].

3. Biodegradable Functional Inorganic Nanoparticles of Carbonate Apatite: Synthesis and Molecular Characterization

Carbonate apatite is a natural component of the body and usually found in the hard tissue, such as bone and teeth. The newly developed carbonate apatite like hydroxyapatite adsorbs DNA, but, unlike the later, it can prevent the growth of its crystals to a significant extent, enabling the synthesis of nano-sized crystals to effectively carry the associated DNA, siRNA/miRNA, proteins or drug across the cell membrane [5-6, 20-24]. It also possesses a high dissolution rate in endosomal acidic pH, leading to the rapid release of the bound therapeutics for a subsequent high level of protein expression.

3.1. Molecular Aspects of DNA Delivery with Nano-Crystals of Carbonate Apatite

Being inspired by the need to develop a superior non-viral approach in terms of both efficiency and biocompatibility, the development of an inorganic DNA nanocarrier based on in vitro-synthesized carbonate apatite, a major component of hard tissues in the body, has recently been reported. Particles are generated in the presence of DNA in a bicarbonate-buffered solution or medium containing appropriate concentrations of phosphate and calcium salts through incubation at a high temperature. The chemical reaction required for particle formation takes place with Ca^{2+}, PO_4^{3-} and HCO_3^-, and DNA that is negatively charged can be electrostatically associated with the cationic (Ca^{2+} rich) domains of the particles as shown in Figure 2 [25].

The number of particles as well as their size, two crucial factors for determining transfection potency, are dramatically influenced by the concentrations of calcium, phosphate and bicarbonate and, additionally, by the pH of the medium (or solution), incubation temperature and time. Increasing the concentration of any of the reactants (calcium, phosphate and bicarbonate), in general, accelerates particle formation by providing a stronger driving force for the chemical reaction while the other parameters (pH, incubation temperature and time) are constant. On the other hand, an increase in pH and temperature (or incubation time) shifts the ionization equilibrium of phosphate and bicarbonate towards the forward direction and consequently favors particle generation by increasing the rate of the reaction.

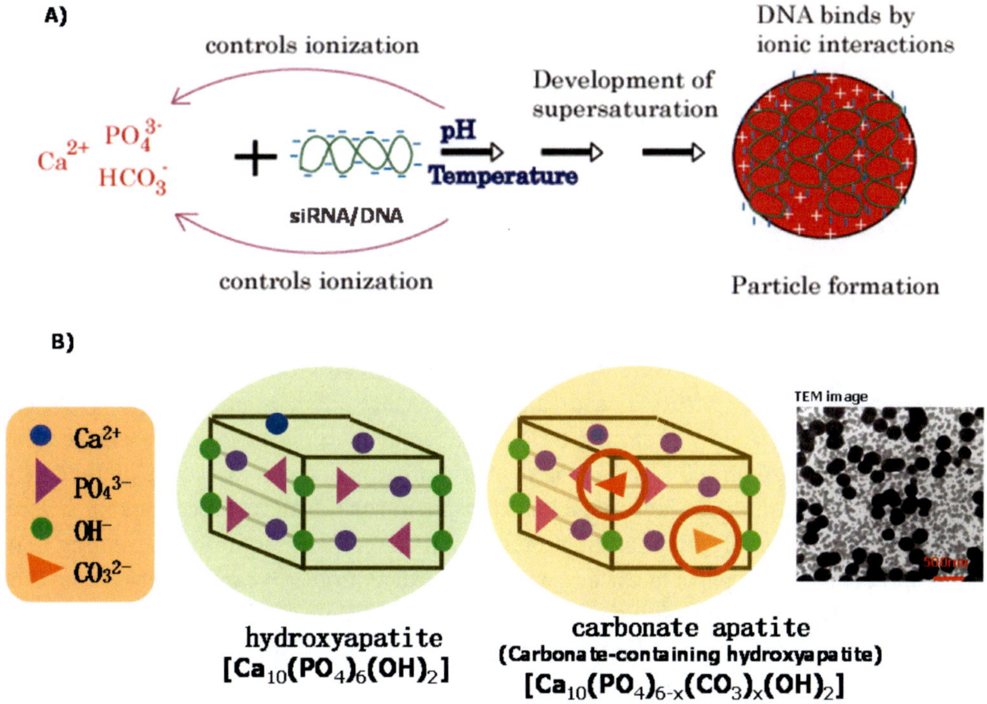

Figure 2. Molecular aspects of carbonate apatite nanoparticles. A) Generation of DNA or siRNA/apatite complex based on ionic interaction, B) Molecular structure of hydroxyapatite and carbonate apatite. Owing to lower crystallinity, carbonate apatite derives lower size of crystalline particle resulting in higher solubility in acidic solution and consequently suitable for drug delivery. TEM image: Carbonate apatite.

The induction of particle formation is generally accompanied by growth and aggregation of the particles, leading to larger-sized crystals. However, bicarbonate, which is a minor component of the final apatite structure, $Ca_{10}(PO4)_{6-X}(CO3)_X(OH)_2$, prevents particle aggregation and generates small-sized crystals in a dose-dependent manner. Due to the small size of the crystal, which has an average diameter of 50- 300 nm and strong binding with DNA (almost 100% efficiency), carbonate apatite can be rapidly internalized with the associated DNA into the cell by endocytosis, following close contact with the cell surface through electrostatic forces. The efficiency of cellular uptake of DNA has been estimated to be at least 10-times more rapid than that achieved by classical CaP method [5, 25].

3.2. Transfection Efficiency and the Role of Crystal Dissolution

Transgene expression results only after multiple barriers are overcome, such as endocytosis, DNA release from the particles and the endosomes, and translocation to the nucleus. Once inside the endosomes, exposure to an increasingly acidic environment results in the consumption of the excess H^+ by phosphate and carbonate ions in the particle, leading to particle dissolution and consequential DNA release. Therefore, inorganic crystals with high acid solubility or low crystallinity would enable quicker DNA release in endosomes, compared with crystals with low solubility or high crystallinity. In addition, the high dissolution rate may contribute to the destabilization of endosomes, to achieve DNA release into the cytoplasm, as vacuolar proton-pumping ATPase-driven massive proton accumulation that causes crystal dissolution could also lead to passive chloride influx to endosomes and subsequent endosome swelling and rupture. Finally, released DNA can enter the nucleus either through the nuclear pore or during cell division and be heavily expressed both in primary and cancer cell lines with an efficiency 5- to 100-times higher than with existing techniques.

3.3. Bio-Recognition Devices Associated with the Inorganic Crystals for Cell-Targeted Delivery

As mentioned before, a gene delivery system should have a precise cell-targeting capability for many therapeutic applications. To achieve this with the inorganic nanocarrier now under discussion, which delivers transgenes in a non-specific manner depending on ionic interactions between the particles and the cell surface, an organic-inorganic hybrid nanocarrier was synthesized by coating the DNA-embedded particle surface. The surface was coated sequentially with a cell-recognizable protein, such as asialofetuin for asialoglycoprotein receptors on hepatocytes or transferrin for transferrin receptors on several cancer cell lines and a highly hydrophilic protein, such as serum albumin for preventing non-specific interactions of the particle surface with cell-surface proteoglycans, other receptors or serum proteins [26]. The composite particles, with dual surface properties, have been shown to accelerate transgene delivery and expression only in the corresponding receptor-bearing cells. Moreover, mimicking the natural mineralization process, extracellular matrix proteins, such as collagen or fibronectin, have been successfully attached to the nano-crystals for integrin-specific efficient delivery of DNA [27]. A remarkable success has been achieved for transfecting embryonic stem cells, where particles initially did not show any significant transfection activity owing to the lack of particle interactions with the cell surface. However, when the particles were complexed with a naturally occurring fibronection and genetically engineered E-cadherin-Fc together with DNA, a synergistic effect resulted in a dramatic enhancement both in transgene delivery and expression in mouse embryonic stem cells that possessed both transmembrane fibronectin-specific integrin and E-cadherin [28].

4. Potential of Carbonate Apatite as Novel Nanocarrier for Therapeutics Delivery

Extra and intracellular fascinating properties of carbonate apatite like biocompatibility, pH sensitivity, high uptake efficiency, quick dissolution and endosomal escaping capacity, and high efficiency remarkably expected to make it highly promising for therapeutics (DNA, siRNA/miRNA, protein, anticancer drug) transportation for targeted delivery to different primary, cancer and embryonic/induced pluripotent stem (ES)/iPS cell as shown in Figure 3.

4.1. Carbonate Apatite Mediated Delivery of Small Interfering RNA As Novel Candidate for Gene Therapy

RNA interference is one of the most powerful and specific mechanisms for gene silencing by degradation of mRNA. Owing to this property, siRNAs are emerging as promising therapeutic agents for the treatment of inherited and acquired diseases, as well as research tools for the elucidation of gene function in both health and disease. Approaches to transfer desirable siRNA into living cells enable various applications both in basic researches as well as in clinics through effective knockdown of desirable gene, thus offering a suitable potential to gene therapy. siRNAs are short stretch (19–30 nucleotides) dsRNAs capable of degrading the complementary mRNA in the cytoplasm. In the cytoplasm, long dsRNAs are cleaved by the endoribonuclease Dicer into short dsRNA duplexes or siRNA. siRNA are loaded onto RNA-induced silencing complex (RISC). RISC contains Argonaute 2 (Ago-2) which cleaves and releases one strand from the dsRNA, resulting in an activated form of RISC with a single-strand RNA (guide siRNA) that directs the specificity of the target mRNA recognition through complementary base pairing. Ago-2 then cleaves the target mRNA between bases 10 and 11 relative to the 5′ end of the siRNA antisense strand, thereby causing mRNA degradation and gene silencing. The site of action of siRNA therapeutics is the cytosol. The barriers to siRNA delivery are multiple and depend on the targeted organs and the administration routes. In general, locoregional delivery of siRNA has fewer barriers compared to systemic delivery. For example, intranasal inhalation of siRNA against respiratory syncytial virus, either naked or encapsulated in polycationic liposomes, was about equally effective in reducing the viral infection.

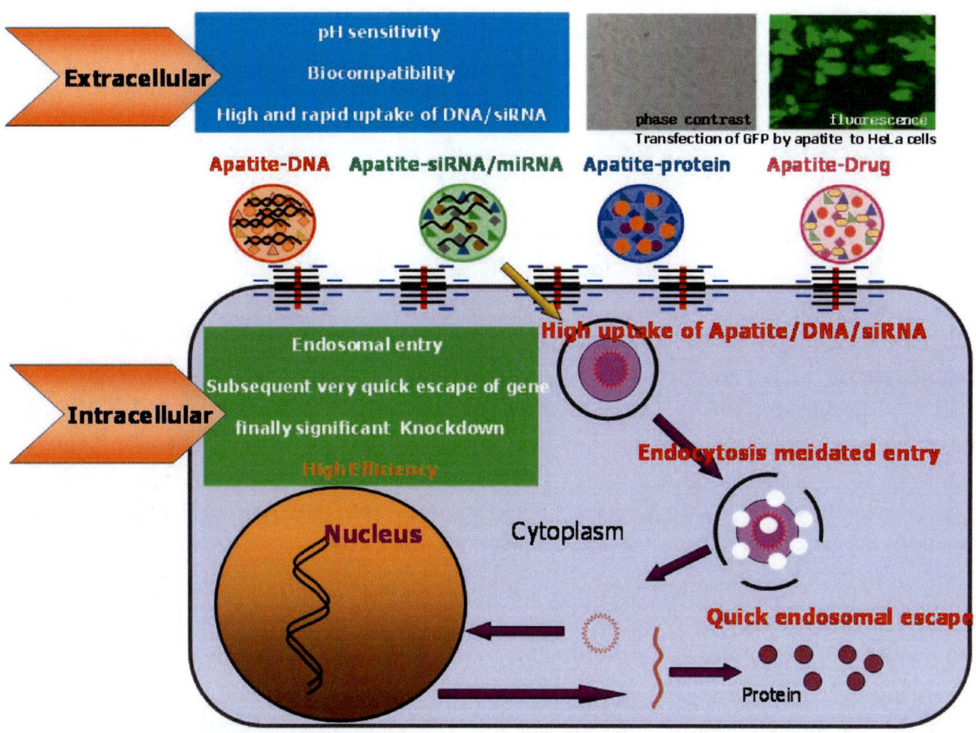

Figure 3. Extra and intracellular impressive findings of super effective pH sensitive carbonate apatite nanoparticle as nucleic acids/protein/drug transporter.

Systemic delivery, e.g., intravenous injection, of siRNA poses significantly greater challenges. After intravenous injection, siRNA is distributed to organs via the blood circulation and at the same time undergoes elimination. Within an organ, siRNA leaves the intravascular space within a blood vessel to enter the interstitium (i.e., extravasation). After entering the tissue interstitium, siRNA is transported across the interstitial space to the target cells. After reaching the target cell, siRNA undergoes internalization via endocytosis, a process that involves siRNA being encapsulated in endocytic vesicles that fuse with endosomes. After entering the cell, siRNA must escape from the endosomes and be released from its carrier to the cytosol in order to be loaded onto RISC [29]. Based on proper understanding on gene silencing mechanism, gene therapy through intracellular delivery of a gene-silencing element is a promising approach to treat critical diseases; therefore, RNA interference is one of the most powerful and specific mechanisms for gene silencing by degradation of mRNA. Although RNA interference is highly promising, the success of RNAi is highly dependent on the delivery of exogenous siRNA and clinical trials of therapeutic siRNA are prohibited, because the efficiency of siRNA or DNA delivery into cytosol of target cells using a nonviral delivery system is still inadequate. Efficient delivery of siRNA to the cytosol of target cells depends on both the endosomal entry and their subsequent escape from endosomal/lysosomal compartments ensuring high gene knockdown efficiency. Thus intracellular siRNA or DNA delivery approach offering high efficiency is highly expected. To fully exploit the potential of siRNA as therapeutics, complex nanoparticle fabricated between the nano apatite and siRNA by manipulating surface charge of carbonate apatite. Since siRNA is negatively charged due to its phosphate backbone while the carbonate apatite particles are positively charged due to the presence of calcium ions in the apatite structure, it was presumed that siRNA would bind to the apatite by ionic interactions. The highly compact spherical siRNA/apatite complex (50~200nm) showed excellent biocompatibility and significant uptake (More than 3 times) of siRNA in Hela cell than Lipofectamine 2000. Dissolution study of siRNA/apatite nanoparticle in different pH conditions revealed that siRNA/apatite particles were highly stable at physiological pH 7.4 but at pH 7.0, the particles began to dissolute and at endosomal pH 6.0, particles are totally dissolved which is beneficial for in vivo drug delivery. This quick dissolution in endosomal pH is a very unique phenomenon for drug delivery system so that siRNA can be released easily from the carrier and escaped from the endosome. It is, therefore, important for a delivery system to avoid the fate of lysosomal degradation as crucial step and major obstacle of siRNA delivery by facilitating release of the internalized siRNA into the cytoplasm. A confocal microscopic analysis was performed on HeLa cells after the delivery of fluorescein siRNA/apatite complexes and observed the cellular localization of siRNA in a time dependent manner. Endosomes/lysosomes were stained with LysoTracker Red. Most of the fluorescein siRNAs (Green) were co-localized in stained endosomes after 1h incubation of HeLa cells with siRNA/apatite complexes which indicated the pathway of siRNA internalization was endocytosis. However, interestingly after 2h, the co- localized siRNA partially escaped and after 4h, most of the siRNAs escaped the endocytic vesicles. This is the first system that ensured the prominent siRNA escape with least time (2-4h only). Therefore, apatite mediated gene silencing efficiency of siRNA against gene coding for GFP and luciferase were found to be highly significant than cytotoxic Lipofectamine 2000 (Figure 4). Moreover, this research aimed to silence cyclin B1 gene whose expression is essentially needed during the onset of mitosis and thus, knock-down of the gene might result in apoptosis of cancer cells.

Knockdown of cyclin B1 gene with only 10 nM of siRNA delivered by carbonate apatite resulted in the significant death of cancer cells. Taken together, all the impressive results like pH sensitivity and biocompatibility of vector, high and rapid uptake of siRNA to cell facilitating endosomal entry, subsequent very quick escape of siRNAs from endo/lysosome and finally significant knockdown of both reporter and functional genes like cyclin B1 and other functional genes, suggesting that the new method of siRNA delivery is highly promising for pre-clinical and clinical cancer therapy [20, 30-32]. Apart from siRNA, it was found that bile acids increase levels of microRNAs 221 and 222, leading to degradation of CDX2 during esophageal carcinogenesis where Pre-MIR precursors and antisense MIR inhibitors were transfected by carbonate apatite nanoparticles [22].

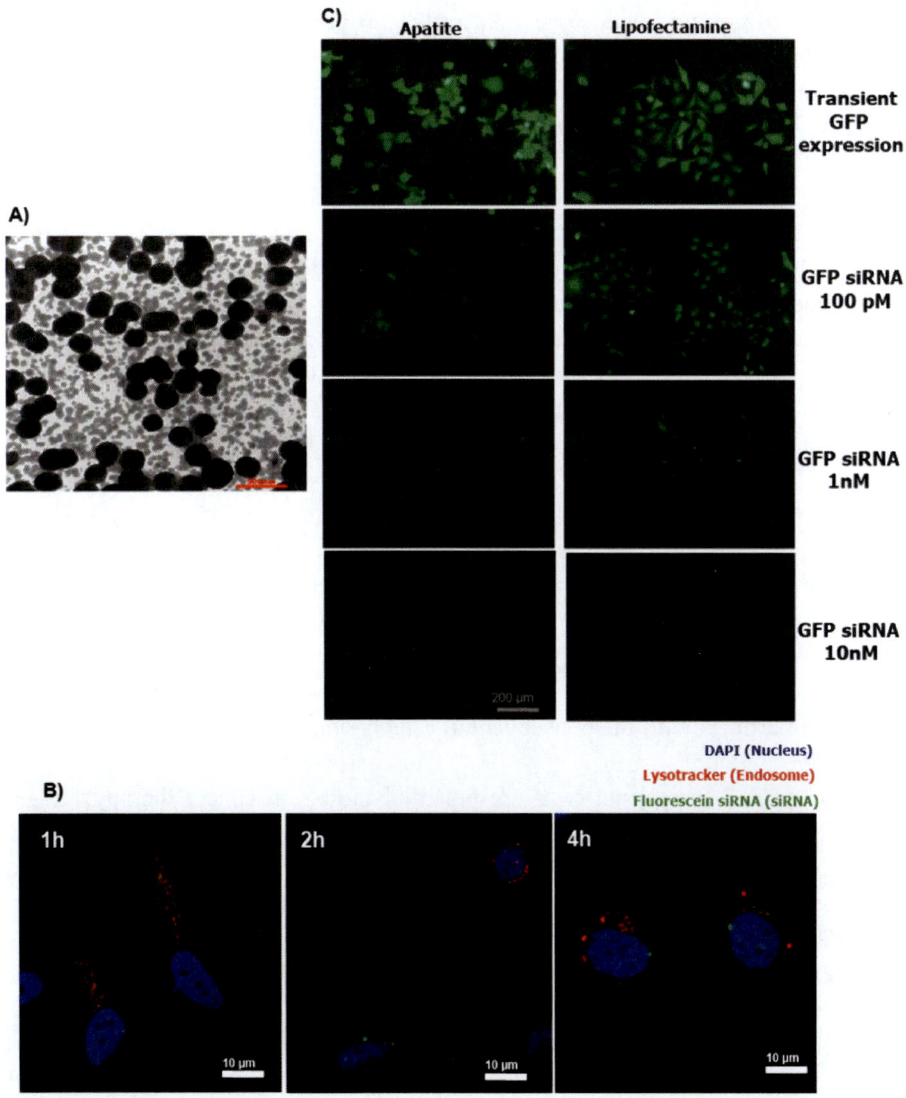

Figure 4. Carbonate apatite-facilitated intracellularly delivered siRNA for efficient knockdown of functional genes. A) Desirable nano size, B) endosomal entry, time dependent exit of siRNA and C) finally effective knockdown of reporter gene (GFP) mediated by carbonate apatite nanoparticles [20]. Scale bar in TEM photograph 500 nm.

4.2. DNA-Immobilized Inorganic Nano-Crystals for Efficient Transgene Expression in Human Lymphocytes and Cancer Cells: Improvement and Mechanism Elucidation

Gene therapy is a revolutionary and promising approach for treating critical human disease at the genetic level. Genes encode proteins through messenger RNA (mRNA) to perform the major functions in a biological system. Delivery of exogenous genes is, therefore, essentially needed to confer a new function in the system that could assist in treating a genetic or acquired disease. Considering the potential of gene delivery, an efficient pH sensitive DNA delivery and expression system have been developed for many primary, cancer and embryonic stem cell lines based on some fascinating properties of carbonate apatite—the ability to prevent crystal growth for generation of nanoscale particles as needed for efficient endocytosis and a fast dissolution rate in endosomal acidic compartments to facilitate DNA release from the particles and endosomes. According to previous protocol direct mixing of DNA in a bicarbonate-buffered medium (DMEM) containing appropriate concentrations of phosphate and calcium salts ($CaCl_2$), followed by incubation at 37°C for 30 min, results in microscopically visible carbonate apatite particles with major components of carbonate, phosphate, and Ca^{2+} and with high transfection activity in many cancer and primary cell types. Because cell culture medium contains high concentrations of NaCl and glucose (in addition to different types of amino acids, vitamins, and other salts at very low concentrations), it is necessary to explore their roles in particle growth, DNA adsorption to the particles, and final transfection activity. Moreover, to implement carbonate apatite-mediated transfection in some cancer cells and primary cells being cultured in special medium or for *in vivo* delivery, DMEM should be replaced with a solution containing all of the components for efficient synthesis of DNA/carbonate apatite particles. Particle formation or synthesis is the result of the growth of particles as induced by the optimal concentrations of the reactants (Ca^{2+}, PO_4^{3}–, and HCO^{3} $^{-}$) at a defined pH and temperature, and an increase in the concentration of any of the reactants leads to the further growth or synthesis of the particles while simultaneously increasing the number and sizes of the growing particles. Therefore, recently a more improved new synthetic approach was reported where carbonate apatite crystals synthesized in bicarbonate buffered, inorganic, phosphate-containing solution electrostatically associate a significantly greater amount of DNA than in cell culture medium. In addition, the existence of $CaCl_2$ and NaCl as the major electrolytes dramatically influences the growth of the particles with the consequential effects on DNA uptake, transgene delivery and expression [6].

Since leukemia and lymphoma cells are potential targets in cancer therapy for genetic manipulation either by transgene expression or silencing of endogenous gene expression and genetically engineered autologous lymphocytes expressing a chimeric antigen against a receptor overexpressed in tumor or tumor vasculature are promising cell based therapeutics for cancer in addition. However, the successful implementation of these attractive approaches is hindered due to the lack of a smart device for efficient transgene delivery and expression in the lymphocytes. Therefore, recently an efficient nanocarrier of carbonate apatite for intracellular delivery and release of DNA molecules, achieving very high level of transgene expression in primary cell like human leukemia cells as well as cancer cell lines [33]. However, its efficacy in human T leukemia cells is comparatively low. Here, it is revealed that simultaneous stimulation of human T leukemia cells by the most commonly used phorbol ester-based protein kinase C (PKC) activator and cytocalasin D, an actin filament disrupting agent, dramatically enhanced carbonate apatite-mediated transgene delivery and expression in the cells by synergistically activating protein kinase C (PKC), while rapidly extruding Ca^{2+} of intracellularly dissolved particles through plasma membrane-associated Ca^{2+} ATPase (Figure 5) [23].

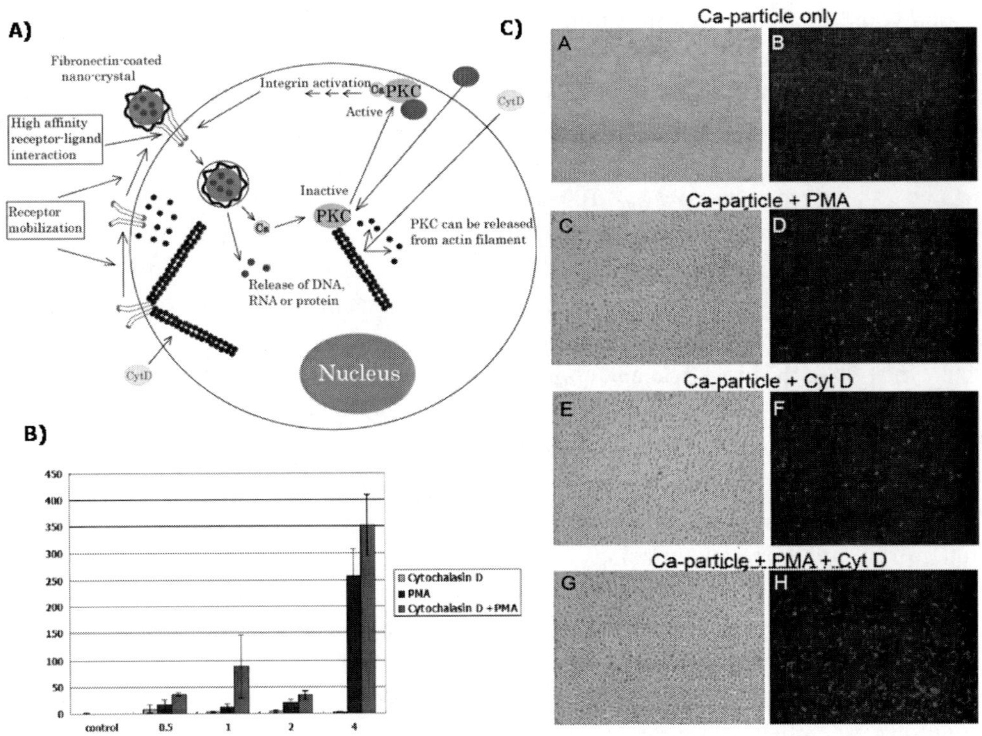

Figure 5. Efficient transgene expression in human lymphocytes mediated by carbonate apatite nanoparticles. A) A proposed mechanism of synergistic effects of nano-crystals, PKC activator and actin disrupting agent on lymphocyte transfection, B) PMA and cytochalasin D treatment enhanced carbonate-apatite mediated cellular DNA uptake and C) resulting in enhanced transgene expression in human T-lymphocytes [23].

This research also elucidated the intracellular fate of elevated calcium which acts as a vital second messenger. Moreover, endocytosis of the DNA-associated particles across the cell membrane was found to follow the clathrin-dependent route in both normal and activated cells. The findings thus offer significant insights for pre-clinical and clinical ex-vivo trials of cancer utilizing autologous T cells [23].

4.3. Carbonate Apatite As an Intracellular Protein Transporter

The transfer of specific proteins into living cells to enable the regulation of cell function or the tracking of the intracellular distribution of proteins is a desirable objective for offering a potential alternative to gene therapy. Generally, there are three kinds of methods to deliver proteins into cells. Physical methods such as microinjection and electroporation have difficulty to be applied in vivo. Another way is the use of carrier molecules such as cationic lipids and polymers that are able to enter cells through endocytosis via ionic interactions. The protein loaded carriers are internalized into cells, escape the endo-lysosomes and finally enter the cytoplasm. However, very small amount of internalized protein loaded carriers escape into the cytosolic compartment and result in denaturation of proteins. Therefore, smart carriers should be used for the rapid escape of proteins from endo-lysosomes. Therefore, Protein/carbonate apatite complexes were successfully fabricated for intracellular delivery of functional proteins since the carbonate apatite being highly water solubility under an acidic condition could easily be dissolved in endosomes following endocytosis, thus releasing the electrostatically associated proteins in cytoplasm [34].

In this study, protein/carbonate apatite complexes were characterized as an intracellular protein delivery system and checked the intracellular delivery of proteins by carbonate apatite nanoparticles *in vitro*. Fluorescently-labeled bovine serum albumin as a model protein was effectively delivered into nearly 100% of HeLa cells by the simple addition of protein/carbonate apatite complexes to the cells. Confocal microscopic imaging suggested the endosomal release of protein delivered with carbonate apatite. Intracellularly delivered β-galactosidase did not lose its enzymatic activity. The intracellular delivery system of protein using pH-sensitive carbonate apatite carrier with a very simple procedure will be a highly effective method to the biological and clinical researches.

4.4. Intracellular Delivery of Anticancer Drugs Based on Carbonate Apatite Nanocomposites for Growth Retardation of Cancer Cell And Solid Tumor

Contemporary cancer therapy, particularly with respect to drug delivery, has begun an evolution from traditional methodology. Part of this change is based on the need to increase the therapeutic index of chemotherapy drugs. Since the central problem in conventional cancer chemotherapy is the severe toxic side effects of anticancer drugs on healthy tissues and invariably the side effects impose dose reduction, treatment delay, or discontinuation of therapy, therefore delivery of specific drugs into the cancer cells to achieve high cytotoxicity with minimal side effect is a desirable objective for offering a potential approach for cancer therapy.

To achieve these goals, the current focus is on the development of novel carriers with prolonged blood circulation time for both existing and incoming new drugs in order to achieve high efficacy in killing cancer cells and minimizing the side effects, while defining better therapeutic targets in various cancer cells. Consequently, in the past decade, nanoparticle-based drug delivery systems have shown exciting efficacy for cancer treatments due to their improved pharmacokinetics and biodistribution profiles via the enhanced permeability and retention (EPR) effect. However, the EPR effect can only enhance the accumulation of nanoparticles in tumor tissues; the poor cellular internalization as well as insufficient intracellular drug release always limits the dosages of anticancer drugs to the level below the therapeutic window, which hampers the efficacy of cancer chemotherapy. To address the challenges, environment-responsive delivery systems have been attempted to improve the drug bioavailability. Of these stimuli, pH-responsiveness is the most frequently used, as pH values in different tissues and cellular compartments vary tremendously. Doxorubicin (DOX), one of the most commonly used antitumor drugs, has a broad spectrum of antitumor properties in therapy against hematopoietic malignancies and solid tumors. However, DOX may produce free radicals; leading to serious side effects especially cardiomyopathy en route to congestive heart failure. Among the synthetic inorganic carriers, pH-sensitive nanoparticles of carbonate apatite being formed in a supersaturated solution of calcium, phosphate and carbonate ions have been getting much attention by virtue of their biodegradability and resemblance to body hard tissue components. Moreover, owing to its heterogeneous charge distribution and several fascinating properties such as ability of preventing crystal growth for generation of nanoscale particles as needed for efficient endocytosis, and fast dissolution kinetics in endosomal acidic compartments to facilitate the release of therapeutics from the particles and endosomes, carbonate apatite has been proven as the potential tool for the delivery of different therapeutics. Therefore, recently pH sensitive carbonate apatite-DOX complexes were successfully fabricated for the first time for intracellular delivery of anticancer drug DOX since the carbonate apatite being highly stable in typical physiological pH could easily be dissolved in endosomal acidic pH following endocytosis, thus quickly releasing the associated drugs in cytoplasm for effective therapeutic action [21]. It was found that the positively charged anticancer drugs like Cisplatin or Doxorubicin can electrostatically bind with the negatively charged carbonate or phosphate-rich domain of the apatite particle and after effective cellular interaction the loaded drugs can effectively be released in the acidic compartment of endosome for subsequent anticancer action [30]. The apatite-DOX formation is based on electrostatic interactions where the positively charged DOX binds to negatively charged. Detailed information on particle size was obtained by AFM and an enlarged image focusing on the central area showed single spherical particles ranging from 30 to 50 nm as well as the combined bodies consisting of several single particles. The maximum height of individual single particle of carbonate apatite was found to be 69.8 nm.

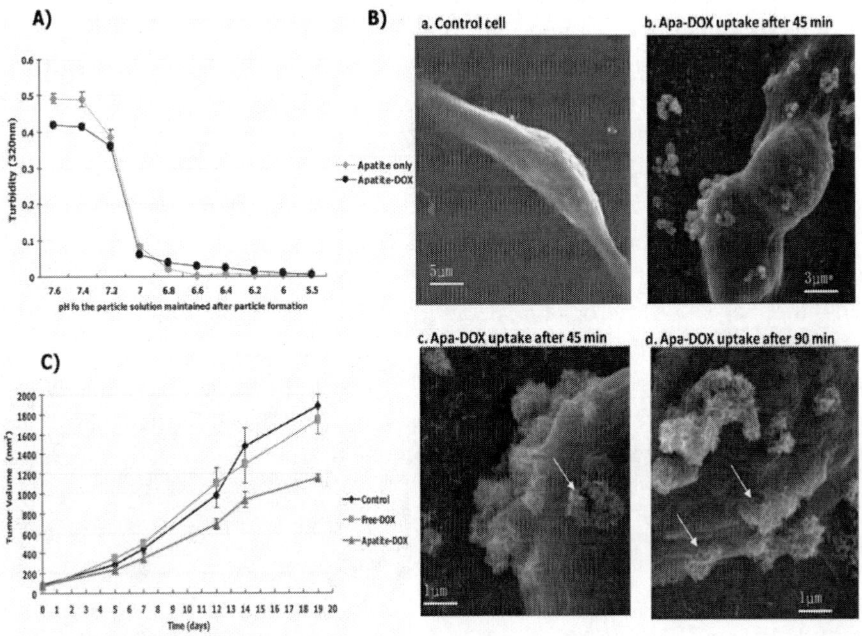

Figure 6. Intracellular Delivery of Doxorubicin/Carbonate Apatite Nanocomposites. A) Dissolution profiles of apatite-DOX particles under different pH conditions, B) Observation of cellular uptake behavior of carbonate apatite into HCT116 colon cancer cells by SEM and C) Inhibition of colorectal tumor growth in BALB/cA nude mice by the carbonate apatite-mediated doxorubicin delivery [21].

Therefore, the inorganic crystal had characteristics of nano-scale size which is effective especially for passive targeting. Moreover, enhanced cellular uptake and quick release of

DOX in response to endosomal low pH resulted in enhanced colorectal tumor inhibitory effects both in vitro and in vivo as shown in Figure 6. The extraordinary proliferation inhibitory effect in a number of cancer cell lines was achieved by virtue of its pH sensitivity resulting in quick release of the drugs from the carrier and probably overcoming the MDR and getting advantages over different slow release carrier systems. Thus, we invented a novel nano-based pH targeting therapeutic strategy against malignancies, which is highly promising for preclinical and clinical cancer therapy [21].

Conclusion

Nonviral approaches were developed to facilitate transfer of exogenous genes into target cells without the complication of immunogenicity or insertion mutation commonly seen in viral vectors. These methods differ widely in their transfection efficiency and toxicity. In the past few years, the work continued in developing new nonviral methods, particularly in the area of chemical vectors. Despite intensive efforts for the last three decades, there has been a lack of proper understanding of the molecular and cellular barriers to gene delivery, which could assist in developing a superior non-viral technique. Therefore, an increasing number of inorganic nanoparticles have been studied as drug or gene delivery carriers, because of their versatile physicochemical properties. Inorganic nanoparticles are readily available, can be easily functionalized, and possess good biocompatibility and low cytotoxicity. It is evident that the versatility of inorganic nanoparticles makes them very suitable as potential delivery carriers. However, the transfection efficiency of these inorganic nanohybrids is relatively low at this stage. Therefore an advanced technology for gene delivery to mammalian cells based on the properties of carbonate apatite has been developed which prevent crystal growth for the generation of nano-scale particles. These properties are beneficial for efficient endocytosis and the high-affinity binding of nucleic acids/proteic molecules for delivering DNA, siRNA/miRNA, protein and classical anticancer drugs to different primary, cancer and ES/iPS cells in either a specific or non-specific manner, and allow a fast dissolution rate in endosomal acidic compartments to facilitate the therapeutic release from the particles and endosomes. These novel approaches could pave the way to the wide potential applications of carbonate apatite, from laboratories to clinical medicine, and provide insights to create a new era for inorganic crystal-based gene and drug delivery.

References

[1] Luo D, Saltzman WM. Synthetic DNA delivery systems. *Nature Biotechnology* 2000; 18: 33–37.

[2] Chowdhury EH, Akaike T. Bio-functional inorganic materials: An attractive branch of gene-based nano-medicine delivery for 21st century. *Curr. Gene. Ther.* 2005; 5: 669–676.

[3] Reischl D, Zimmer A. Drug delivery of siRNA therapeutics: potentials and limits of nanosystems. *Nanomedicine* 2009; 5: 8–20.

[4] Pouton CW, Seymour LW. Key issues in non-viral gene delivery. *Adv. Drug Deliv. Rev.* 1998; 34: 3–19.
[5] Chowdhury EH, Maruyama A, Kano A, Nagaoka M, Kotaka M, Hirose S, et al. pH-sensing nano-crystals of carbonate apatite: Effects on intracellular delivery and release of DNA for efficient expression into mammaliancells. *Gene* 2006; 376: 87–94.
[6] Hossain S, Tada S, Akaike T, Chowdhury EH. Influences of electrolytes and glucose on formulation of carbonate apatite nanocrystals for efficient gene delivery to mammalian cells. *Anal. Biochem.* 2010; 397: 156–161.
[7] Duncan R. The dawning era of polymer therapeutics. *Nat. Rev. Drug Discov.* 2003; 2: 347–360.
[8] Kamimura K, Suda T, Zhang G, Liu D. Advances in Gene Delivery Systems. *Pharmaceut Med.* 2011; 25: 293-306.
[9] Prokop A, Davidson JM. Gene Delivery into Cells and Tissues. In: Lanza R, Langer R, Vacanti J, editors. *Princeples of Tissue Engineering*. Elsevier Academic Press, ABD; 2007. P. 493-515.
[10] Moghimi SM, Hunter AC, Murray JC. Long-circulating and target-specific nanoparticles: Theory to practice. *Pharmacol. Rev.* 2001; 53: 283–318.
[11] Witlox MA, Lamfers ML, Wuisman PI, Curiel DT, Siegal GP. Evolving gene therapy approaches for osteosarcoma using viral vectors: review. *Bone* 2007; 40: 797-812.
[12] Yang NS, Sun WH. Gene gun and other non-viral approaches for cancer gene therapy. *Nat. Med.* 1995; 1: 481-483.
[13] Aris A, Villlaverde A. Modular protein engineering for non-viral gene therapy. *Trends Biotechnol.* 2004; 22: 371–377.
[14] Medina-Kauwe LK, Xie J, Hamm-Alvarez S. Intracellular trafficking of nonviral vectors. *Gene. Ther.* 2005; 12: 1734–1751.
[15] Li W, Nicol F, Szoka FC Jr. GALA: a designed synthetic pH responsive amphipathic peptide with applications in drug and gene delivery. *Adv. Drug Deliv. Rev.* 2004; 23: 967–985.
[16] Akinc A, Thomas M, Klibanov AM, Langer R. Exploring polyethylenimine-mediated DNA transfection and the proton sponge hypothesis. *J. Gene. Med.* 2005; 7: 657–63.
[17] Lukacs GL, Haggie P, Seksek O, Lechardeur D, Freedman N, Verkman AS. Size-dependent DNA mobility in cytoplasm and nucleus. *J. Biol. Chem.* 2000; 275: 1625–1629.
[18] Hossain S, Chowdhury EH, Akaike T. Nanoparticles and toxicity in therapeutic delivery: The ongoing debate. *Ther. Deliv.* 2011; 2: 125–132.
[19] Hossain S, Akaike T, Chowdhury EH. Current approaches for drug delivery to central nervous system. *Curr. Drug Deliv.* 2010; 7: 389–397.
[20] Hossain S, Stanislaus A, Chua MJ, Tada S, Tagawa Y, Chowdhury EH, et al. Carbonate apatite-facilitated intracellularly delivered siRNA for efficient knockdown of functional genes. *J. Control. Release* 2010; 147: 101–108.
[21] Hossain S, Yamamoto H, Chowdhury EH, Wu X, Hirose H, Haque A, et al. Fabrication and intracellular delivery of doxorubicin/carbonate apatite nanocomposites: effect on growth retardation of established colon tumor. *PLoS One* 2013; 8: e60428.
[22] Matsuzaki J, Suzuki H, Tsugawa H, Watanabe M, Hossain S, Arai E, et al. Bile acids increase levels of microRNAs 221 and 222, leading to degradation of CDX2 during esophageal carcinogenesis. *Gastroenterology* 2013; 145: 1300-1311.

[23] Nag K, Hossain S, Naznin S, Chowdhury EH. PKC activation promotes internalization of DNA-immobilized inorganic nano-crystals by clathrin-dependent endocytosis for efficient transgene expression in human lymphocytes. *J. Nanomed. Nanotech*. 2013; 4: 174.

[24] Haque A, Hexig B, Meng Q, Hossain S, Nagaoka M, Akaike T. The effect of recombinant E-cadherin substratum on the differentiation of endoderm-derived hepatocyte-like cells from embryonic stem cells. *Biomaterials* 2011; 32: 2032–2042.

[25] Chowdhury EH, Akaike T. High performance DNA nano-carriers of carbonate apatite: Multiple factors in regulation of particle synthesis and transfection efficiency. I*nt. J. Nanomedicine* 2007; 2: 101–106.

[26] Chowdhury EH, Akaike T. A bio-recognition device developed onto nano-crystals of carbonate apatite for cell-targeted gene delivery. *Biotechnol. Bioeng*. 2005; 90: 414–421.

[27] Chowdhury EH, Nagaoka M, Ogiwara K, Zohra FT, Kutsuzawa K, Tada S, et al. Integrin-supported fast rate intracellular delivery of plasmid DNA by extracellular matrix protein embedded calcium phosphate complexes. *Biochemistry* 2005; 44: 12273–12278.

[28] Kutsuzawa K, Chowdhury EH, Nagaoka M, Maruyama K, Akiyama Y, Akaike T. Surface functionalization of inorganic nano-crystals with fibronectin and E-cadherin chimera synergistically accelerates trans-gene delivery into embryonic stem cells. *Biochem. Biophys. Res. Commun*. 2006; 350: 514–520.

[29] Grünweller A, Hartmann RK. RNA interference as a gene-specific approach for molecular medicine. *Curr. Med. Chem*. 2005; 12: 3143–3161.

[30] Stanislaus A, Bakhtiar A, Salleh D, Tiash S, Fatemian T, Hossain S, et al. Knockdown of PLC-gamma-2 and calmodulin 1 genes sensitizes human cervical adenocarcinoma cells to doxorubicin and paclitaxel. *Cancer Cell. Int*. 2012; 12: 30.

[31] Stanislaus A, Hossain S, Chua MJ, Kunnath AP, Wen QC, Syakira SN, et al. Fabrication and intracellular delivery of siRNA/carbonate apatite nano-composites for effective knockdown of cyclin B1 gene. *Drugs and Therapy Studies* 2011; 1: e8.

[32] Stanislaus A, Kunnath AP, Tiash S, Fatemian T, Kamaruzman NI, Bakhtiar A, et al. Intracellular delivery of NF-κB small interfering RNA for modulating therapeutic activities of classical anti-cancer drugs in human cervical cancer cells. *Drugs and Therapy Studies 2013*; 3: e7.

[33] Kutsuzawa K, Tada S, Hossain S, Fukuda K, Maruyama K, Akiyama Y, et al. Disrupting actin filaments promotes efficient transfection of a leukemia cell line using cell adhesive protein-embedded carbonate apatite particles. *Anal. Biochem*. 2009; 388: 164-166.

[34] Tada S, Chowdhury EH, Cho CS, Akaike T. pH-sensitive carbonate apatite as an intracellular protein transporter. *Biomaterials* 2010; 31: 1453–1459.

In: Apatite
Editors: M. Iafisco and J. M. Delgado-López
ISBN: 978-1-63321-500-9

Chapter 10

APATITE ENRICHMENT OF HYDROGEL BIOMATERIALS FOR BONE REGENERATION

Timothy E. L. Douglas*
Department of Molecular Biotechnology, Ghent University,
Gent, Belgium

ABSTRACT

Hydrogels are gaining ground as materials for bone regeneration due to advantages including injectability and ease of incorporation of bone-forming cells and bioactive substances, such as growth factors and mineralization-promoting enzymes. The incorporation of a mineral phase is considered desirable for a number of reasons. Firstly, the presence of mineral promotes bioactivity, i.e. the formation of a direct bond with surrounding bone. Secondly, the mineral phase can improve binding of bioactive substances (e.g. growth factors) or itself serve as a delivery vehicle for such substances. Thirdly, the increase in mechanical strength and roughness due to the presence of mineral can stimulate osteogenic differentiation of bone-forming cells. Fourthly, the presence of mineral can improve cell adhesion, proliferation, differentiation and further, cell-mediated mineralization. This book chapter aims to review the literature concerning apatite-enriched hydrogel composite biomaterials, with special emphasis on strategies to synthesize or incorporate apatite within hydrogels. The strategies can be subdivided into the following general categories, which may be used alone or in combination: a) incorporation of pre-formed apatite particles; b) creation of nucleation sites for apatite growth by incorporation of solid particles, calcium-binding biomolecules or functional chemical groups; c) delivery of calcium and phosphate ions through incubation in physiological mineralization solutions, pre-, alternate or post-soaking in superphysiological solutions, double diffusion or assisted diffusion strategies or formation of apatite particles during hydrogel formation and d) elevation of calcium and phosphate concentrations within the hydrogel through enzymatic or cellular action. Studies concerning the influence of the presence of the apatite component on biological properties, both in vitro and in vivo, will also be discussed.

* Corresponding author: Email: Timothy.Douglas@UGent.be.

Keywords: Hydrogel, composite, bone, apatite, mineralization

INTRODUCTION

Hydrogels, or crosslinked polymer networks containing entrapped water, are widely applied as materials for drug delivery and as tissue engineering scaffold materials for regeneration of certain tissues, including cartilage, interverterbral discs and soft tissues. Recently, hydrogels have also been gaining ground as materials for bone regeneration due to a number of inherent advantages.

One advantage is injectability, allowing minimally-invasive implantation with limited scarring, patient discomfort and hospitalization time. In addition, injectable materials can fit a defect site perfectly. Other advantages are connected with the fact that the main component of a hydrogel is water and that hydrogels are formed through a sol-gel transition from polymer solution to hydrogel polymer network. Bone-forming cells (e.g. osteoblasts, mesenchymal stem cells (MSC)) can be easily incorporated in a polymer solution before gelation. After gelation, cells remain entrapped within the hydrogel polymer network with uniform distribution. Nutrients, oxygen and dissolved biochemical cues (e.g. drugs, growth factors) can reach cells entrapped in the hydrogel polymer network by diffusion. Waste products of cell metabolism and biochemical cues produced by the entrapped cells can diffuse out of the hydrogel polymer network.

It is straightforward to incorporate not only cells, but also nano- and microparticles. These can improve mechanical properties and promote survival, vitality and osteogenic differentiation of bone-forming cells. Furthermore, soluble bioactive substances, such as growth factors, drugs and mineralization-promoting enzymes, can also be easily incorporated by addition to the polymer solution prior to gelation. Such bioactive substances are entrapped within the hydrogel polymer network. Their rate of release from the hydrogel polymer network is determined by a range of factors. These include the size and conformation of the entrapped substance, and the pore size of the hydrogel polymer network. These factors affect diffusion of the substance out of the hydrogel. Diffusion is also affected by interactions between the substance and the hydrogel polymer network (e.g. ionic, hydrophobic).

Current ceramic materials for bone tissue regeneration suffer from certain shortcomings. They are difficult to handle, brittle, do not fit defect sites exactly, and certain types are poorly resorbable. The use of hydrogels offers the possibility of overcoming these drawbacks.

However, the main drawback of hydrogels, their low mechanical strength, is also connected with the fact that the main component is water. This can lead to loss of integrity and failure if excessive loads are applied. Furthermore, cells incorporated in hydrogels may bind to the polymer network and exert tensile forces, which can lead to hydrogel shrinkage.

The advantages and disadvantages of hydrogel biomaterials are summarized in Table 1.

Numerous studies in the last 10 years have involved development of composite materials consisting of a hydrogel and a mineral phase, most often a form of calcium phosphate (CaP). The presence of a mineral phase is intended to endow hydrogel-mineral composites with some of the advantages of CaP-based ceramic materials while retaining the advantages of hydrogel materials. The mineral phase is considered desirable to promote bioactivity, or the formation of a direct bond with surrounding bone [1]. Furthermore, the mineral phase can encourage binding of external bioactive substances such as growth factors [2] and can itself serve as a delivery vehicle for bioactive substances [3, 4]. In addition, the mineral phase can reinforce the hydrogel, increasing stiffness and thus promoting osteogenic differentiation of cells [5]. It is conceivable that increased surface roughness caused by the presence of mineral may also stimulate osteogenic differentiation [6, 7]. Finally, cell adhesion, proliferation, differentiation and cell-mediated mineralization may be improved by the presence of mineral in the hydrogel [8-11].

One of the most popular mineral phases, if not the most popular, used to enrich hydrogels is apatite. Calcium-deficient hydroxyapatite (CDHA) is the form of CaP present in bone tissue. This article aims to review the literature concerning apatite-enriched hydrogel composite biomaterials, with special emphasis on:

1) Biological effects of apatite enrichment of hydrogels
    i) in vitro
    ii) in vivo
2) Strategies to synthesize or incorporate apatite within hydrogels

Table 1. Advantages and disadvantages of hydrogel biomaterials

Advantages	Disadvantages

• Injectability, allowing minimally-invasive implantation
• Ease of incorporation of
  • Cells
  • Bioactive substances
  • Growth factors
  • Enzymes
  • Micro-, nanoparticles
• Highly hydrated environment
  • Delivery of bioactive substances by diffusion
  • Easy diffusion to incorporated cells
    • Nutrients
    • Oxygen
    • Biochemical signals
  • Easy diffusion from incorporated cells
    • Metabolic waste products
    • Cell-produced biochemical signals
• Not brittle
• Mechanical stimulation of cells possible
• Good resorbability
• Certain hydrogels of mammalian origin:
  • Binding sites for cells
	• Lack of mechanical strength
	  • Non-load-bearing
	  • Stiffness not suitable for osteogenic cell differentiation
	  • If cells incorporated:
	    • shrinkage
	• If of mammalian origin:
	  • Concerns connected with disease transmission
	• If of non-mammalian origin
	  • Lack of binding sites for cells
	• Poor protein adsorption due to hydrophilicity
	  • Poorer cell adhesion

These can be subdivided into the following general categories, which may be used alone or in combination:

i) incorporation of pre-formed apatite particles
ii) creation of nucleation sites for apatite growth by:
    iia) incorporation of solid particles
    iib) incorporation of calcium-binding biomolecules
    iic) functionalization with chemical groups with a high affinity for apatite
iii) delivery of calcium and phosphate ions through:
    iiia) incubation in physiological mineralization solutions
    iiib) pre-soaking, alternate soaking and post-soaking
    iiic) double diffusion and assisted diffusion methods
    iiid) simultaneous apatite and hydrogel formation
iv) elevation of calcium and phosphate concentrations within the hydrogel through:
    iva) enzymatic action
    ivb) cellular action

A graphical overview of the aforementioned strategies is given in Figure 1.

Biological Effects of Apatite Enrichment of Hydrogels

Incorporation of apatite into hydrogels has positively impacted behavior of bone-forming cells in vitro and promoted bone regeneration in vivo. Although there are numerous strategies to enrich hydrogels by apatite, there are relatively few biological studies involving biological characterization of apatite-hydrogel composites in vitro or in vivo. In the majority of these studies, hydrogels were enriched with apatite by addition of pre-formed apatite particles. Hence, biological assessment of hydrogels enriched by apatite by other methods remains an unexplored area.

In Vitro

Studies on bone marrow-derived MSC have demonstrated the ability of incorporated apatite nanoparticles to stimulate MSC proliferation, osteogenic differentiation and MSC-induced hydrogel mineralization. Bongio et al. incorporated 21% w/v apatite into 15% w/v hydrogels of oligo(poly(ethylene glycol) fumarate) (OPF), which were loaded with rat bone marrow-derived MSC at an initial cell concentration of 10^6 cells/ml hydrogel [9]. Significantly higher numbers of rat bone marrow-derived MSC were observed after 8, 12, 16 and 24 days of culture, while significantly greater cell-mediated mineral production was detected after 16 and 24 days of culture.

Patel et al. incorporated rat bone marrow-derived MSC into cyclic acetal hydrogels synthesized from poly(ethylene glycol) diacrylate (PEGDA) and 5-ethyl-5-(hydroxymethyl)-b, b-di- methyl-1,3-dioxane-2-ethanol diacrylate (EHD) at an initial cell concentration of 10^6 cells/ml hydrogel [10]. Incorporation of 5 and 50 ng apatite/ml hydrogel increased gene expression of bone morphogenetic protein-2 (BMP-2) and osteocalcin (OC) significantly after 1, 4 and 8 days of culture, while 50 ng apatite/ml hydrogel significantly raised alkaline phosphatase (ALP) expression after 8 days of culture, suggesting enhanced osteoblastic differentiation. 50 ng apatite/ml hydrogel also increased immunohistochemical staining for OC after 1 day of culture.

Anderson et al. incorporated 50% w/v apatite into 2% w/v peptide amphiphile (PA) hydrogels functionalized with the RGDS (Arg-Gly-Asp-Ser) ligand. Human bone marrow MSC were also incorporated (50000 cells/hydrogel). Apatite incorporation promoted gene expression of collagen type I and OC after 14 days and collagen type I after 28 days compared to apatite-free hydrogels [12].

In Vivo

Anderson et al. implanted the aforementioned PA hydrogels into critical-size (6 mm) rat femoral defects. Hydrogels were immobilized at the defect site by means of an intramedullary nail. Apatite-free hydrogels promoted callus formation, leading to narrowing of the defect. Enrichment of the hydrogel with 50% apatite promoted new bone formation to a greater extent, allowing bridging of the defect and closure [12].

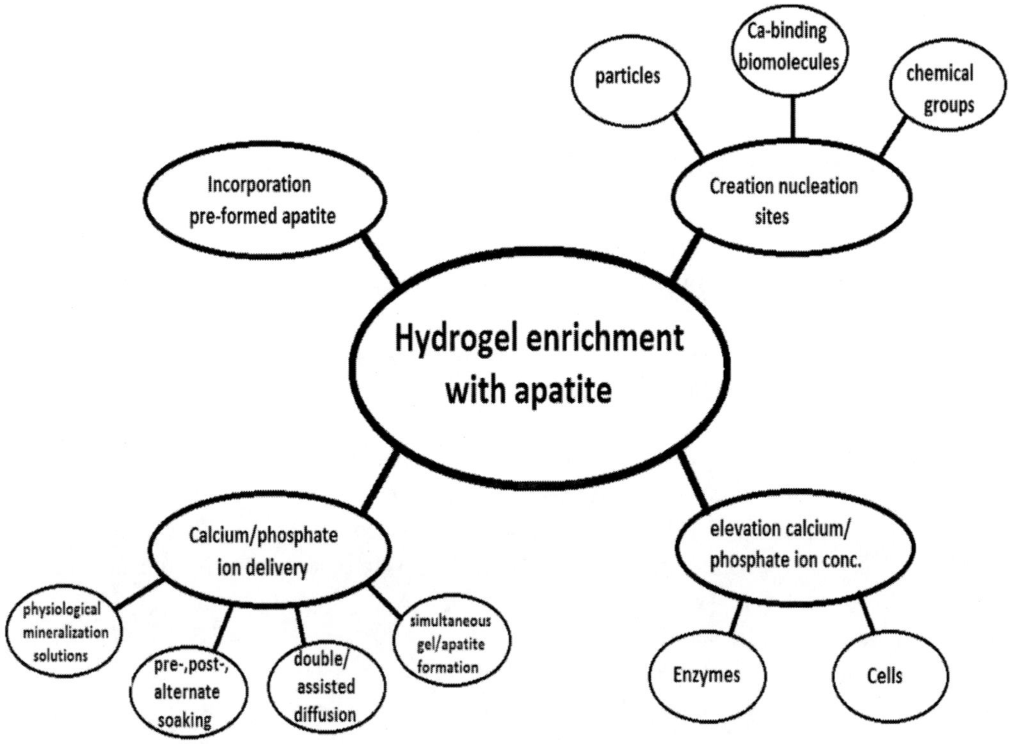

Figure 1. Graphical overview of strategies to enrich hydrogel biomaterials with apatite.

Suzawa et al. induced apatite formation in agarose hydrogels by an alternative soaking method. Agarose hydrogels were alternately soaked in solutions of 0.2 M $CaCl_2$ at pH 7.4 and 0.12 M Na_2HPO_4 at 4 °C. Each immersion lasted 1 hour and the alternate soaking cycle was repeated 12 times. This lead to formation of 1.5 mg apatite/hydrogel (diameter 4 mm, thickness 0.5 mm), equivalent to 23.9 % w/v. Hydrogels, both pure and apatite-enriched, were implanted into cranial defects of diameter 4 mm in six-week old male Wistar rats. Two implantation methods were compared. The first method involved homogenization of mineralized discs to produce a substance with a consistency similar to that of toothpaste, which was subsequently implanted in the defect site. The second method involved direct implantation of hydrogels in the defect site without homogenization. Both homogenized and non-homogenized apatite-enriched hydrogels promoted new bone formation and defect closure to a greater extent than pure hydrogels after 8 weeks implantation. This was demonstrated by micro-computer tomography analysis [13].

Federovich et al. implanted Matrigel hydrogels enriched with apatite nanoparticles subcutaneously into immunodeficient mice. After 6 weeks, apatite nanoparticles acted osteoinductively and activated osteoclasts [14].

Hulsart-Billstrom et al. used aldehyde-modified hyaluronic acid hydrogels enriched with the osteogenic growth factor bone morphogenetic protein 2 (BMP-2) at a concentration of 30 µg/0.2 ml hydrogel, or 150 µg/ml hydrogel. These were enriched with apatite particles of different sizes at concentrations of 50 µg/0.2 ml hydrogel, or 250 µg/ml hydrogel, and implanted at ectopic sites in rats, more specifically the quadriceps muscle [15]. Computerised tomography studies revealed that apatite nanoparticles of size 20 nm induced formation of denser bone after 4 weeks than apatite microparticles in the range 1-50 µm and greater than 100 µm.

STRATEGIES TO SYNTHESIZE/INCORPORATE APATITE IN HYDROGELS

Incorporation of Pre-Formed Apatite Particles

One of the most simple and widely applied methods to incorporate apatite in hydrogels is simple addition of apatite particles. These have been shown to serve as nucleation sites for further mineral growth in hydrogels such as OPF [16] and alginate [17].

There are numerous publications reporting mechanical reinforcement of hydrogels by incorporation of apatite particles. Hydrogel materials thus reinforced include benzyl ester of hyaluronic acid (HYAFF11) [18], poly(lactide ethylene oxide fumarate) (PLEOF) [19], acrylamide [20], poly(ethylene glycol) (PEG) [21], poly(ethylene glycol) maleate citrate (PEGMC) [22], chitosan/β-glycerophosphate [23], poly(vinyl alcohol) (PVA) [24], poly(lactic acid-co-glycolic acid)-g-poly(ethylene glycol) (PLGA-g-PEG) [25], poly(2-hydroxyethyl methacrylate) (pHEMA) [26] and OPF [9].

In most of the aforementioned cases, increasing apatite content also increased mechanical strength. However, there have been reports of composite instability above certain apatite contents, e.g. 21% w/v in the case of OPF [9, 27] and 10% in the case of PEG [21]. Conceivably, particles may hinder formation of the hydrogel polymer network, leading to mechanical weakness of the polymer network. There may be an optimal apatite particle content for each hydrogel type. For example, the mechanical strength of PA hydrogel-apatite composites was highest at an apatite content of 50% w/v, higher than at 37 % w/v and 66% w/v [12]. The effect of apatite particle content on mechanical strength may be influenced by particle size, size distribution and polymer concentration in the hydrogel polymer network.

It should be mentioned that a further application of apatite particles is as a slow release system for calcium ions (in combination with glycerol and the acidifier glucono-delta-lactone (GDL)) to crosslink alginate and induce alginate hydrogel formation [17].

Creation of Nucleation Sites for Apatite Growth

Incorporation of Solid Particles

An alternative to direct incorporation of pre-formed apatite particles is incorporation of other ceramic particles which serve as templates for apatite growth in a suitable mineralization solution, e.g. simulated body fluid (SBF), which contains mineral ions at concentrations similar to those found in blood plasma. Addition of bioglass particles to collagen hydrogels and chitosan/β-glycerophosphate hydrogels resulted in apatite formation after incubation in SBF [28-30].

Incorporation of Calcium-Binding Biomolecules

Another alternative is the incorporation of calcium-binding biomolecules in the hydrogel network. One biopolymer with a high affinity for calcium is alginate, which has been incorporated into gelatin hydrogels to promote apatite formation [31]. Methacrylated alginate grafted onto PEG-monomethacrylate and poly(propylene glycol)-monomethacrylate has promoted apatite formation [32]. Another such biopolymer is bone sialoprotein (BSP), which has promoted apatite formation in collagen hydrogels [33].

Functionalization with Chemical Groups with a High Affinity for Apatite

Yet another strategy is the use of peptide sequences which display an affinity for apatite, such as those rich in glutamic acid (Glu). Short Glu-groups covalently crosslinked to PLEOF hydrogels have improved interactions with apatite nanoparticles dispersed in the hydrogel, leading to increased mechanical strength [19]. Hydrogels based on crosslinked polyglutamic acid (PGA) have shown high apatite-forming ability in SBF, which was ascribed to the high number of carboxyl groups [34]. It was found that increasing molecular weight of PGA leads to higher mineralizability, which the authors claimed was due to a higher availability of carboxyl groups after crosslinking. Previous work by the same group on PGA hydrogels revealed that pre-treatment with $CaCl_2$ before incubation in SBF favored formation of apatite over amorphous CaP [35]. Other peptides derived from silk fibroin have promoted apatite formation in dense collagen hydrogels [36].

Delivery of Calcium and Phosphate Ions

Incubation in Physiological Mineralization Solutions

The most commonly used mineralization solution is SBF. Immersion in SBF is considered to simulate mineralization behavior in vivo and thus serves as a test of a material's mineralizability under physiological conditions. Several hydrogels have been mineralized with apatite through immersion in SBF, including crosslinked pectin [37], poly(ethylene imine) (PEI) [38], chitosan [39] and poly(N-isopro-pylacryamide)-co-poly(ethylene glycol) dimethacrylate with added tri-methacryloxypropyltrimethoxysilane (MPS) [40].

Pre-soaking, Alternate Soaking and Post-soaking

Some authors have presoaked hydrogels in calcium solutions so that calcium ions bind to the hydrogel. These bound calcium ions subsequently act as nucleation sites for CaP growth upon incubation in SBF. In this manner, apatite formation has been promoted in PGA hydrogels [35, 41].

Alternate soaking strategies involve alternate immersion of a hydrogel in solutions containing calcium ions and phosphate ions. This immersion cycle can be repeated to increase the amount of CaP formed. Using this strategy, apatite has been formed in hydrogels of PVA [42, 43] chitosan [44] and cellulose [45].

Poorly crystalline CaP formed in hydrogels during soaking can be converted into apatite by post-soaking in an alkali solution. In this way, amorphous CaP formed in chitosan hydrogels by an alternate soaking method was transformed into apatite by post-soaking in 5% NaOH [46, 47].

Another post-treatment strategy, only suitable for hydrogels which can withstand high temperatures, is hydrothermal treatment to convert another mineral phase into apatite. Goto et al. loaded PVA hydrogels with alpha- en beta-TCP particles, which transformed into apatite upon hydrothermal treatment with water vapor at 160ºC [48].

Double Diffusion and Assisted Diffusion

Certain groups have used double-diffusion systems, where calcium and phosphate ions diffuse from different reservoirs into the hydrogel. For a review of such systems, the reader is referred to the review by Dorvee et al. [49]. As an example, Azami at al. employed a double diffusion system to induce brushite formation in gelatin hydrogels [50], which was converted into apatite by subsequent post-soaking in SBF.

Other groups have taken the strategy of double diffusion a step further by using current-assisted diffusion techniques such as electrophoresis. Agarose and polyacrylamide hydrogels have been mineralized with apatite in this manner [51-53].

Simultaneous Apatite and Hydrogel Formation

One potential problem with the addition of pre-formed apatite particles to hydrogels (see above) is aggregation and non-homogeneous distribution. One strategy is the addition of dispersants such as citrate, which has improved distribution of apatite nanoparticles in gelatin hydrogels [54]. Another strategy is simultaneous apatite and hydrogel formation, also referred to as "chemical mixing" [27], where the inorganic reagents for apatite formation are added during gelation and react to form apatite particles within the hydrogel. Advantages of chemical mixing are more uniform distribution of apatite particles and less agglomeration [27]. However, if cell encapsulation is desired, the reagents added may prove toxic to cells. Chemical mixing has been used to generate apatite in OPF [27], alginate [55, 56], chitosan [57] and chitosan-collagen hydrogels [58]. Another technique is re-swelling of a dried hydrogel in a mineralization solution. PEG hydrogels containing N-acrloyl-6-aminocaproic acid (A6ACA) moieties have been mineralized with apatite by re-swelling in SBF [11].

Elevation of Calcium and Phosphate Concentrations through Enzymatic or Cellular Action

Enzymatic Action

The enzyme ALP is responsible for mineralization of bone tissue in vivo and acts by cleaving phosphate groups from organophosphates. Formation of a metal phosphate inside hydrogels can be induced by incorporation of ALP into the polymer network during the hydrogel formation process, followed by incubation of the hydrogel in a solution containing divalent metal ions, e.g. Ca^{2+}, Mg^{2+}, and an organophosphate substrate for ALP, e.g. glycerophosphate [8, 59-63]. Soluble metal ions and organophosphate enter the hydrogel by diffusion. Subsequently free phosphate ions are cleaved from organophosphate under the action of ALP. These phosphate ions are then free to react with metal ions to form an insoluble metal phosphate which precipitates and remains entrapped within the hydrogel. Byproducts of this reaction, such as the organic part of organophosphate, are able to exit the hydrogel by diffusion. By far the most popular metal ion used in literature is Ca^{2+}, while the most popular organophosphate used as a substrate for ALP is glycerophosphate.

A schematic overview of enzymatic mineralization of hydrogels under the action of ALP is given in Figure 2.

This approach is applicable to the vast majority of hydrogel materials. ALP can be entrapped physically in hydrogels without the need for an extra crosslinking step, which avoids the use of potentially toxic crosslinkers (e.g. glutaraldehyde). ALP has a higher molecular weight (185 kD) [64] than growth factors widely used in regenerative medicine. For comparison, the molecular weight of transforming growth factor beta 1 (TGF-beta 1) is 25 kD [65]. Hence, the rate of release of ALP is expected to be slower due to physical entrapment. Release studies of hydrogels containing 2.5 mg ALP/ml hydrogel have shown negligible release after 6 hours from catechol-PEG (cPEG) (200 mg cPEG/ml hydrogel) and OPF (100 mg OPF/ ml hydrogel), approximately 30% release from both collagen (3 mg collagen/ml hydrogel) and gellan gum (GG) (7 mg GG/ml hydrogel) [8, 60]. ALP has also been enclosed in microparticles of sodium alginate/hydroxypropylcellulose [66]. ALP was released slowly upon immersion of microparticles in a solution of calcium glycerophosphate (CaGP) and induced apatite formation.

The type of CaP formed in hydrogels is dependent on mineralization medium and hydrogel type. Mineralization of ALP-loaded chitosan hydrogels [61] and ALP-soaked collagen sheets [67-69] led to formation of a combination of apatite and amorphous CaP. Predominantly apatite formed in ALP-loaded cPEG and GG hydrogels [8, 60], as well as polyHEMA hydrogels [62]. Two hydrogel types based on fibrils formed from artificial peptides differed in their ability to promote apatite formation [70]. CaP but not apatite was not detected in platelet-rich fibrin (PRF) containing entrapped ALP [59, 71].

It should be mentioned that the concentrations of calcium ions, glycerophosphate and ALP differed greatly in these studies. Certain studies use superphysiological concentrations (0.1 M CaGP, 2.5 mg ALP/ml hydrogel or higher) in order to achieve faster mineralization [8, 59-61, 72]. Other researchers have chosen concentrations more similar to physiological values in order to mimic the mineralization process in vivo more closely [62, 63, 70].

The vast majority of studies on ALP-mediated mineralization of hydrogels aim at the formation of CaP within the hydrogel. However, in principle, other metal ions (e.g. Mg^{2+},

Zn^{2+}, Sr^{2+}) can be similarly incorporated into the CaP phase of mineralized hydrogels by addition of the ions to the mineralization solution.

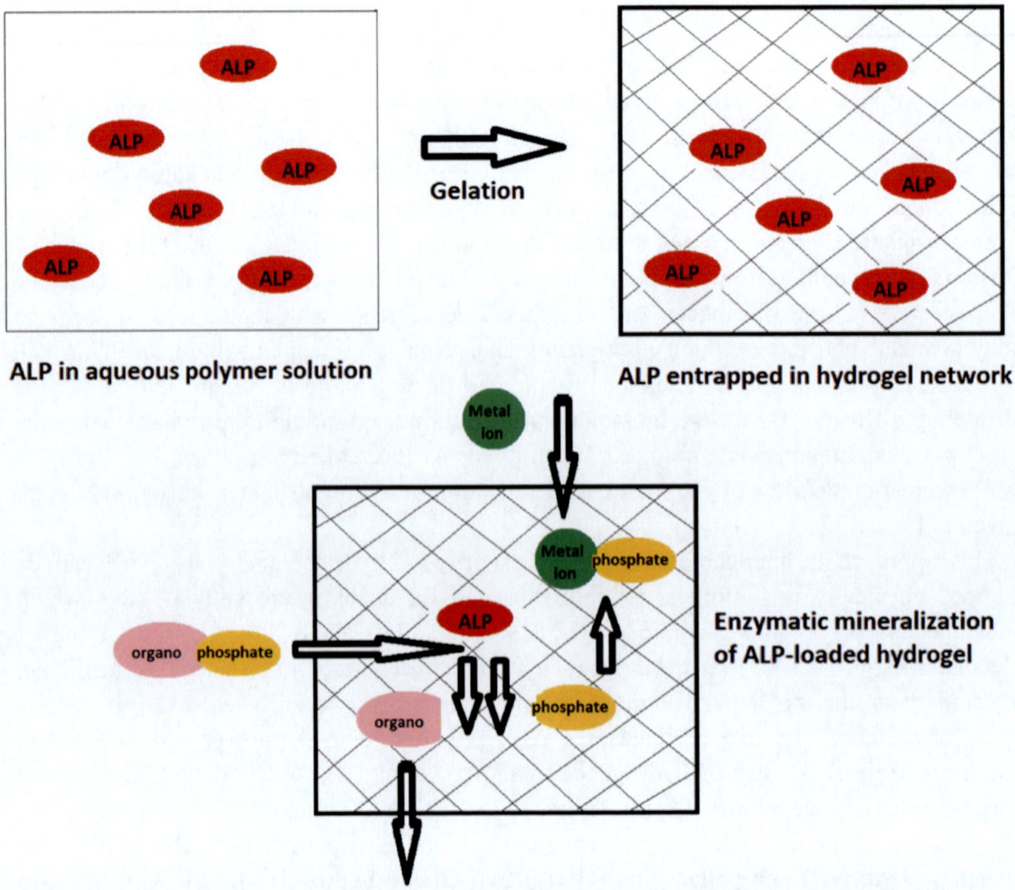

Figure 2. Schematic representation of enzymatic mineralization of hydrogels via the action of alkaline phosphatase (ALP). Soluble ALP can be added to a polymer solution before gelation (top left). After gelation, the ALP remains physically entrapped to a certain extent within the hydrogel polymer network (top right). Upon incubation in a mineralization solution (bottom) an ALP substrate, namely soluble organophosphate, enters the hydrogel by diffusion. Under the enzymatic action of ALP, phosphate is cleaved from the organic part. The organic part is then free to exit the hydrogel by diffusion. Metal ions (e.g. Ca^{2+}, Mg^{2+}) diffuse into the hydrogel and react with the cleaved phosphate ions to form insoluble metal phosphates. These precipitate and remain entrapped within the hydrogel network.

The presence of other metal ions, e.g. Mg^{2+}, can favor formation of amorphous CaP if the concentration is sufficiently high. A CaGP:magnesium glycerophosphate (MgGP) ratio of 0.075 M:0.025 M in mineralization medium led to formation of magnesium-doped apatite, however ratios of 0.05 M:0.05 M and 0.025 M:0.075 M led to formation of amorphous CaP containing magnesium [73]. In the same study, the presence of magnesium in mineral, both apatite and amorphous CaP, led to improved adhesion of osteoblast-like MC3T3-E1 cells. Cell proliferation was also improved after 11 days of culture. Hydrogels mineralized with apatite and magnesium-doped apatite supported formation of multinucleated osteoclast cells from RAW 264.7 osteoclast precursor cells.

It is obvious from the abovementioned literature that judicious choice of hydrogel material, polymer concentration within the hydrogel, ALP concentration in the hydrogel, and concentrations of calcium, ALP substrate and other metal ions are necessary to tailor the type and amount of CaP formed and the reaction of cells cultivated on the hydrogel-CaP composites. Variation in the aforementioned parameters hinders comparison of results from different studies.

Cellular Action

There are numerous publications on the ability of bone-forming cells to generate bone tissue, of which apatite is the main component. For reasons of brevity, only publications dealing with the formation and detection of apatite in hydrogels by cellular action will be discussed here.

DeVolder et al. incorporated poly(lactic- co-glycolic acid) (PLGA) microparticles within collagen hydrogels with the aim of raising rigidity, thus stimulating mineral formation by mouse-derived MSC [74]. Incorporation of 1 % w/v PLGA microparticles in a 0.3% w/v collagen hydrogel increased storage modulus by an order of magnitude at measuring frequencies of 1 and 5 Hz. Mouse-derived MSC were incorporated at an initial density of 10^6 cells/ml hydrogel and cultured in osteogenic differentiation medium. An increase in elastic modulus thanks to the presence of 10% w/v PLGA microparticles after 5 and 7 days of culture was observed. Enhanced mineral formation due to the presence of PLGA microparticles was also demonstrated by histological analysis and von Kossa staining for phosphate deposits, which were claimed to be apatite.

Coyac et al. incorporated dental pulp cells in dense collagen hydrogels [75]. These dense collagen scaffolds were produced by a plastic compression method involving compression of a 0.2% w/v collagen hydrogel with simultaneous absorption of water which was pressed out of the hydrogel as a result of compression. The result was a dense collagen hydrogel with a collagen concentration of over 10% w/v. Dental pulp cells at a seeding density of 1.5×10^5 cells/ml hydrogel were incorporated into 0.2 % w/v collagen hydrogels prior to the plastic compression procedure. Cells were cultured for 24 days. Osteogenic differentiation of pulp cells was confirmed by increased production of ALP after 8 and 16 days of culture, as well as by gene expression of the osteogenic markers ALP and osteopontin (OPN). Mineralization was proven by scanning electron microscopy (SEM) and transmission electron microscopy (TEM) of mineral deposits. Furthermore, biochemical assays to measure levels of calcium and phosphate revealed significantly higher levels of both elements within cell-loaded dense collagen hydrogels compared to acellular hydrogels after 8, 16 and 24 days of culture. Formation of apatite within dense collagen hydrogels after 24 days of culture was confirmed by Infrared Spectroscopy (IR), SEM imaging of mineral deposits characteristic for apatite, TEM imaging of nanocrystals with a needle-like shape typical for apatite as well as TEM-assisted selected area electron diffraction (SAED), which revealed a diffraction pattern characteristic for apatite.

Conclusion

A wide range of strategies have been applied to incorporate an apatite phase in hydrogels in vitro. The most popular method remains the incorporation of pre-formed apatite particles. More in vitro and in vivo work is required to elucidate the biological effects of incorporation of an apatite phase in hydrogels and the potential of hydrogel-apatite composites for bone regeneration. In particular, in vitro and in vivo characterization of hydrogels enriched with apatite by a method other than incorporation of pre-formed apatite particles is lacking.

Acknowledgment

The author Timothy E.L. Douglas thanks the Research Foundation Flanders (FWO), Belgium for financial support in the framework of a postdoctoral fellowship.

References

[1] LeGeros, RZ. *Calcium Phosphates in Oral Biology and Medicine.* San Francisco: Karger, 1991.

[2] Ruhe, PQ; Boerman, OC; Russel, FG; Spauwen, PH; Mikos, AG; Jansen, JA. Controlled release of rhBMP-2 loaded poly(dl-lactic-co-glycolic acid)/calcium phosphate cement composites in vivo. *J Control Release*. 2005 Aug 18, 106(1-2), 162-71.

[3] Iafisco, M; Di, Foggia, M; Bonora, S; Prat, M; Roveri, N. Adsorption and spectroscopic characterization of lactoferrin on hydroxyapatite nanocrystals. *Dalton Trans*. 2012 Jan 28, 40(4), 820-7.

[4] Tang, Y; Du, Y; Li, Y; Wang, X; Hu, X. A thermosensitive chitosan/poly(vinyl alcohol) hydrogel containing hydroxyapatite for protein delivery. *J Biomed Mater Res A*. 2009 Dec 15, 91(4), 953-63.

[5] Discher, DE; Janmey, P; Wang, YL. Tissue cells feel and respond to the stiffness of their substrate. *Science*. 2005 Nov 18, 310(5751), 1139-43.

[6] Boyan, BD; Lossdorfer, S; Wang, L; Zhao, G; Lohmann, CH; Cochran, DL; et al. Osteoblasts generate an osteogenic microenvironment when grown on surfaces with rough microtopographies. *Eur Cell Mater*. 2003 Oct 24, 6, 22-7.

[7] Evans, ND; Minelli, C; Gentleman, E; LaPointe, V; Patankar, SN; Kallivretaki, M; et al. Substrate stiffness affects early differentiation events in embryonic stem cells. *Eur Cell Mater*. 2009, 18,1-13, discussion -4.

[8] Douglas, T; Wlodarczyk, M; Pamula, E; Declercq, H; de, Mulder, E; Bucko, M; et al. Enzymatic mineralization of gellan gum hydrogel for bone tissue-engineering applications and its enhancement by polydopamine. *J Tissue Eng Regen Med*. 2012 Oct 5.

[9] Bongio, M; van, den, Beucken, JJ; Nejadnik, MR; Leeuwenburgh, SC; Kinard, LA; Kasper, FK; et al. Biomimetic modification of synthetic hydrogels by incorporation of

adhesive peptides and calcium phosphate nanoparticles: in vitro evaluation of cell behavior. *Eur Cell Mater*. 2011, 22, 359-76.

[10] Patel, M; Patel, KJ; Caccamese, JF; Coletti, DP; Sauk, JJ; Fisher, JP. Characterization of cyclic acetal hydroxyapatite nanocomposites for craniofacial tissue engineering. *J Biomed Mater Res A*. 2010 Aug, 94(2), 408-18.

[11] Phadke, A; Shih, YR; Varghese, S. Mineralized synthetic matrices as an instructive microenvironment for osteogenic differentiation of human mesenchymal stem cells. *Macromol Biosci*. 2012 Aug, 12(8), 1022-32.

[12] Anderson, JM; Patterson, JL; Vines, JB; Javed, A; Gilbert, SR; Jun, HW. Biphasic peptide amphiphile nanomatrix embedded with hydroxyapatite nanoparticles for stimulated osteoinductive response. *ACS Nano*. 2011 Dec 27, 5(12), 9463-79.

[13] Suzawa, Y; Funaki, T; Watanabe, J; Iwai, S; Yura, Y; Nakano, T; et al. Regenerative behavior of biomineral/agarose composite gels as bone grafting materials in rat cranial defects. *J Biomed Mater Res A*. 2010 Jun 1, 93(3), 965-75.

[14] Fedorovich, NE; Leeuwenburgh, SC; van der Helm, YJ; Alblas, J; Dhert WJ. The osteoinductive potential of printable, cell-laden hydrogel-ceramic composites. *J Biomed Mater Res A*. 2012 Sep, 100(9), 2412-20.

[15] Hulsart-Billstrom, G; Hu, Q; Bergman, K; Jonsson, KB; Aberg, J; Tang, R; et al. Calcium phosphates compounds in conjunction with hydrogel as carrier for BMP-2: a study on ectopic bone formation in rats. *Acta Biomater*. 2011 Aug, 7(8), 3042-9.

[16] Nejadnik, MR; Mikos, AG; Jansen, JA; Leeuwenburgh, SC. Facilitating the mineralization of oligo(poly(ethylene glycol) fumarate) hydrogel by incorporation of hydroxyapatite nanoparticles. *J Biomed Mater Res A*. 2012 May, 100(5), 1316-23.

[17] Cardoso, DA; van den Beucken, JJ; Both, LL; Bender, J; Jansen, JA; Leeuwenburgh, SC. Gelation and biocompatibility of injectable alginate-calcium phosphate gels for bone regeneration. *J Biomed Mater Res A*. 2014 Mar, 102(3), 808-17.

[18] Sanginario, V; Ginebra, MP; Tanner, KE; Planell, JA; Ambrosio, L. Biodegradable and semi-biodegradable composite hydrogels as bone substitutes: morphology and mechanical characterization. *J Mater Sci Mater Med*. 2006 May, 17(5), 447-54.

[19] Sarvestani, AS; He, X; Jabbari, E. Effect of osteonectin-derived peptide on the viscoelasticity of hydrogel/apatite nanocomposite scaffolds. *Biopolymers*. 2007 Mar, 85(4), 370-8.

[20] Tolga, Demirtas, T; Karakecili, AG; Gumusderelioglu, M. Hydroxyapatite containing superporous hydrogel composites: synthesis and in-vitro characterization. *J Mater Sci Mater Med*. 2008 Feb, 19(2), 729-35.

[21] Paxton, JZ; Donnelly, K; Keatch, RP; Baar, K. Engineering the bone-ligament interface using polyethylene glycol diacrylate incorporated with hydroxyapatite. *Tissue Eng Part A*. 2009 Jun, 15(6), 1201-9.

[22] Gyawali, D; Nair, P; Kim, HK; Yang, J. Citrate-based Biodegradable Injectable hydrogel Composites for Orthopedic Applications. *Biomater Sci*. 2013 Jan 1, 1(1), 52-64.

[23] Huang, Z; Tian, J; Yu, B; Xu, Y; Feng, Q. A bone-like nano-hydroxyapatite/collagen loaded injectable scaffold. *Biomed Mater*. 2009 Oct, 4(5), 055005.

[24] Zheng, Y; Lv, H; Wang, Y; Lu, H; Qing, L; Xi, T. Performance of novel bioactive hybrid hydrogels in vitro and in vivo used for artificial cartilage. *Biomed Mater*. 2009 Feb, 4(1), 015015.

[25] Lin, G; Cosimbescu, L; Karin, NJ; Tarasevich, BJ. Injectable and thermosensitive PLGA-g-PEG hydrogels containing hydroxyapatite: preparation, characterization and in vitro release behavior. *Biomed Mater*. 2012 Apr, 7(2), 024107.

[26] Song, J; Xu, J; Filion, T; Saiz, E; Tomsia, AP; Lian, JB; et al. Elastomeric high-mineral content hydrogel-hydroxyapatite composites for orthopedic applications. *J Biomed Mater Res A*. 2009 Jun 15, 89(4), 1098-107.

[27] Leeuwenburgh, SC; Jansen, JA; Mikos, AG. Functionalization of oligo(poly(ethylene glycol)fumarate) hydrogels with finely dispersed calcium phosphate nanocrystals for bone-substituting purposes. *J Biomater Sci Polym Ed*. 2007, 18(12), 1547-64.

[28] Eglin, D; Maalheem, S; Livage, J; Coradin, T. In vitro apatite forming ability of type I collagen hydrogels containing bioactive glass and silica sol-gel particles. *J Mater Sci Mater Med*. 2006 Feb, 17(2), 161-7.

[29] Couto, DS; Hong, Z; Mano, JF. Development of bioactive and biodegradable chitosan-based injectable systems containing bioactive glass nanoparticles. *Acta Biomater*. 2009 Jan, 5(1), 115-23.

[30] Marelli, B; Ghezzi, CE; Mohn, D; Stark, WJ; Barralet, JE; Boccaccini, AR; et al. Accelerated mineralization of dense collagen-nano bioactive glass hybrid gels increases scaffold stiffness and regulates osteoblastic function. *Biomaterials*. 2011 Dec, 32(34), 8915-26.

[31] Stancu, IC; Dragusin, DM; Vasile, E; Trusca, R; Antoniac, I; Vasilescu, DS. Porous calcium alginate-gelatin interpenetrated matrix and its biomineralization potential. *J Mater Sci Mater Med*. 2011 Mar, 22(3), 451-60.

[32] Cha, C; Kim, ES; Kim, IW; Kong, H. Integrative design of a poly(ethylene glycol)-poly(propylene glycol)-alginate hydrogel to control three dimensional biomineralization. *Biomaterials*. 2011 Apr, 32(11), 2695-703.

[33] Hunter, GK; Poitras, MS; Underhill, TM; Grynpas, MD; Goldberg, HA. Induction of collagen mineralization by a bone sialoprotein--decorin chimeric protein. *J Biomed Mater Res*. 2001 Jun 15, 55(4), 496-502.

[34] Miyazaki, T; Mukai, J; Ishida, E; Ohtsuki C. Apatite mineralization behavior on polyglutamic acid hydrogels in aqueous condition: effects of molecular weight. *Biomed Mater Eng*. 2013, 23(5), 339-47.

[35] Sugino, A; Miyazaki, T; Ohtsuki, C. Apatite-forming ability of polyglutamic acid hydrogels in a body-simulating environment. *J Mater Sci Mater Med*. 2008 Jun, 19(6), 2269-74.

[36] Marelli, B; Ghezzi, CE; Alessandrino, A; Barralet, JE; Freddi, G; Nazhat, SN. Silk fibroin derived polypeptide-induced biomineralization of collagen. *Biomaterials*. 2012 Jan, 33(1), 102-8.

[37] Ichibouji, T; Miyazaki, T; Ishida, E; Sugino, A; Ohtsuki, C. Apatite mineralization abilities and mechanical properties of covalently cross-linked pectin hydrogels. *Materials Science and Engineering C*. 2009, 29, 1765–9.

[38] Shkilnyy, A; Graf, R; Hiebl, B; Neffe, AT; Friedrich, A; Hartmann, J, et al. Unprecedented, low cytotoxicity of spongelike calcium phosphate/poly(ethylene imine) hydrogel composites. *Macromol Biosci*. 2009 Feb 11, 9(2), 179-86.

[39] Chesnutt, BM; Viano, AM; Yuan, Y; Yang, Y; Guda, T; Appleford, MR, et al. Design and characterization of a novel chitosan/nanocrystalline calcium phosphate composite scaffold for bone regeneration. *J Biomed Mater Res A*. 2009 Feb, 88(2), 491-502.

[40] Ho, E; Lowman, A; Marcolongo, M. In situ apatite forming injectable hydrogel. *J Biomed Mater Res A*. 2007 Oct, 83(1), 249-56.
[41] Koh, MY; Ohtsuki, C; Miyazaki, T. Modification of polyglutamic acid with silanol groups and calcium salts to induce calcification in a simulated body fluid. *J Biomater Appl*. 2011 Feb, 25(6), 581-94.
[42] Taguchi, T; Kishida, A; Akashi, M. Apatite formation on/in hydrogel matrices using an alternate soaking process (III), effect of physico-chemical factors on apatite formation on/in poly(vinyl alcohol) hydrogel matrices. *J Biomater Sci Polym Ed*. 1999, 10(8), 795-804.
[43] Taguchi, T; Kishida, A; Akashi, M. Apatite formation on/in hydrogel matrices using an alternate soaking process: II. Effect of swelling ratios of poly(vinyl alcohol) hydrogel matrices on apatite formation. *J Biomater Sci Polym Ed*. 1999, 10(3), 331-9.
[44] Madhumathi, K; Shalumon, KT; Rani, VV; Tamura, H; Furuike, T; Selvamurugan, N, et al. Wet chemical synthesis of chitosan hydrogel-hydroxyapatite composite membranes for tissue engineering applications. *Int J Biol Macromol*. 2009 Jul 1, 45(1), 12-5.
[45] Hutchens, SA; Benson, RS; Evans, BR; O'Neill, HM; Rawn, CJ. Biomimetic synthesis of calcium-deficient hydroxyapatite in a natural hydrogel. *Biomaterials*. 2006 Sep, 27(26), 4661-70.
[46] Li, B; Wang, Y; Jia, D; Zhou, Y. Gradient structural bone-like apatite induced by chitosan hydrogel via ion assembly. *J Biomater Sci Polym Ed*. 2011, 22(4-6), 505-17.
[47] Li, B; Wang, Y; Jia, D; Zhou, Y; Cai, W. Mineralization of chitosan rods with concentric layered structure induced by chitosan hydrogel. *Biomed Mater*. 2009 Feb, 4(1), 015011.
[48] Goto, T; Kim, IY; Kikuta, K; Ohtsuki, C. Hydrothermal synthesis of composites of well-crystallized hydroxyapatite and poly(vinyl alcohol) hydrogel. *Materials Science & Engineering C-Materials for Biological Applications*. 2012 Apr 1, 32(3), 397-403.
[49] Dorvee, JR; Boskey, AL; Estroff, LA. Rediscovering hydrogel-based double-diffusion systems for studying biomineralization. *Crystengcomm*. 2012, 14(18), 5681-700.
[50] Azami, M; Moosavifar, MJ; Baheiraei, N; Moztarzadeh, F; Ai, J. Preparation of a biomimetic nanocomposite scaffold for bone tissue engineering via mineralization of gelatin hydrogel and study of mineral transformation in simulated body fluid. *J Biomed Mater Res A*. 2012 May, 100(5), 1347-55.
[51] Watanabe, J; Akashi, M. Integration approach for developing a high-performance biointerface: Sequential formation of hydroxyapatite and calcium carbonate by an improved alternate soaking process. *Applied Surface Science*. 2008, 255, 344–9.
[52] Liu, G; Zhao, DC; Tomsia, AP; Minor, AM; Song, XY; Saiz, E. Three-Dimensional Biomimetic Mineralization of Dense Hydrogel Templates. *Journal of the American Chemical Society*. 2009 Jul 29, 131(29), 9937-+.
[53] Kamitakahara, M; Kimura, K; Ioku, K. Synthesis of nanosized porous hydroxyapatite granules in hydrogel by electrophoresis. *Colloids and Surfaces B-Biointerfaces*. 2012 Sep 1, 97, 236-9.
[54] Leeuwenburgh, SC; Ana, ID; Jansen, JA. Sodium citrate as an effective dispersant for the synthesis of inorganic-organic composites with a nanodispersed mineral phase. *Acta Biomater*. 2010 Mar, 6(3), 836-44.

[55] Rajkumar, M; Meenakshisundaram, N; Rajendran, V. Development of nanocomposites based on hydroxyapatite/sodium alginate: Synthesis and characterisation. *Materials Characterization*. 2011 May, 62(5), 469-79.

[56] Xie, M; Olderoy, MO; Andreassen, JP; Selbach, SM; Strand, BL; Sikorski, P. Alginate-controlled formation of nanoscale calcium carbonate and hydroxyapatite mineral phase within hydrogel networks. *Acta Biomater*. 2010 Sep, 6(9), 3665-75.

[57] Chen, J; Nan, K; Yin, S; Wang, Y; Wu, T; Zhang, Q. Characterization and biocompatibility of nanohybrid scaffold prepared via in situ crystallization of hydroxyapatite in chitosan matrix. *Colloids Surf B Biointerfaces*. 2010 Dec 1, 81(2), 640-7.

[58] Luo, D; Sang, L; Wang, X; Xu, S; Li, X. Low temperature, pH-triggered synthesis of collagen–chitosan–hydroxyapatite nanocomposites as potential bone grafting substitutes. *Materials Letters*, 2011, 65, 2395–7.

[59] Douglas, TE; Gassling, V; Declercq, HA; Purcz, N; Pamula, E; Haugen, HJ; et al. Enzymatically induced mineralization of platelet-rich fibrin. *J Biomed Mater Res A*. 2012 May, 100(5), 1335-46.

[60] Douglas, TE; Messersmith, PB; Chasan, S; Mikos, AG; de, Mulder, EL; Dickson, G; et al. Enzymatic mineralization of hydrogels for bone tissue engineering by incorporation of alkaline phosphatase. *Macromol Biosci*. 2012 Aug, 12(8), 1077-89.

[61] Douglas, TE; Skwarczynska, A; Modrzejewska, Z; Balcaen, L; Schaubroeck, D; Lycke, S; et al. Acceleration of gelation and promotion of mineralization of chitosan hydrogels by alkaline phosphatase. *Int J Biol Macromol*. 2013 Feb 8, 56C, 122-32.

[62] Filmon, R; Basle, MF; Barbier, A; Chappard, D. Poly(2-hydroxy ethyl methacrylate)-alkaline phosphatase: a composite biomaterial allowing in vitro studies of bisphosphonates on the mineralization process. *J Biomater Sci Polym Ed*. 2000, 11(8), 849-68.

[63] Spoerke, ED; Anthony, SG; Stupp, SI. Enzyme Directed Templating of Artificial Bone Mineral. *Adv Mater*. 2009, 21, 425–30.

[64] Bruder, SP; Caplan, AI. A monoclonal antibody against the surface of osteoblasts recognizes alkaline phosphatase isoenzymes in bone, liver, kidney, and intestine. *Bone*. 1990, 11(2), 133-9.

[65] Pfeilschifter, J; Bonewald, L; Mundy, GR. Characterization of the latent transforming growth factor beta complex in bone. *J Bone Miner Res*. 1990 Jan, 5(1), 49-58.

[66] Karewicz, A; Zasada, K; Bielska, D; Douglas, TE; Jansen, JA; Leeuwenburgh, SC; et al. Alginate-hydroxypropylcellulose hydrogel microbeads for alkaline phosphatase encapsulation. *J Microencapsul*. 2014, 31(1), 68-76.

[67] Beertsen, W; van, den, Bos, T. Alkaline phosphatase induces the mineralization of sheets of collagen implanted subcutaneously in the rat. *J Clin Invest*. 1992 Jun, 89(6), 1974-80.

[68] Doi, Y; Horiguchi, T; Moriwaki, Y; Kitago, H; Kajimoto, T; Iwayama, Y. Formation of apatite-collagen complexes. *Journal of Biomedical Materials Research*. 1996, 31, 43-9.

[69] Yamauchi, K; Goda, T; Takeuchi, N; Einaga, H; Tanabe, T. Preparation of collagen/calcium phosphate multilayer sheet using enzymatic mineralization. *Biomaterials*. 2004 Nov, 25(24), 5481-9.

[70] Gungormus, M; Branco, M; Fong, H; Schneider, JP; Tamerler, C; Sarikaya, M. Self assembled bi-functional peptide hydrogels with biomineralization-directing peptides. *Biomaterials*. 2010 Oct, 31(28), 7266-74.

[71] Gassling, V; Douglas, TE; Purcz, N; Schaubroeck, D; Balcaen, L; Bliznuk, V; et al. Magnesium-enhanced enzymatically mineralized platelet-rich fibrin for bone regeneration applications. *Biomed Mater*. 2013 Oct, 8(5), 055001.

[72] Olderoy, MO; Xie, M; Andreassen, JP; Strand, BL; Zhang, Z; Sikorski, P. Viscoelastic properties of mineralized alginate hydrogel beads. *J Mater Sci Mater Med*. 2012 Jul, 23(7), 1619-27.

[73] Douglas, TEL; Krawczyk, G; Pamula, E; Declercq, H; Schaubroeck, D; Bucko, MM; et al. Generation of composites for bone tissue engineering applications consisting of gellan gum hydrogels mineralized with calcium and magnesium phosphate phases by enzymatic means. *J Tissue Eng Regen Med*. 2014 accepted 7.1.2014.

[74] DeVolder, RJ; Kim, IW; Kim, ES; Kong, H. Modulating the Rigidity and Mineralization of Collagen Gels Using Poly(Lactic-Co-Glycolic Acid) Microparticles. *Tissue Eng Pt A*. 2012 Aug, 18(15-16), 1642-51.

[75] Coyac, BR; Chicatun, F; Hoac, B; Nelea, V; Chaussain, C; Nazhat, SN; et al. Mineralization of Dense Collagen Hydrogel Scaffolds by Human Pulp Cells. *J Dent Res*., 2013 Jul, 92(7), 648-54.

In: Apatite
Editors: M. Iafisco and J. M. Delgado-López

ISBN: 978-1-63321-500-9

Chapter 11

FUNCTIONALIZED APATITE NANOCRYSTALS FOR BIOMEDICAL APPLICATIONS

Luis M. Rodríguez-Lorenzo*

Polymeric Nanomaterials and Biomaterials, Institute of Polymer Science and Technology-Spanish Research Council (CSIC), Spain

ABSTRACT

Biological nanocrystalline apatites are the main inorganic components of hard tissues in mammals. The most important characteristics conferring the special properties to nanocrystalline apatites are the nanosized dimensions, their large surface-to-volume ratio and the existence of a surface hydrated layer, nonapatitic in nature. This hydrated layer exhibits a great ionic mobility, as well as ion exchange and adsorption capacities. The idea of functionalizing hydroxyapatite may have arisen from the observation that some glass compositions in the Na_2O–CaO–SiO_2–P_2O_5 system show a greater bioactivity rate than calcium phosphate materials. The possibilities of introducing ionic substitutions are greater for nanoapatites than for microparticles. Biological ocurrence, positions within the structure, induced properties and synthesis routes are discussed for several anionic and cationic functionalization including SiO_4^{2-}, CO_3^{2-}, F^-, Mg^{2+}, Sr^{2+}, Zn^{2+}, Mn^{2+}, Na^+, K^+ and Ag^+. Apatite nanoparticles can also be surface functionalized with organic molecules including polymers, peptides, proteins or nucleic acids. Organic functionalization proceeds mostly in a non covalent mode. The biomimetic approach has been the most common method used to obtain biodegradable polymers functionalized apatites. However, grafting of degradable polymers onto nano-OHAp surfaces can also be achieved by using radio-frequency plasma polymerization or miniemulsion methods. In addition, hydroxyapatites offer a tunable composition and dissolution rate that make them appropriate for controlled drug release. The most common interaction between apatites and active drugs is hydrogen bonding. The loading and release capacity of apatites can be modulated by modifying the surface features of the nanoparticles. Nanosized apatite particles are one of the most viable options for gene delivery systems.

Keywords: Functionalized apatites, inorganic functionalization, organic functionalization

* Corresponding Author. Email: luis.rodriguez-lorenzo@ictp.csic.es.

INTRODUCTION

As it is frequently repeated, biological nanocrystalline apatites are the main inorganic components of hard tissues in mammals (bone and tooth, with the exception of enamel which is closer to stoichiometric hydroxyapatite) [1]. The most important characteristics conferring the special properties to nanocrystalline apatites are the nanosized dimensions, their large surface-to-volume ratio and the existence of a surface hydrated layer, nonapatitic in nature. This hydrated layer exhibits a great ionic mobility, as well as ion exchange and adsorption capacities. The composition of this layer depends on the pH and ionic content of the medium where the apatite crystal has been synthesized, either in nature or in the lab. The ionic exchange ability is a key factor for the process of bone remodeling and bone maturity has been related with the loss of this ability [2].

Several inorganic ions have been recognized to have important funtions within the organism: Sr^{2+}, Mg^{2+}, Mn^{2+}, Zn^{2+} and others [3-6]. Nanocrystalline apatites are used in nature as ion reservoirs for these ions thanks to their exchange ability. Thus, bones are essential not only as mechanical support for the body of vertebrates but also as regulators of numerous functions within the body [7]. Also, proteins have a great binding ability to apatites. Natural bone can be described as a tissue made of apatite nanoparticles embebed in a collagen matrix. However, protein adsorption on nanocrystalline apatites is a complex phenomenon and it has been explained in different ways, by hydrophobic or by electrostatic forces, i.e. through the interaction of the positively charged calcium ions from the apatite to the negatively charged carboxylate groups of the amino acid side chains of the proteins [8].

Synthetic apatite nanoparticles can be surface functionalized in a similar way with inorganic ions and organic species. They include anionic and cationic ions and groups and also organic molecules such as peptides, proteins, or nucleic acids, therapeutic agents and fluorescent molecules [7]. Functionalized apatites nanocrystals can be considered not only as structural materials but also as functional materials. Frequently, apatite nanoparticles only fulfill their function if they release the incorporated species, thus, functionalized apatite nanocrystals can also be considered as controlled delivery systems. Figure 1 is an esquematic summary of the interactions of the nanocrystal surface with the environment.

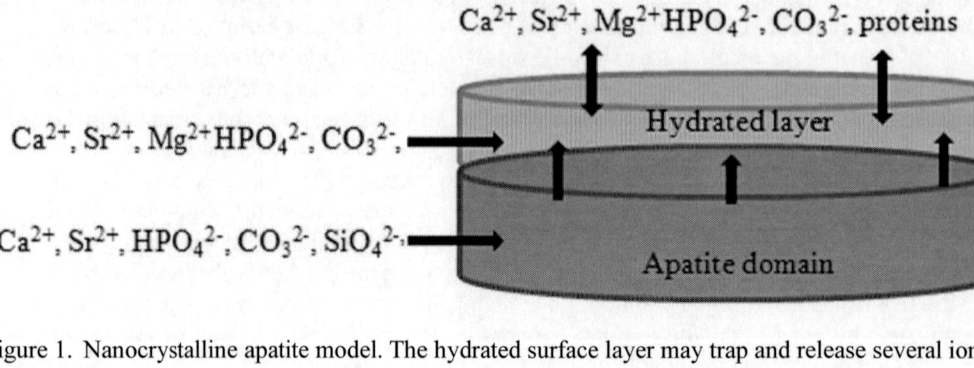

Figure 1. Nanocrystalline apatite model. The hydrated surface layer may trap and release several ions from solution.

This chapter provides with a general vision of the functionalization with different species that have been assayed on apatite nanocrystals. First, how the idea of functionalizing apatites with SiO_4^{4-} may have arisen from the observation than silica based glasses offer a higher bioactivity rate than synthetic crystalline apatites. Then, the most important anionic functionalizations, SiO_4^{4-}, CO_3^{2-} and F^- are treated with special attention of the possible substitution mechanism that are involved. The cationic functionalization section focuses in the different locations than monovalent and divalent cations can occupy within the apatite structure. Finally, the organic functionalization section describes the possible interacions between macromolecules, proteins and drugs with the inorganic surface. This chapter is not a comprehensive review thus some aspect that may be relevant for some readers have been left out. The introduction of magnetic species or trivalent cations such as Eu^{3+} or Gd^{3+} that can be used in theranostic are between those subjects that are not treated in this chapter [9, 10].

Ionic Functionalization in Apatites and Bioglass Bioactivity

Biological apatites are carbonate hydroxyapatites with reduced contents of structural hydroxyl groups and small amounts of several ions. The list of ions that can be found is rather variable depending of the examined tissue, health and age of the individual [11]. Other than carbonate; magnesium, potassium, sodium, chloride, fluoride and hydrogen phosphate are the main ions that can be found in natural tissues [12].

Synthetic stoichiometric hydroxyapatite $Ca_{10}(PO_4)_6(OH)_2$ (OHAp) has been historically used for bone augmentation due to the similarity of its structure with the natural apatites. It has been applied both as a block and in a particulate form [13]. It has shown excellent biocompatibility [14] and, in recent studies, to bypass a host foreign body response system and ability to integrate with the surrounding tissues, unlike other artificial materials [15]. Also it has been shown to alter growth and expression profiles of bone metastatic tumors [16]. However, the capacity to enhance normal bone formation is still controversial with some researches reporting negative results [17].

Hydroxyapatite is a bioactive material. Bioactive materials are defined as materials that stimulate a beneficial response from the body, particularly, bonding to the host tissue [18]. Their success depends upon achieving a stable attachment to connective tissue. Hydroxyapatite is capable of achieving a mechanically stable attachment to bone when applied as a coating over metallic materials or to attach directly by chemical bonding with the bone when applied as a granulate [19]. However, some glass compositions in the Na_2O–CaO–SiO_2–P_2O_5 system have shown a greater bioactivity rate than calcium phosphate materials [20, 21]. The greater bioactivity of glasses has been related with the dissolution products, i.e. soluble silica and calcium ions that stimulate osteogenic cells to produce bone matrix [22]. In a brief summary, bone bonding on bioglass occurs as a result of a rapid sequence of chemical reactions on the surface of the implant when it gets in contact with living tissues. Initially, a rapid release of soluble ionic species occurs that leads to the formation of a high surface area hydrated silica layer and a polycrystalline carbonate hydroxyapatite bi-layer on the glass surface. This layer enhances adsorption and desorption of growth factors reducing the time required for the macrophages to initiate the tissue repair and the attachment, proliferation and differentiation of osteoblasts. Mineralization of the matrix follows soon thereafter and mature osteocytes, encased in a collagen- carbonate hydroxyapatite matrix, are the final product. All these stages take from 6 to 12 days [23].

Hydroxyapatite bioactivity seems to follow a dissolution-reprecipitation mechanism as well [24]. The introduction of silicate groups into the hydroxyapatite lattice may provide the media with the Si ions required to obtain greater bioactivity kinetics. Following this thought, great efforts have been made during the last few decades to produce silicate substituted apatites and to study the effects of this substitution on the biological response of the material [25]. As an example, HR-TEM was used to compare the rates of dissolution of phase pure hydroxyapatite and silicate substituted hydroxyapatite in an ovine model. The dissolution of the ceramic grains followed the order 1.5wt% Si-HA>0.8 wt% Si-HA>pure OHAp at 6 and 12 weeks in vivo [26]. The morphology of the apatite deposits and timing of events at the interfaces of bone with pure HA and with Si-HA implants, were different. A more rapid remodeling of the bone surrounding Si-HA was observed when compared to pure OHAp [27]. In general, silicate substituted apatites enhanced bone apposition, bone in-growth and cell-mediated degradation [28] when compared with OHAp. These biological effects of Si substitution in apatites have been related with the greater solubility, the more electronegative surface and the finer microstructure of silicate substituted apatites in comparison with OHAp. They have been also related to direct effects of Si ions in the physiological processes of bone and connective tissues [25].

However, there are concerns about the Si ion release mechanism from silicate substituted apatites and the mechanism for the biological action of the Si ions. Whether the biological behavior of silicate substituted apatites are due to the Si ion release or to other features of the silicate substituted apatites, such as topography effects or greater release of Ca ions, is currently in discussion [29].

Inorganic Functionalization

Si functionalized Apatites

When the characteristics of the bulk material are considered only and the surface effects are dismissed, the introduction of Si ions into the calcium phosphate apatite structure has been achieved through the substitution of the PO_4^{3-} groups by SiO_4^{4-} groups [30]. The phosphate group forms a tetrahedron with the P atom located at the tetrahedron center, in a T-site, and the four oxygen atoms at the corners. The amount of phosphate groups that can be substituted by silicate groups is then limited by the effective ionic radii of the substituting Si^{4+} in comparison to the effective ionic radii of the P^{5+} and the distortion that the difference may introduce in the structure. And also, by the capacity of the apatite structure to accept the counter ion needed for maintaining charge neutrality. The nature of the counter ion depends on the synthesis procedure and on the availability of water during the reaction [25]. Possibilities include the introduction of O^{2-} or OH^- vacancies and the introduction of excess Ca^{2+} or H^+. Studies such as Rietveld refinements on neutron powder diffractograms have shown a reduction at the hydroxyl site occupancy for Si-functionalized apatites in comparison to hydroxyapatite [31] revealing the introduction OH- vacancies as the prevalent mechanism to maintain charge neutrality. The assumed formulation for Si functionalized apatites is then: $Ca_5(PO_4)_{3-x}(SiO_4)_xOH_{1-x}$. Below a limiting Si concentration, single-phase Si functionalized apatite can be achieved. Whereas different phase mixtures of Si functionalized apatite plus other phases such as Si-tricalcium phosphates, silicocarnitite and amorphous compounds are produced over that limit. The value of the limit depends on the sintering temperature, the Ca/P and the Ca/(P+Si) ratios of the reaction mixture. Single-phase Si functionalized apatite can be produced by maintaining a constant Ca/(P+Si) ratio of 1.67–1.75. Computational studies also show that there is not a unique Si functionalized apatite material but different types depending on the details of the preparation conditions [25].

Reduction to Nanoscale

The possibilities of introducing ionic substitutions are greater for nanoapatites. In general, nanoparticles bear a larger number of grain boundaries and defects, an altered electronic structure and a huge surface area in comparison with microparticles [32]. For hydroxyapatites, whether it is assumed that there are two surface phosphate ions per unit cell, and that the {100} form provides with all the surface faces, the fraction of surface phosphate ions is ~***A***/1200 where ***A*** is the specific surface area in m^2/g [11]. As an example, reported ***A*** for hydroxyapatites synthesized by a wet chemistry route range from 42 to 100 m^2/g [33] indicating that between ~3.5 and ~8% of the phosphates are on the surface and could be replaced without the restrictions impose by the stoichiometry of the lattice. Roughly, the equivalent spherical diameter (assuming a 3.0g/cc density for all of them) for these particles would be between 20 and 47 nm. In addition, nanoapatites have hydroxyl groups at the surface that make them "prone" to develop a hydrated surface layer containing labile ionic species [34, 35]. This layer has been defined as "non apatitic" and can be easily exchanged with ions and/or macromolecules from the surrounding fluids [36]. The exact structure and chemical composition of the hydrated layer depends on several factors such as Ca/P ratio or crystallinity. Some NMR studies have shown that the surface chemistry is governed by the presence of hydrogen phosphate units for specimens with a Ca/ ratio 1.48[37] whereas others have detected the presence of OCP-like defects as well, in samples with an ***A*** $93m^2/g$ [35].

Also, nanoapatites may possess other surface features such as a large number of nanoscale pores that may affect the type and amount of adsorption of selective species, which could modify the biological response to these materials. This variability is the key factor to understand the variability of biological responses reported for similar compositions. Protein–surface interaction is the first event in the process of integration of an implanted device with a hosting tissue but only a few rules of thumb are known about the surface parameters of a particular material that modulates and eventually determines the adsorption and activity of a layer of proteins. For hydroxyapatites, only a general rule that a small crystal size favors bone regeneration while a high crystallinity favors bone resorption is generally accepted [8].

Synthesis Methods and Biological Response of Si-OHAp

Several methods for the synthesis of Si functionalized apatite nanoparticles and films have been reported. A few examples are collected next. Aqueous precipitation from $CaOH_2$, H_3PO_4 and silicon acetate was probably the first reported synthesis method [30], however studies on remodeling processes at the bone–implant interface were made with sintered particles at 1200ºC having 0.45 (0.17) μm grain size and 0.45 g/m^2 ***A*** [27], thus, after losing their nano character. 600 nm thickness Si-OHAp films with a Si content of approximately 0.8 wt% have been produced on titanium substrates by magnetron co-sputtering. The as-deposited films required a 700 °C heat treatment for 3 h to yield the Si-OHAp thin films. The in vitro cell culture study with human osteoblast-like (HOB) cells demonstrated that these coatings promoted the growth of cells [38]. Nano-sized silicon functionalized hydroxyapatite (nSiOHAp) particles have also been deposited on glass and titanium substrates by electrostatic spray deposition. The surface roughness induced by the Si-OHAp islands was in the range 60 to 80 nm and scanning electron microscopy (SEM) studies revealed that extracellular matrix (ECM) produced by the HOB cells on these nSiOHAp deposits was well organized [39]. Si-OHAp coatings with 0.14 to 1.14 % Si have been prepared by a biomimetic process on pure titanium substrates. The atomic ratio of Ca to (P + Si) in the Si-OHAp coatings was in a range of 1.61–1.73, increasing slightly with an increase in the Si content. However, no difference was observed in the cell response between Si-OHAp coatings and OHAp coatings studied comparatively [40]. (Si-CaPs) coatings have been prepared by a sol–gel technique on biodegradable magnesium yttrium alloy (Mg_4Y) substrates. A heat treatment of the dip coated substrates at 450 °C was necessary to obtain the nano-sized particles. *In vitro* biocompatibility experiments only demonstrate that the Si-CaPs coated substrates are more cytocompatible than the uncoated alloy surfaces but not significant differences were reported in relation with the Si content on the apatites [41]. Nano-SiOHAp containing 0.8 wt% Si coatings have been prepared on glass substrates *via* the Drop-on-Demand (DoD) micro-dispensing technique. A uniform, phase-pure coating 34.5 ± 1.0 μm thick without any undesirable extra phases was achieved [42]. Finally, Nano sized hydroxyapatites with silicon substitution of three different silicon concentrations were successfully prepared by a rapid microwave assisted synthesis method. The crystallite size and particle size was observed to decrease with increases in silicon content and the particle morphology shift from spheroidal for pure hydroxyapatite to elongated ellipsoidal crystals for silicon substituted powders [43]. However there are concerns about the Si ion release mechanism from silicate functionalized apatites and the mechanism for the biological action of the Si ions. As stated above, the greater bioactivity of silicate functionalized apatites can be attributed to a greater calcium release, to a change of the surface chemistry, or to a change of surface topography caused by the introduction of the Si^{4+} into the apatite structure other than directly to the Si^{4+} presence [29]. Studies that help to discriminate the effect of each of these phenomena are still needed.

Carbonate Functionalized Apatites

Carbonate is the most frequent substituting group in biological apatites. The quantity depends on the species, age and health of the individual. Typical compositions range from 3.5% for enamel to 5.6% in dentine or 7.4% in bone [44]. As a consequence, the synthesis and biological aspects of carbonate functionalized apatites have been the focus of extensive investigations. These works have been reviewed somewhere else [11, 45-47].

The complexity of the carbonate apatite structure proceeds from the different position where the carbonate can be located within the structure. Hydroxyapatite may host carbonate groups in two different positions plus the non apatitic layer positions. The location in the hydroxyl ion position produces the type A carbonate apatite and the location in the phosphate ion position forms the type B carbonate apatites. For both positions when carbonate gets into the apatite structure the charge balance has to be maintaining by introducing a further ion or a vacancy. The possibilities are then various.

The location of carbonate groups within the apatite structure has been frequently studied by vibrational spectroscopies [48, 49]. IR assignments calculated for carbonate bands in A-type and B-type substitutions are collected in table 1. However, registered spectra from synthesized samples are usually more complex. Both synthetic and biological apatites may present both substitutions simultaneously and also, adsorbed carbonate in the non apatitic layer; these are called labile carbonates in older literature [11]. Carbonate location in synthetic apatites depends on the synthesis method. The presence of counter ions frequently used to facilitate the incorporation of the carbonate group into the structure causes a shift on the wavenumber of the vibrational modes and changes in the multiplicity that increases the complexity of the analysis.

A - Carbonate apatites (A-CO_3-Ap) can be prepared by solid state reactions at high temperatures. Heating OHAp in dry CO_2 atmosphere at 900ºc yields this type of CO_3-Ap [11]. The introduction of one CO_3^{2-} group in the OH^- position must be accompanied by the loss of two OH^- groups that compensates for the difference in charge. Then, A-CO_3-Ap can be described with the model:

$$Ca_{10}(PO_4)_6(OH)_{2-2x}(CO_3)_x \quad (0 \leq x \leq 1)$$

A second synthesis option is to introduce a monovalent cation that compensates the charge valance. However, reactions between $CaHPO_4$, $CaCO_3$ and $KHCO_3$ at 650ºC in a CO_2 atmosphere have yielded a mixture of A and B carbonate substitutions. The amount of B substitution increases for greater carbonate contents [50]. Assignments made for the carbonate vibration modes in one of these products have also been collected in table 1.

B-carbonate apatites (B-CO_3Ap) are usually obtained by precipitation reactions [11]. A type substitution is also possible but only to a limited extend. Following discussion considers only the B-type environment for simplification. Charge balance in alkali and fluoride free precipitated apatites is achieved by the introduction of vacancies in the hydroxyl and calcium positions. For every negative charge loss due to the introduction of CO_3^{2-} in the PO_4^{3-} position, a further negative charge is lost by the loss of an OH^- group and then the charge compensation is achieved by losing a Ca ion. They have been described with the formula:

$$Ca_{10-x+y}(CO_3)_x(PO_4)_{6-x}(OH)_{2-x+2y} \quad (0 \leq x \leq 2 \text{ and } 2y \leq x)$$

IR spectra assignation for this composition can also be found in table 1. Another possibility of achieving charge balance is by the simultaneous incorporation of CO_3^{2-} and OH^- into the PO_4^{3-}position, thus the introduction of vacancies is not needed. Assignments that correspond to this sort of substitution can also be found in the literature and are collected in table 1.

However, the most frequent method to achieve charge compensation in precipitated apatites is by the simultaneous addition of CO_3^{2-} and alkali cations. For precipitated sodium containing carbonate apatite, the next model seems to fit with the reported characterization:

$$Ca_{10-x}Na_{2x/3}(CO_3)_x(PO_4)_{6-x}(OH)_{2-x/3}(H_2O)_x \quad (0 \leq x \leq 3)$$

In this model, Na ions replace Ca^{2+} and 1/3 of OH^- is lost to create vacancies. Thus, Hydroxyl groups are always present. The IR absorption frequencies may also vary depending of the amount of carbonate and Ca/P ratio of the sample. An example for this functionalization is also found in table 1. Fluoride containing carbonate apatites have also been obtained by precipitation. The F^- partially replaces the OH^- along the C axis channel not affecting the introduction of carbonate. They are described with a similar model:

$$Ca_{10-x}Na_{2x/3}(CO_3)_x(PO_4)_{6-x}(OH,F)_{2-x/3}(H_2O)_x \quad (0 \leq x \leq 3)$$

Sodium free fluoride carbonate apatites have been also prepared. They follow the same charge balance mechanism than B-CO_3-Ap but the number of vacancies is reduced by the incorporation of F^- in both, the C axis site and adjacent to the CO_3^{2-} groups.

The labile carbonate environment has been identified by a weak shoulder at 866 cm^{-1} in deconvoluted IR spectra of synthetic and natural apatites [11]. Also the appearance of several carbonate environments in solid state NMR spectra has been used to identify adsorbed carbonate in several carbonate apatites synthesized by precipitation. A strong resonance around 168-169 ppm that cannot be attributed to neither a A or a B environment was assigned to a carbonated amorphous calcium phosphate [52] and related with the non apatitic surface hydrated layer [53] mentioned above.

As it can be deduced from the former paragraphs the analysis of carbonated nano apatite in “real” samples is complex for either synthetic products [54] and natural tissues [48] and due to the key role in biomineralization processes and in biomimetic materials, it remains as an open subject of discussion.

Fluoride Functionalized Apatites

Fluoride substitution has been already mentioned in relation with carbonate apatites. Fluoride is found in natural tissues and the fluoride apatite structure has been used as the starting point and model to study the structure of other apatites. Hydroxy-fluorapatites occur naturally in the body, in both bone and teeth [55]. In particular, they are found in the enamel of teeth, where the degree of fluoridation may be as high as 50% of the hydroxyl positions. The presence of fluorine provides the low solubility and good acid resistance that is necessary for teeth [56].

The wide use of fluoride in drinking water, at a concentration level of approximately 1 ppm, and in dental products has been proven beneficial for reducing tooth decay. In bone, the fluorine content is approximately 1wt% (10 000 ppm). Fluoride has also been administered in bone to prevent the reduction in bone density associated with osteoporosis even though results are controversial [57].

Table 1. Assignments for carbonate bands in infrared spectra from different carbonate apatites

	A-CO_3-Ap[a] (CO_3^{2-} for OH^-)	B-CO_3-Ap[a] (CO_3^{2-} for PO_4^{3-})	AB-CO_3-Ap[a]	A-CO_3 (plus K^+)	B-CO_3-Ap[b] (CO_3^{2-} for PO_4^{3-})	$Ca_{10-x}Na_{2x/3}(CO_3)_x(PO_4)_{6-x}(OH)_{2-x/3}(H_2O)_x$ [b]	$Ca_{10-x}Na_{2x/3}(CO_3)_x(PO_4)_{6-x}(OH,F)_{2-x/3}(H_2O)_x$ [b]	Non apatitic layer[b]
υ_3 (cm^{-1})	1518 1413	1450	1518-A 1442-B 1410-A 1364-B	1550 1470	1452 1416	1464 1419	1450 1435	
υ_2 (cm^{-1})	839	831	851-B 840-A		873	875	862	
υ_4 (cm^{-1})	736 654	681 666	725-B 659-A					
Labile								866
Ref	[49]	[49]	[49]	[51]	[33]	[49]	[11]	[11]

[a] Calculated wavenumber

[b] Registered wavenumber

The fluoride ion is found in the OH lattice position. Complete solubility can occur for the full range of possible compositions due to the similarity in the size of the fluoride ion, 1.32Å, and the hydroxide ion, 1.68Å. This location explains the lack of change in cell parameter dimensions in the *c* axis direction, while a contraction can be observed for the *a* cell parameter. This contraction can be attributed to the smaller size of the fluoride ion, relative to the hydroxide ion. Partial substitution of a hydroxide ion with a fluoride ion will produce solid solutions of OH-FAp and not mixtures of OHAp plus FAp. The composition can be explained with the following formula:

$$Ca_{10}(PO_4)_6F_x(OH)_{2-x} \qquad (0 \leq x \leq 2)$$

The absorption peak intensity changes observed in IR spectra provides an indication of the OH^- and F^- presence in a homogeneous F-OHAp solid solution. However, the spectrum of fully substituted FAp may shows a weak band at 740cm^{-1} as well, indicating that some OH^- groups remain from the reaction media and being an indication of the difficulties of producing pure FAp through a wet reaction route. Table 2 provides with an example of bands assigned to OH, OH-F-HO and OH-F in the IR spectra of a solid solution family of F-OHAp covering the full range of substitution [58].

Table 2. Bands assigned to OH, OH-F-HO and OH-F in the IR spectra of $Ca_{10}(PO_4)_6F_x(OH)_{2-x}$ from ref [58]

x	OH	OH···F	OH···F···OH
0.0	3572, 630		
0.4		3528	3556, 712, 658
0.8		3532, 738	3558, 688
1.2		3534, 752	662
1.6		3534, 746	-
2.0		740	-

When nanoparticles are considered, an interesting result frequently observed is the absence of hydroxyl bands in the IR spectra. Some researchers have neglected the existence of hydroxyl groups for mineral bone based on this absence [59] while others tend to believe that, in poorly crystalline apatites, hydroxyl groups are simply not resolved for several reasons [60]. A possible explanation is that there is enough coupling to the fluoride to mask the characteristic hydroxyl band. The unit cell contains four available (0, 0, z) sites, if a fluorine ion occupies one of these sites, an OH group will preferentially occupy the next occupied site. Also there is a dependence of the specific surface area on the degree of fluoridation. When a family of compounds is synthesized in the same conditions modifying the fluoride content only, the specific surface area is greater for small fluoride contents than for OHAp. Also, ***A*** decreases for greater fluoride contents. This founding has been also related with the interaction between the fluoride and the hydroxyl ion that slows down the growth rate during the synthesis reaction. The slowest growing rate has been found to be for values around 50% substitution for which the smaller particles have been synthesized [58].

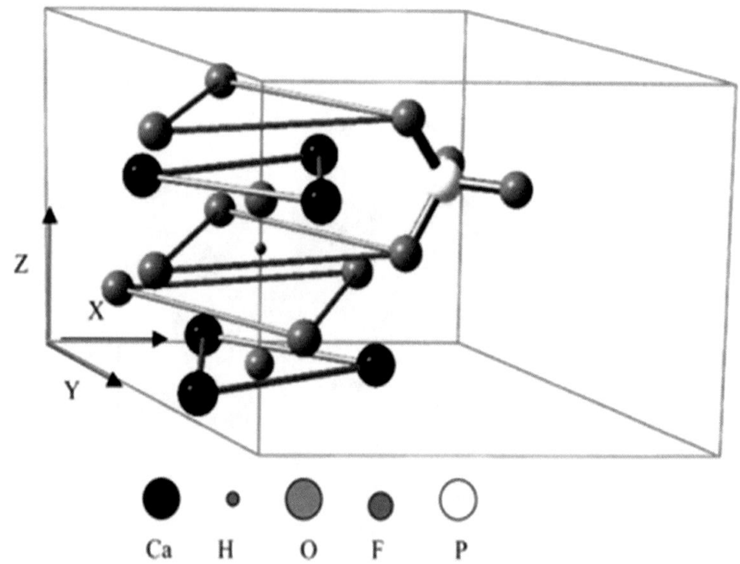

Figure 2. Arrangement of F^- and OH^- ions around the vertical axes at the corner of the apatite unit cell. Reprinted with permission from [61] Rodriguez-Lorenzo LM et al. Structural and chemical analysis of well crystallized hydroxyfluorapatites. *J phys chem B*. 2003; 107: 8316-20 Copyright (2003) American Chemical Society.

Cationic Functionalized Apatites

As mentioned in the introduction, biological apatites contain several cationic substitutions. Ca^{2+} can be partially or completely replaced in the apatite structure by any bivalent ion with the only condition of having a comparable ionic radius. This possibility of substitution explains the presence of these ions in vertebrate bones and provides them with invaluable physical and biological properties. Taking advantage of this ability to accept substitutions, synthetic apatites has been produced with many ions such as Pb^{2+}, Sr^{2+}, Ba^{2+}, Cd^{2+} etc. [62-66]. In this chapter only cations with a marked biological interest are discussed.

Calcium ions occupy two different crystallographic sites in the apatite structure. Ca(1) is found in a octahedral site at 1/3, 2/3, 0 with a coordination number of 9 and multiplicity 4 in f position in Wyckoff notation and Ca(2) is found at the mirror plane at z ¼ and ¾ with a coordination number of 7, multiplicity 6 and h in Wyckoff notation. Whether the substituting ion charge is different from 2, the charge compensation mechanism will rule in the distribution of the substituting ion between the two possible positions. Considering first divalent cations and in a first approach, the distribution between the two sites will be determined by the relative size of the substituting and the host cation. Since site (2) is greater than site (1), the introduction of cations larger than Ca^{2+}, i.e. Sr^{2+} or Ba^{2+} will led to a preferential occupation of the Ca(2) site. The degree of the preferential occupation of the Ca(2) site by a larger cation increases with an increase in the difference in radii between the substituting and the Ca ion. However, for any substituting cation, many considerations have to be made. The polarizability and ability to form covalent bonds can be important for some cations. Introduction of substituting cations may be accompanied with an anion, i.e. F^-, Cl^- or CO_3^{2-} and the accommodation of the "whole package" must be considered. Also, cation substitution may induce a lowering of the space-group symmetry. The synthesis method plays also an important role, precipitation reactions containing Cd^{2+} and Ca^{2+} have shown preferential occupation of the Ca(2) site by Cd^{2+} but also partial introduction in the Ca(1) site whereas exchange reactions led to substitutions only in the Ca(2) site [63].

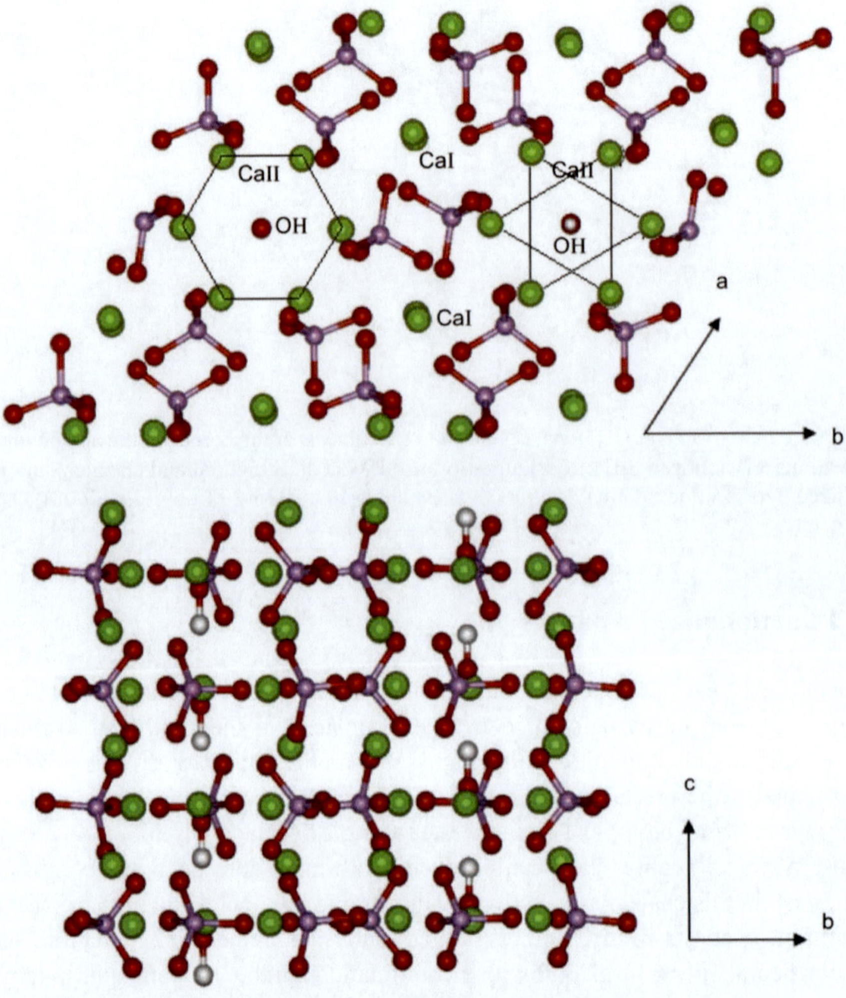

Figure 3. Representation of the hydroxyapatite structure perpendicular to the crystallographic *c* and *a* axes, showing the OH^- channels and the different types of Ca ions. Reprinted with permission from [67] Laurencin D et al. Magnesium incorporation into hydroxyapatite. *Biomaterials*. 2011; 32: 1826-37. Copyright (2011) Elsevier.

Divalent Cation Functionalization-Magnesium

Magnesium is one of the most important bivalent ions associated with biological apatite [68]. Enamel, dentin and bone contain, respectively, 0.44, 1.23, 0.72 wt.% of Mg [4]. Despite its low concentration it plays a key role in bone metabolism, in particular during the early stages of osteogenesis where it stimulates osteoblast proliferation [67]. Magnesium depletion adversely affects all stages of skeletal metabolism, causing cessation of bone growth, decreases in osteoblast activity, osteopenia and bone fragility [69]. Also, relationships have been suggested between the magnesium content in enamel and the development of dental caries [70].

The size difference between Mg^{2+} and Ca^{2+} is 0.28Å (in Pauling scale) and this rather large difference may hinder the replacement of Ca^{2+} by Mg^{2+} and alter the properties of Mg-OHAp. The crystallinity of Mg-OHAp is reduced and as a consequence, the solubility and resorbability are greater than for OHAp. Although some authors claim that Mg^{2+} can be found in the Ca(1) position [68], most of the recent reports based on solid state NMR spectroscopy (^{43}Ca NMR), X-ray absorption spectroscopy and theoretical calculations based on electronic structure techniques and interatomic potential methods seem to indicate that Mg enters preferentially in the Ca(2) position [67]. It is necessary to remark here that these reported results do not follow the reasoning discussed above about the preferential occupation of larger ions for site 2.

Magnesium-substituted OHAp (Ca,MgOHAp) has been prepared by precipitation, hydrolysis and mechanochemical methods. Samples containing up to 1.5 wt.% of Mg without simultaneous carbonate substitution have been obtained by precipitation [71]. Samples with up to 7.5 wt.% of Mg have been synthesized at 120°C in hydrothermal conditions [72]. And a stable, phase pure magnesium-substituted crystalline OHAp containing up to 29 wt.% Mg has been reported using the mechanochemical–hydrothermal technique. In this preparation, at least 75 wt.% Mg substituted for calcium ions in the apatite lattice structure [73].

Strontium

The presence of Sr^{2+} in bone influences bone mineral density and bone mineral content responsible for the strength [74]. The strontium levels in bone vary according to the anatomical site. Greater amounts of strontium are found in cancellous bone than in cortical bone and greater concentrations have been observed in newly formed bone than in old bone [75]. There is also dependence between age and concentration of Ca and Sr in bones, observed during the examination of fractured bones. There is a significant decrease on the concentration of Ca and Sr after 55 and 45 years of age for male and female and demineralization seems to preferentially occur in the bones forming the hip [76]. Strontium was also found to have antiresorptive and anabolic activities and has been thought to have potential interest for the treatment of osteoporosis [77-80]. Nano-Sr-OHAp supports increased osteoblast adhesion, proliferation and viability in culture when compared with nano-OHAp [81]. The dissolution products, apparently strontium (Sr^{2+}) from Sr-OHAp-coated implants, enhanced the ALP activity and *in vitro* mineralization ability [82]. More recently, Sr^{2+} has shown to promote angiogenic expression. However, the mechanism of promoted angiogenic behaviors initiated by Sr components remains unknown [83]. The introduction of strontium into hydroxyapatite changes the crystallinity and dissolution rate of the calcium phosphate as well as the net positive charge on calcium phosphate nanoparticles. For this reason, nano-Sr-OHAps have been thought to be candidates as vectors for gene delivery [84]. The functionalization of apatites for gene delivery will be discussed later.

The difference between magnesium substitution and strontium substitution in hydroxyapatite can be attributed to the size of the ions. Sr^{2+} radius is larger than ca^{2+} radius whereas Mg^{2+} radius is smaller as discussed above. The difference is 0.14Å in the Pauling scale, a 114% larger than Ca^{2+}. Since site (2) is greater than site (1), Sr^{2+} will preferentially occupy site (2) position as confirmed from X-ray structure refinements [85]. However, the degree of ordering of the Sr^{2+} ions in Ca,SrOHAp decreases with an increase in the strontium content [11]. Nanoparticles of SrOHAp and CaSrOHAp have been synthesized by aqueous precipitation [81, 85] and by a sol gel method [84]. Microspheres have been manufactured in a subsequent step from precipitated powders by a hydrothermal treatment [83]. Coatings on metallic surfaces, to induce bioactivity to these surfaces, have been assayed by plasma spray [82] or by electrodeposition in combination with fluoride [86].

Zinc

Zinc is another important trace element in all biological tissues. The Zn ion content ranges from 0.0126 wt.% to 0.0217 wt.% in human bone. Zn ion takes part in the activity of more than 300 types of metallo-enzymes and proteins, including, alkaline phosphatase and can promote osteoblasts cells proliferation, differentiation and bone formation [87]. It has been claimed that Zn ions directly stimulate DNA synthesis by either enzyme stimulation or by altering the binding of F1 and F3 histones to DNA [88]. Thus, the use of Zn ions as active drugs has been largely investigated. In vitro, concentrations of zinc ion that favor the growth of most micro-organisms are in the range of 10^{-5} to 10^{-7} M [89]. Higher ion concentrations start to produce cytotoxic effects. This toxicity has been used for attacking cancerous cells [90] and bacteria through Zn containing nanoparticles [89]. The observed inferior limit for non-inducing cytotoxic effects is 120 μM zinc –equivalent and whatever the particles used to dose the Zn ion into the body, i.e. ZnO, $ZnCl_2$ or Zn-OHAp, the solubility of these particles will play a pivotal role in the dosification rate and biological activity [91, 92].

Zn is incorporated into the apatite structure as other cations discussed before. The incorporation of zinc in the OHAp structure yields a decrease in the particles size and increases the solubility. Zn-OHAp presents a greater resorbability and bioactivity character than OHAp [93]. Zn-OHAp also has shown size-dependent cytotoxicity in cell cultures using HepG2 cells. Upon increasing the amount of zinc in the OHAp, a decrease in the particles size can be observed and with the decrease in particle size a decrease in the IC50 has been observed that it has been attributed to the effect of the Zn ion [94]. Zn-OHAp has been used as an anticancer treatment as well. Briefly, the Zn^{2+} containing nanoparticles trigger the changes of expression of the apoptotic proteins in cancerous cells.

EXAFS experiments on Zn-OHAp containing 800ppm of Zn^{2+} have shown that the incorporation of Zn^{2+} into the apatite structure occurs in the Ca (1) site in agreement with the theory that states that larger ions go to the larger site in the apatite and since Ca^{2+} is larger than Zn^{2+}, 0.99 vs 0.74Å in the Pauling scale, Ca^{2+} will occupy site (2) and Zn^{2+} site (1). Also Zn^{2+} can be found in the interstice between the two hydroxides [93]. However there are also reports that show that Zn^{2+} incorporates on Ca (2) site. The differences may be originated in the synthesis method and Zn^{2+} content [95]. Zn-OHAp has been synthesized mainly by precipitation by using $Ca(NO_3)_2.4H_2O$ and $Zn(NO_3)_2.6H_2O$ as sources for Ca and Zn ions and $(NH_4)_2HPO_4$ solutions as source for phosphate ions at 10 pH [87, 96, 97]. Sol-gel methods have been used for producing nanoparticles using $Ca(NO_3)_2.4H_2O$, ZnO and triethyl phosphate as precursors for nanopowders [98] and coatings using $Ca(NO_3)_2.4H_2O$ and $Zn(NO_3)_2.6H_2O$ solved in ethanol and P_2O_5 as source for phosphate [89].

Manganese

Manganese is a microelement that plays an important role in osteogenesis. Significant amounts of this element are found in bones. Manganese is also an essential agent in the synthesis of mucopolysaccharides, and its deficiency leads to disturbed synthesis of organic components of bone matrix. Manganese deficiency decreases also thickness and length of growing bones [99]. In addition, manganese present in tissues activates integrins (cellular receptors involved in intracellular adhesion) and increases their affinity to appropriate ligands. This feature is especially important considering the integration of bone substitutes, i.e. implants. Addition of manganese ions to hydroxyapatite promotes the interaction of an implant with the host tissue and hence, it reduces the time required for the osteointegration of the implant [5]. Some studies have shown that titanium implants coated with manganese doped calcium phosphate form a larger number of stable links with highly branched osteoblasts, as compared to the same type of implant coated with conventional calcium phosphate. Thus, it can be stated that manganese strongly promotes differentiation of osteoblasts and formation of new links between the implant and the tissue.

Substitution of manganese for calcium in hydroxyapatite has been a subject of several studies [100, 101]. These studies have focused on the impact of manganese on the firing process of apatite bioceramics. It has been demonstrated that the presence of manganese does not produce any alterations in terms of possible formation of secondary phases (impurities) such as TCP or CaO. And, it has been found that manganese doped OHAp bears a greater stability during the firing process than pure hydroxyapatite. Hardness and density of the manganese-doped apatite were also greater than those of pure OHAp.

Monovalent Cation Functionalization

The introduction of alkali cations has been the most frequent method to achieve charge compensation in carbonated apatites. Biological apatites are carbonateapatites and they bear small amounts of Na^+ and K^+. Na ion introduction has been discussed in the section dedicated to carbonate functionalization. The differences between introducing Na^+ and introducing K^+ in the apatite structure are originated in the size difference between them. K^+ is larger than Ca^{2+} and Ca^{2+} is larger than Na^+. As a consequence, it is easier to introduce Na^+ in the Ca ion positions than K^+. And, when the purpose is to introduce a large amount of carbonate in an apatite synthesis, it would be preferable to use Na^+ instead of K^+. However, the introduction of both cations follows the same charge balance mechanisms described before. Small differences in physical properties have been observed between Na^+ and K^+ containing apatites [50]. These differences can be attributed to the differences in size and also to the different positions where the alkali cations can be located in the structure. In substitution experiments carried out in Sr –apatites, Na^+ are introduced only in site (1) whereas K^+ are introduced only in site (2) [85].

Although not found in natural bones, silver cations can be introduced into the apatite structure yielding interesting antibacterial properties. Material-related infection is a major clinical problem, which complicates the bone healing process, resulting in localized bone destruction. Bacteria are capable of adhering onto implant surfaces, forming a biofilm and when bacteria have developed resistance to antibiotic treatment, removal of the infected implant is the only effective solution [42]. Silver and silver ions have long been known to have strong inhibitory and bactericidal effects and a broad spectrum of anti-microbial activities [102]. Ag ions are known to interfere with the transmembranous energy transfer and electron transport of bacterial microorganisms, they block the respiratory chain at the cytochromes and cause metabolite efflux, inducing massive proton leakage through the bacterial membrane which, results in complete deenergization and ultimately cell death. It also binds to the bacteria DNA increasing the stability of the bacterial double helix and thus inhibiting DNA replication [103]. However, it is known that different forms of Ag^+ confer different degrees of bactericidal properties [104]. Amorphous and polycrystalline hydroxyapatite and Ag^+ doped polycrystalline coatings are non-toxic but silver doped amorphous apatite coatings are mildly toxic from 1.2 wt.% Ag. The minimal level of silver ions necessary to clear infections is not known yet [102]. It also appears that the introduction of Ag^+ does not have a strong influence in the osteoconductivity properties of the apatites [103]. Ag ions can be incorporated into the apatite structure through an ion-exchange method in either sol–gel or coprecipitation routes [105]. No studies have been found that show the distribution of Ag^+ between sites (1) and (2) neither about the capacity of Ag for Ca substitution in the structure or how the distribution may affect to the resulting bactericidal properties, but since Ag^+ is larger than Ca^{2+}, and a ion exchange method have been proposed, a preferential occupation of the Ca (2) site can be expected.

ORGANIC FUNCTIONALIZATION

Apatite nanoparticles can also be surface functionalized with organic molecules including amphiphilic molecules, fluorescent molecules, polymers, peptides, proteins or nucleic acids. Organic functionalization proceeds mostly in a non-covalent mode. It is an adsorption driven by hydrophobic or van der Waals forces. Organic functionalization is usually achieved during the synthesis procedure and usually the molecules are incorporated after the hydroxyapatite has already begun the nucleating process [7].

Amphiphilic molecules are used to stabilize suspensions, prevent particle agglomeration and design particles morphology [8, 106, 107]. As an example, Bis(2-ethylhexyl)sulfo-succinate has been used for the purpose of assembling apatite particles into enamel-like bundles [108]. But this sort of interaction is out of the scope of the topic of functionalization.

Polymer Functionalization

The biomimetic approach has been most common method used to obtain biodegradable polymers functionalized apatites. First, the polymers are treated with strong alkali to generate carboxylate groups that can be bound to Ca^{2+}. Then, the apatite is grown by soaking the "active" polymer in a calcium and phosphate containing solution such as a simulated body fluid (SBF) [109] or by soaking in successive solutions, first in a calcium rich solution, second in a phosphate rich solution [110]. Studies made on self-assembled monolayers (SAMs) have shown that -COOH but also, $-PO_4H_2$ and $–SO_3H$ [111] functional groups are able to nucleate apatite. poly(εcaprolactone), poly(lactide-co-glycolide) and poly(l-lactic acid) have been mineralized this way [112-114]. However, some polymers such as PCL did not calcify in SBF even after such modification. Successive treatments in $CaCl_2$ and K_2HPO_4 solutions were necessary prior to the treatment in SBF in order to promote the deposition of apatitic phases [112]. It is assumed that the irregular accumulation of phosphate groups on the surface of certain polymers, caused by unfavorable reactivity ratios of specific pairs of monomers, leads regionally to multiple bonding of Ca ions. Consequently, they are no longer available to bind to the phosphate ions present in SBF and therefore unable to participate in further nucleation and crystallization [115].

Other alternatives tested include the surface functionalization with –SiOH groups [116] or with oxyphosphorus groups. Alginate or collagen-like polypeptide have been modified with -SiOH groups and Ca^{2+} ions [117] then mineralized by soaking in SBF [118]. poly(l, 4-phenylene terephthalamide) or poly(1,3-phenylene isophthalamide), commercially known as Kevlar and Nomex have been copolymerized with dimethylphosphonomethyl group, $–CH_2–P{=}O)(OCH_3)_2$ in order to induce biomineralization. While the modified Kevlar copolymers induced significant deposition of hydroxyapatite in a supersaturated calcifying medium, no deposits were formed on the Nomex copolymers. This was attributed to a stereochemical situation in the latter polymer whereby the dimethylphosphonomethyl group is ''buried" due to formation of hydrogen bonds between –NH– and –P=O moieties, rendering the group inaccessible to Ca ions. The nucleation induced by the modified Kevlar copolymers was explained by the chelation of Ca ions to the negative end of the P=O bond, which in principle can trigger ion accumulation events [119]. In conclusion, although the cation chelation mechanism may have purportedly accounted for the calcification of some synthetic polymers, such a mechanism appeared inconsistent with the calcification of certain acrylic hydrogels [120]

However, and applying the carboxylation mechanism, surface functionalized polymer nanoparticles can be used as templates to obtain hybrid materials. As an example, OHAp-coated poly(styrene-co-acrylic acid) copolymer particles have been synthesized [121]. In a different approach, surface grafting of degradable polymers onto nano-OHAp surface can be achieved by using radio-frequency plasma polymerization or miniemulsion methods [110]. The compatibility between surfaces is critical. And, since the nanoapatite crystals are hydrophilic when the polymer is hydrophobic, a block polymeric surfactant has to be added [122].

Drug Delivery Systems

Several apatite based drug delivery systems have been developed [123-127]. Many of them are based on the preparation of porous reservoirs filled with the intended drug [123]. The rate of drug release is controlled by diffusion, osmosis or degradation of the matrix [128]. These systems are not considered here. Only those with a chemical interaction between the drug and the calcium phosphate are of interest in the present chapter. Hydroxyapatites offer a tunable composition and dissolution rates that offers a controlled drug release over time. They enable a reduced dosage and frequency of administration that yields fewer side effects. Hydroxyapatites are less toxic than silica, quantum dots, carbon nanotubes, or magnetic particles [129] and they have a greater biocompatibility than polymers [130]. The dissolution of hydroxyapatite is greater at low pH which it can be found in endolysosomes and in the vicinity of tumors. Thus, it provides an advantage in the delivery of drugs into malign zones or cell organelles. A couple of examples of drug delivery from nanoapatites are given next. Vancomycin is a glycopeptide antibiotic used to treat infections caused by Gram-positive bacteria including methicillin-resistant staphylococcus [131]. Vancomycin can be adsorbed on OHAp through the hydrogen bonding interactions between the OH group, which existed in both the drug molecules and on the OHAp surfaces [132]. Since the physical interaction of hydrogen bonding is relatively weak comparing with the chemical bonding, the loaded drug will be released into the medium due to the dissociation of the hydrogen bonding [133]. By increasing the specific surface area of the apatite particles, a greater loading of vancomycin can be achieved. Ionic modification of the apatite frequently induces a greater specific surface area. As an example, SrHAp porous microspheres with higher amounts of OH groups have shown a greater loading and release efficiency compared with other nanoparticles [83]. Another example would be Ibuprofen which is a non-steroidal anti-inflammatory drug. Adsorption and release of ibuprofen on hydroxyapatite particles for short periods has been modulated by modifying the crystallinity and surface area of the particles [134].

Bisphosphonates (Bps) possess a strong affinity to hydroxyapatite under physiological conditions [135]. Bisphosphonates are the drug of choice in the treatment of osteoporosis [136]. The local release of Bps has been presented as an alternative to oral administration to overcome the low bioavailability and the gastrointestinal disturbance generally caused by oral administration [137]. Alendronate and zolendronate are between the most known and clinically used Bps but several other molecules have been synthesized and tested [137-141]. The interaction between apatites and bisphosphonates has been described in several ways but it does not consist in the simple precipitation of the Bp on the surface of the apatite [139]. For alendronate, an interaction between the hydroxyl group of the apatite and the amine group of alendronate resulting in the formation of N–H–O associations has been found by IR spectroscopy [142]. 3-D computational modeling has predicted an interaction at the calcium Ca(1) site where the bidentate and tridentate coordination of the P–C–P and R_1–OH domains of the bisphosphonate interact [143]. Zolendronate interactions with apatites have shown concentration dependence. For concentrations <0.05 mol/l calcium deficient apatites seem to undergo chemisorption of the drug through a surface adsorption process, due to a PO_3 for PO_4 exchange, described by Freundlich equations. At concentrations >0.05 mol/l, crystalline needles of a zoledronate complex have been found onto the calcium deficient apatite surfaces [144].

Protein and Peptide Functionalization and Gene Delivery

OHAp shows a high affinity for proteins [145]. The thorough understanding of the protein–apatite interactions should help to understand not only the different response from host tissues to implanted OHAp but also it should help in the design of tailored medical devices for specific needs [146]. The adsorption of proteins on solid surfaces is a complicated process consisting in many events. Factors affecting protein adsorption on material surfaces include, protein concentration [147], molecular weight (MW) of the protein [148], isoelectrical point (PI) of the protein [147], hydrophilicity or hydrophobicity of the material [149], presence of functional groups [146], surface grain/particle size (PS), crystallite size (CS) [150],etc. Plenty of literature can be found on this subject. Check review [151] for a wide perspective on this subject. The main interactions between materials and proteins have been generally accepted to be hydrophobic and electrostatic [152, 153]. However, OHAp present a hydrophilic character ($\theta < 60$) [150] and in hydrophilic surfaces, repulsive solvation forces arising from strongly bound water occurs and they should avoid protein binding [154, 155]. Negatively charge proteins at physiological pH such as fibrinogen or fibronectin have been proved to be adsorbed on apatites surfaces by Raman spectroscopy. The interaction occurs between the positively charged calcium ions of the apatite and the negatively charged carboxylate groups of the amino acid side chains or the terminal carboxylate groups of the proteins [8]. In the other side, PO_4^{3-} groups have been described to interact with basic protein residues [156]. In addition, OHAp has two types of crystal planes, which carry different net charges: positive on *a* planes and negative on *c* planes, and there are speculations that the *a* planes tend to adsorb acidic proteins, whereas the *c* planes tend to attract the basic ones [157]. Thus, the synthesis of elongated particles, i.e. crystals, would lead to a greater adsorption of negatively charged acidic proteins. Also, a combination of a large specific surface area with a small crystallite size seems to be factors that favor the adsorption of acidic proteins [8].

Surface functionalization with cell-binding peptides has been also assayed to attain control over cell-biomaterial interactions and promote a more efficient healing process upon the introduction of OHAp in the body [158]. The most common peptide for biomimetic surface modification is Arg-Gly-Asp (RGD), a ubiquitous peptide present on ECM proteins that have shown to promote cell adhesion in multiple cell types [159].

The same type of interaction occurs between DNA and OHAp particles. OHAp and other calcium phosphate particles have been intensively used as smart DNA-carriers for more than 30 years [160]. The encapsulation of DNA into OHAp nanoparticles protects the nucleic acid from cytoplasmatic environment and enables its efficient delivery into the cell nucleus. The degradation of the apatite does not produce any toxic chemicals apart from the release of Ca ions (which may in the worst case scenario initiate protein aggregation), and these reactions are also scalable in size and imply low manufacturing costs. In that sense, nanosized apatite particles present one of the most viable options for gene delivery systems [161]. The possibility of multiple functionalization enhances the potential of hydroxyapatite as vehicles for gene transfection [123]. As an example, DNA containing Mg-OHAp particles have been functionalized with ECM proteins (fibronectin and collagen). The particle-bound DNA can be then internalized by specific cell surface integrin receptors. The rapid cellular endocytosis resulted in significantly high DNA delivery in the cytoplasm and eventually in a significant enhancement in transgene expression [162]. Apatites may also interact with cell membranes without protein functionalization. Cell membrane lipids are amphiphilic molecules. The interaction between apatites and lipids usually happen between the OH groups of the apatite and the hydrophilic part of the lipid. As an example, Cholesterol monohydrate develop (001) faces similar to the apatite. The two crystal structures show a close superposition of hydroxyl groups that result in hydrogen bonding. A crystalline layer of either of the two structures can thus serve as a quasi-epitaxial nucleation surface for the growth of the other [163]. One final consideration, the efficiency of OHAP nanoparticles for drug or gene delivery relies not only on the functional groups or molecules that can be attached but also on the size. A few studies have shown that 50 nm nanoparticles penetrate the cell membrane faster than those smaller than 15 nm or larger than 200 nm. Smaller is not necessarily the best option when it comes to choosing the most optimal size for a given drug/gene delivery application [94].

In summary the versatile apatite structure provides with an ideal framework to introduce different types of functional groups, both organic and inorganic. Not much attention have been dedicated in this chapter to multiple functionalization for a question of simplification and space, but the potential to combine functional groups is enormous and as a consequence, apatites are one of the most versatile and powerful instruments that can be found to induce different responses in the human body.

REFERENCES

[1] Gomez-Morales J, Iafisco M, Manuel Delgado-Lopez J, Sarda S, Drouet C. Progress on the preparation of nanocrystalline apatites and surface characterization: Overview of fundamental and applied aspects. *Progress in Crystal Growth and Characterization of Materials.* 2013;59:1-46.

[2] Cazalbou S, Combes C, Eichert D, Rey C. Adaptative physico-chemistry of bio-related calcium phosphates. *Journal of Materials Chemistry.* 2004;14:2148-53.

[3] Canalis E, Hott M, Deloffre P, Tsouderos Y, Marie PJ. The divalent strontium salt S12911 enhances bone cell replication and bone formation in vitro. *Bone.* 1996;18:517-23.

[4] LeGeros R. Calcium phosphates in oral biology and medicine. Basel: *Karger;* 1991.

[5] Bracci B, Torricelli P, Panzavolta S, Boanini E, Giardino R, Bigi A. Effect of Mg2+, Sr2+, and Mn2+ on the chemico-physical and in vitro biological properties of calcium phosphate biomimetic coatings. *Journal of Inorganic Biochemistry.* 2009;103:1666-74.

[6] Place ES, Rojo L, Gentleman E, Sardinha JP, Stevens MM. Strontium- and Zinc-Alginate Hydrogels for Bone Tissue Engineering. *Tissue Engineering Part A.* 2011;17:2713-22.

[7] Uskoković V, Uskoković DP. Nanosized hydroxyapatite and other calcium phosphates: Chemistry of formation and application as drug and gene delivery agents. *Journal of Biomedical Materials Research - Part B Applied Biomaterials.* 2011;96 B:152-91.

[8] Fernandez-Montes Moraleda B, San Roman J, Rodriguez-Lorenzo LM. Influence of surface features of hydroxyapatite on the adsorption of proteins relevant to bone regeneration. *Journal of Biomedical Materials Research Part A.* 2013;101A:2332-9.

[9] Iafisco M, Delgado-Lopez JM, Varoni EM, Tampieri A, Rimondini L, Gomez-Morales J, et al. Cell surface receptor targeted biomimetic apatite nanocrystals for cancer therapy. *Small* (Weinheim an der Bergstrasse, Germany). 2013;9:3834-44.

[10] Liu Z, Wang Q, Yao S, Yang L, Yu S, Feng X, et al. Synthesis and characterization of Tb3+/Gd3+ dual-doped multifunctional hydroxyapatite nanoparticles. *Ceramics International.* 2014;40:2613-7.

[11] Elliott JC. Structure and chemistry of the apatites and other calcium orthophosphates. Amsterdam: *Elsevier;* 1994.

[12] Dorozhkin SV. Calcium orthophosphates. *Journal of Materials Science.* 2007;42:1061-95.

[13] Cabanas MV, Rodriguez-Lorenzo LM, Vallet-Regi M. Setting behavior and in vitro bioactivity of hydroxyapatite/calcium sulfate cements. *Chemistry of materials.* 2002;14:3550-5.

[14] Huang J, Best SM, Bonfield W, Brooks RA, Rushton N, Jayasinghe SN, et al. In vitro assessment of the biological response to nano-sized hydroxyapatite. Journal of Materials Science-Materials in Medicine. 2004;15:441-5.
[15] Oyane A, Ootsuka T, Hayama K, Sogo Y, Ito A. Enhanced immobilization of acidic proteins in the apatite layer via electrostatic interactions in a supersaturated calcium phosphate solution. *Acta Biomaterialia.* 7:2969-76.
[16] Pathi SP, Lin DDW, Dorvee JR, Estroff LA, Fischbach C. Hydroxyapatite nanoparticle-containing scaffolds for the study of breast cancer bone metastasis. *Biomaterials.* 2011;32:5112-22.
[17] Gellynck K, Shah R, Parkar M, Young A, Buxton P, Brett P. Small molecule stimulation enhances bone regeneration but not titanium implant osseointegration. *Bone.* 2013;57:405-12.
[18] Jones JR. Review of bioactive glass: From Hench to hybrids. *Acta Biomaterialia.* 2013;9:4457-86.
[19] Hench L.L. Ceramics, glasses and glass ceramics. In: Ratner B.D., Hoffman A.S., Schoen F.J., J.E. L, editors. Biomaterials science An introduction to materials in medicine. *SanDiego: Academic press;* 1996. p. 73-84.
[20] Ragel CV, Vallet-Regi M, Rodriguez-Lorenzo LM. Preparation and in vitro bioactivity of hydroxyapatite/solgel glass biphasic material. *Biomaterials.* 2002;23:1865-72.
[21] Oonishi H, Hench LL, Wilson J, Sugihara F, Tsuji E, Matsuura M, et al. Quantitative comparison of bone growth behavior in granules of Bioglass (R), A-W glass-ceramic, and hydroxyapatite. *Journal of Biomedical Materials Research.* 2000;51:37-46.
[22] Hench L, JM. P. Third generation biomedical materials. *Science.* 2002;295:1014-7.
[23] Hench LL. The story of Bioglass (R). *Journal of Materials Science-Materials in Medicine.* 2006;17:967-78.
[24] Weng J, Liu Q, Wolke JGC, Zhang XD, deGroot K. Formation and characteristics of the apatite layer on plasma-sprayed hydroxyapatite coatings in simulated body fluid. *Biomaterials.* 1997;18:1027-35.
[25] Pietak AM, Reid JW, Stott MJ, Sayer M. Silicon substitution in the calcium phosphate bioceramics. *Biomaterials.* 2007;28:4023-32.
[26] Porter AE, Patel N, Skepper JN, Best SM, Bonfield W. Comparison of in vivo dissolution processes in hydroxyapatite and silicon-substituted hydroxyapatite bioceramics. *Biomaterials.* 2003;24:4609-20.
[27] Porter AE, Patel N, Skepper JN, Best SM, Bonfield W. Effect of sintered silicate-substituted hydroxyapatite on remodelling processes at the bone-implant interface. *Biomaterials.* 2004;25:3303-14.
[28] Hing KA, Revell PA, Smith N, Buckland T. Effect of silicon level on rate, quality and progression of bone healing within silicate-substituted porous hydroxyapatite scaffolds. *Biomaterials.* 2006;27:5014-26.
[29] Bohner M. Silicon-substituted calcium phosphates - A critical view. *Biomaterials.* 2009;30:6403-6.
[30] Gibson IR, Best SM, Bonfield W. Chemical characterization of silicon-substituted hydroxyapatite. *Journal of Biomedical Materials Research.* 1999;44:422-8.
[31] Leventouri T, Bunaciu CE, Perdikatsis V. Neutron powder diffraction studies of silicon-substituted hydroxyapatite. *Biomaterials.* 2003;24:4205-11.

[32] Traykova T, Aparicio C, Ginebra MP, Planell JA. Bioceramics as nanomaterials. *Nanomedicine.* 2006;1:91-105.

[33] Rodriguez-Lorenzo L.M., Vallet-Regi M. Controlled crystallization of calcium phosphate apatites. *Chemistry of materials.* 2000;12:2460-5.

[34]] Panda RN, Hsieh MF, Chung RJ, Chin TS. FTIR, XRD, SEM and solid state NMR investigations of carbonate-containing hydroxyapatite nano-particles synthesized by hydroxide-gel technique. *Journal of Physics and Chemistry of Solids.* 2003;64:193-9.

[35] Rodriguez-Lorenzo LM. Studies on calcium deficient apatites structure by means of MAS-NMR spectroscopy. *J Mater Sci Mater Med.* 2005;16:393-8.

[36] Rey C, Combes C, Drouet C, Sfihi H, Barroug A. Physico-chemical properties of nanocrystalline apatites: Implications for biominerals and biomaterials. Materials Science & Engineering C-Biomimetic and Supramolecular Systems. 2007;27:198-205.

[37] Jager C, Welzel T, Meyer-Zaika W, Epple M. A solid-state NMR investigation of the structure of nanocrystalline hydroxyapatite. Magnetic Resonance in Chemistry. 2006;44:573-80.

[38] Thian ES, Huang J, Best SM, Barber ZH, Bonfield W. Magnetron co-sputtered silicon-containing hydroxyapatite thin films - an in vitro study. *Biomaterials.* 2005;26:2947-56.

[39] Huang J, Jayasinghe S, Best SM, Edirisinghe MJ, Brooks RA, Rushton N, et al. Novel deposition of nano-sized silicon substituted hydroxyapatite by electrostatic spraying. *Journal of Materials Science-Materials in Medicine.* 2005;16:1137-42.

[40] Zhang E, Zou C, Yu G. Surface microstructure and cell biocompatibility of silicon-substituted hydroxyapatite coating on titanium substrate prepared by a biomimetic process. *Materials Science & Engineering C-Biomimetic and Supramolecular Systems. 2009;29:298-305.*

[41] Roy A, Singh SS, Datta MK, Lee B, Ohodnicki J, Kumta PN. Novel sol-gel derived calcium phosphate coatings on Mg4Y alloy. *Materials Science and Engineering B-Advanced Functional Solid-State Materials.* 2011;176:1679-89.

[42] Chang L, Sun J, Fuh JYH, Thian ES. Deposition and characterization of a dual-layer silicon- and silver-containing hydroxyapatite coating via a drop-on-demand technique. *Rsc Advances.* 2013;3:11162-8.

[43] Padmanabhan SK, Haq EU, Licciulli A. Rapid synthesis and characterization of silicon substituted nano hydroxyapatite using microwave irradiation. *Current Applied Physics.* 2014;14:87-92.

[44] Dorozhkin SV. Calcium Orthophosphates in Nature, Biology and Medicine. *Materials.* 2009;2:399-498.

[45] Cai Y, Tang R. Calcium phosphate nanoparticles in biomineralization and biomaterials. *Journal of Materials Chemistry.* 2008;18:3775-87.

[46] Hong Y, Fan H, Li B, Guo B, Liu M, Zhang X. Fabrication, biological effects, and medical applications of calcium phosphate nanoceramics. *Materials Science & Engineering R-Reports.* 2010;70:225-42.

[47] Liu Q, Huang S, Matinlinna JP, Chen Z, Pan H. Insight into Biological Apatite: Physiochemical Properties and Preparation Approaches. *Biomed Research International.* 2013.

[48] Khan AF, Awais M, Khan AS, Tabassum S, Chaudhry AA, Rehman IU. Raman Spectroscopy of Natural Bone and Synthetic Apatites. *Applied Spectroscopy Reviews.* 2013;48:329-55.

[49] Ren F, Ding Y, Leng Y. Infrared spectroscopic characterization of carbonated apatite: A combined experimental and computational study. *Journal of Biomedical Materials Research Part A.* 2014;102:496-505.

[50] Verbeeck RMH, Demaeyer EAP, Driessens FCM. Stoichiometry of Potassium-Containing and Carbonate-Containing Apatites Synthesized by Solid-State Reactions. *Inorganic Chemistry.* 1995;34:2084-8.

[51] He QJ, Huang ZL, Cheng XK, Yu J. Thermal stability of porous A-type carbonated hydroxyapatite spheres. *Materials Letters.* 2008;62:539-42.

[52] Nassif N, Martineau F, Syzgantseva O, Gobeaux F, Willinger M, Coradin T, et al. In Vivo Inspired Conditions to Synthesize Biomimetic Hydroxyapatite. *Chemistry of materials.* 2010;22:3653-63.

[53] Huang S-J, Tsai Y-L, Lee Y-L, Lin C-P, Chan JCC. Structural Model of Rat Dentin Revisited. *Chemistry of materials.* 2009;21:2583-5.

[54] Fleet ME, Liu XY. Location of type B carbonate ion in type A-B carbonate apatite synthesized at high pressure. *Journal of Solid State Chemistry.* 2004;177:3174-82.

[55] Gross KA, Rodriguez-Lorenzo LM. Sintered hydroxyfluorapatites. Part 1: Sintering ability of precipitated solid solution powders. *Biomaterials.* 2004;25:1375-84.

[56] Gross KA, Rodriguez-Lorenzo LM. Sintered hydroxyfluorapatites. Part II: Mechanical properties of solid solutions determined by microindentation. *Biomaterials.* 2004;25:1385-94.

[57] Rodriguez-Lorenzo LM, Gross KA. Encapsulation of hydroxyapatite microspheres with fluorapatite using a diffusion process. *Journal of the American Ceramic Society.* 2004;87:814-8.

[58] Rodriguez-Lorenzo L.M., Hart J.N., Gross K.A. Influence of fluorine in the synthesis of apatites. Synthesis of solid solutions of hydroxy-fluorapatite. *Biomaterials.* 2003;24:3777-85.

[59] Rey C, Miquel JL, Facchini L, Legrand AP, Glimcher MJ. Hydroxyl-Groups in Bone-Mineral. *Bone.* 1995;16:583-6.

[60] Taylor MG, Parker SF, Simkiss K, Mitchell PCH. Bone mineral: evidence for hydroxy groups by inelastic neutron scattering. *Physical Chemistry Chemical Physics.* 2001;3:1514-7.

[61] Rodriguez-Lorenzo LM, Hart JN, Gross KA. Structural and chemical analysis of well crystallized hydroxyfluorapatites. Journal of physical chemistry B. 2003;107:8316-20.

[62] Lusvardi G, Menabue L, Saladini M, Spaggiari M. Effect of PH and Anions on Hydroxyapatite-Cu2+ Solid-Liquid Interactions. *Journal of Materials Chemistry.* 1995;5:493-7.

[63] Jeanjean J, Vincent V, Fedoroff M. Structural modification of calcium hydroxyapatite induced by sorption of cadmiun ions. *J Solid State Chemistry.* 1994;108:68-72.

[64] Brook I, Freeman C, Grubb S, Cummins N, Curran D, Reidy C, et al. Biological evaluation of nano-hydroxyapatite-zirconia (HA-ZrO2) composites and strontium-hydroxyapatite (Sr-HA) for load-bearing applications. *Journal of Biomaterials Applications.* 2012;27:291-8.

[65] de Lima IR, Alves GG, Soriano CA, Campaneli AP, Gasparoto TH, Ramos Junior ES, et al. Understanding the impact of divalent cation substitution on hydroxyapatite: An in vitro multiparametric study on biocompatibility. *Journal of Biomedical Materials Research Part A.* 2011;98A:351-8.

[66] Baroth S, Bourges X, Fellah B, Daculsi G. Radiopaque strategy for bone injectable substitute. In: Daculsi G, Layrolle P, editors. *Bioceramics,* Vol 20, Pts 1 and 22008. p. 39-42.

[67] Laurencin D, Almora-Barrios N, de Leeuw NH, Gervais C, Bonhomme C, Mauri F, et al. Magnesium incorporation into hydroxyapatite. *Biomaterials*. 2011;32:1826-37.

[68] Ren F, Leng Y, Xin R, Ge X. Synthesis, characterization and ab initio simulation of magnesium-substituted hydroxyapatite. *Acta Biomaterialia.* 2010;6:2787-96.

[69] Rude RK, Gruber HE. Magnesium deficiency and osteoporosis: animal and human observations. *Journal of Nutritional Biochemistry.* 2004;15:710-6.

[70] Takaishi Y. Significance of the Magnesium in the hard tissues such as Bone and Teeth. *Clinical calcium.* 2012;22:1189-96.

[71] Mayer I, Schlam R, Featherstone JDB. Magnesium-containing carbonate apatites. *Journal of Inorganic Biochemistry. 1997*;66:1-6.

[72] Bigi A, Falini G, Foresti E, Gazzano M, Ripamonti A, Roveri N. Magnesium Influence on Hydroxyapatite Crystallization. *Journal of Inorganic Biochemistry.* 1993;49:69-78.

[73] Suchanek WL, Byrappa K, Shuk P, Riman RE, Janas VF, TenHuisen KS. Preparation of magnesium-substituted hydroxyapatite powders by the mechanochemical-hydrothermal method. *Biomaterials.* 2004;25:4647-57.

[74] Nielsen S.P., Slosman D., Sorensen O.H., basse-Cathalinat B., De Cassin P., Roux C., et al. Influence of strontium on bone mineral density and bone mineral content measurements by dual X-ray absorptiometry. *J Clinical densitometry*. 1999;2:371-9.

[75] Dahl S.G., Allain P., marie P.J., mauras Y., Boivin G., Ammann P., et al. Incorporation and distribution of strontium in bone. *Bone.* 2001;28:446-53.

[76] Burguera M., Burguera J.L., Di Bernardo M.L., Alarcon O.M., Nieto E., Salinas J.R., et al. Age and sex-related calcium and strontium concentration in different types of human bones. *Trace elements and electrolytes.* 2002;19:143-51.

[77] Christoffersen J., Christoffersen M.R., Kolthoff N., Barenholdt O. Effect of strontium ions on growth and dissolution of hydroxyapatite and on bone mineral detection. *Bone.* 1997;20:47-54.

[78] Chaouat D. New osteoformative agents for osteoporosis. *Revue de medicine interne.* 2002;23:759-67.

[79] Reginster J.Y., Hoszowski K., Varela A.R., Balogh A., Clemens M., Fiore C., et al. Strontium ranelate: a new effective antiosteoporotic treatment reducing the incidence of vertebral and non vertebral fractures in postmenopausal women with osteoporosis. *Bone.* 2003;32:S94-S.

[80] Marie P.J., Ammann P., Boivin G., Rey C. Mechanisms of action and therapeutic potential of strontium in bone. *Calcif Tissue Int.* 2001;69:121-9.

[81] Hao Y, Yan H, Wang X, Zhu B, Ning C, Ge S. Evaluation of Osteoinduction and Proliferation on Nano-Sr-HAP: A Novel Orthopedic Biomaterial for Bone Tissue Regeneration. *Journal of Nanoscience and Nanotechnology.* 2012;12:207-12.

[82] Zhang W, Wang G, Liu Y, Zhao X, Zou D, Zhu C, et al. The synergistic effect of hierarchical micro/nano-topography and bioactive ions for enhanced osseointegration. *Biomaterials.* 2013;34:3184-95.

[83] Lin K, Liu P, Wei L, Zou Z, Zhang W, Qian Y, et al. Strontium substituted hydroxyapatite porous microspheres: Surfactant-free hydrothermal synthesis, enhanced

biological response and sustained drug release. *Chemical Engineering Journal.* 2013;222:49-59.

[84] Hanifi A, Fathi MH, Sadeghi HMM. Effect of strontium ions substitution on gene delivery related properties of calcium phosphate nanoparticles. *Journal of Materials Science-Materials in Medicine.* 2010;21:2601-9.

[85] Naddari T, Hamdi B, Savariault JM, El Feki H, Ben Salah A. Substitution mechanism of alkali metals for strontium in strontium hydroxyapatite. *Materials Research Bulletin.* 2003;38:221-30.

[86] Huang Y, Ding Q, Pang X, Han S, Yan Y. Corrosion behavior and biocompatibility of strontium and fluorine co-doped electrodeposited hydroxyapatite coatings. *Applied Surface Science.* 2013;282:456-62.

[87] Tank KP, Sharma P, Kanchan DK, Joshi MJ. FTIR, powder XRD, TEM and dielectric studies of pure and zinc doped nano-hydroxyapatite. *Crystal Research and Technology.* 2011;46:1309-16.

[88] Channa HMA, Ashfaq M, Bangash R, Abbasi A, Qureshi MA. Preventive role of zinc chloride against toxicity of ciprofloxacin on the growing cartilage of Wistar albino rat litter. Journal of Ayub Medical College, Abbottabad: *JAMC.* 2008;20:77-81.

[89] Samani S, Hossainalipour SM, Tamizifar M, Rezaie HR. In vitro antibacterial evaluation of sol-gel-derived Zn-, Ag-, and (Zn plus Ag)-doped hydroxyapatite coatings against methicillin-resistant Staphylococcus aureus. *Journal of Biomedical Materials Research Part A.* 2013;101A:222-30.

[90] Thurber A, Wingett DG, Rasmussen JW, Layne J, Johnson L, Tenne DA, et al. Improving the selective cancer killing ability of ZnO nanoparticles using Fe doping. *Nanotoxicology.* 2012;6:440-52.

[91] Roy R, Tripathi A, Das M, Dwivedi PD. Cytotoxicity and Uptake of Zinc Oxide Nanoparticles Leading to Enhanced Inflammatory Cytokines Levels in Murine Macrophages: Comparison with Bulk Zinc Oxide. *Journal of Biomedical Nanotechnology.* 2011;7:110-1.

[92] Yang S-T, Liu J-H, Wang J, Yuan Y, Cao A, Wang H, et al. Cytotoxicity of Zinc Oxide Nanoparticles: Importance of Microenvironment. *Journal of Nanoscience and Nanotechnology.* 2010;10:8638-45.

[93] Chassot E, Oudadesse H, Irigaray J, Curis E, Benazeth S, Nicolis I. Differentiation of biological hydroxyapatite compounds by infrared spectroscopy, x-ray diffraction and extended x-ray absorption fine structure. *Journal of Applied Physics.* 2001;90:6440-6.

[94] Venkatasubbu GD, Ramasamy S, Avadhani GS, Palanikumar L, Kumar J. Size-mediated cytotoxicity of nanocrystalline titanium dioxide, pure and zinc-doped hydroxyapatite nanoparticles in human hepatoma cells. *Journal of Nanoparticle Research.* 2012;14:819.

[95] Tang Y, Chappell HF, Dove MT, Reeder RJ, Lee YJ. Zinc incorporation into hydroxylapatite. *Biomaterials.* 2009;30:2864-72.

[96] Fujii E, Ohkubo M, Tsuru K, Hayakawa S, Osaka A, Kawabata K, et al. Selective protein adsorption property and characterization of nano-crystalline zinc-containing hydroxyapatite. *Acta Biomaterialia.* 2006;2:69-74.

[97] Stanic V, Dimitrijevic S, Antic-Stankovic J, Mitric M, Jokic B, Plecas IB, et al. Synthesis, characterization and antimicrobial activity of copper and zinc-doped hydroxyapatite nanopowders. *Applied Surface Science.* 2010;256:6083-9.

[98] Kalita SJ, Bhatt HA. Nanocrystalline hydroxyapatite doped with magnesium and zinc: Synthesis and characterization. *Materials Science & Engineering C-Biomimetic and Supramolecular Systems.* 2007;27:837-48.

[99] Medvecky L, Stulajterova R, Parilak L, Trpcevska J, Durisin J, Barinov SM. Influence of manganese on stability and particle growth of hydroxyapatite in simulated body fluid. *Colloids and Surfaces a-Physicochemical and Engineering Aspects.* 2006;281:221-9.

[100] Ramesh S, Tan CY, Peralta CL, Teng WD. The effect of manganese oxide on the sinterability of hydroxyapatite. *Science and Technology of Advanced Materials.* 2007;8:257-61.

[101] Paluszkiewicz C, Slosarczyk A, Pijocha D, Sitarz M, Bucko M, Zima A, et al. Synthesis, structural properties and thermal stability of Mn-doped hydroxyapatite. *Journal of Molecular Structure.* 2010;976:301-9.

[102] Jelinek M, Kocourek T, Remsa J, Weiserova M, Jurek K, Miksovsky J, et al. Antibacterial, cytotoxicity and physical properties of laser - Silver doped hydroxyapatite layers. *Materials Science & Engineering C-Materials for Biological Applications.* 2013;33:1242-6.

[103] Chen W, Oh S, Ong AP, Oh N, Liu Y, Courtney HS, et al. Antibacterial and osteogenic properties of silver-containing hydroxyapatite coatings produced using a sol gel process. *Journal of Biomedical Materials Research - Part A.* 2007;82:899-906.

[104] Chen W, Liu Y, Courtney HS, Bettenga M, Agrawal CM, Bumgardner JD, et al. In vitro anti-bacterial and biological properties of magnetron co-sputtered silver-containing hydroxyapatite coating. *Biomaterials.* 2006;27:5512-7.

[105] Liu X, Mou Y, Wu S, Man HC. Synthesis of silver-incorporated hydroxyapatite nanocomposites for antimicrobial implant coatings. *Applied Surface Science.* 2013;273:748-57.

[106] Uskokovic V, Drofenik M. Synthesis of materials within reverse micelles. *Surface Review and Letters.* 2005;12:239-77.

[107] Uskokovic V, Drofenik M. Reverse micelles: Inert nano-reactors or physico-chemically active guides of the capped reactions. *Advances in Colloid and Interface Science.* 2007;133:23-34.

[108] Chen HF, Clarkson BH, Sun K, Mansfield JF. Self-assembly of synthetic hydroxyapatite nanorods into an enamel prism-like structure. *Journal of Colloid and Interface Science.* 2005;288:97-103.

[109] Kokubo T., Kushitani H., Sakka S., Kitsugi T., T. Y. Solutions able to reproduce in vivo surface structurechanges in bioactive glass-ceramic A-W. *J Biomed Mater Res.* 1990;24:721-34.

[110] Ethirajan A, Landfester K. Functional Hybrid Materials with Polymer Nanoparticles as Templates. *Chemistry-a European Journal.* 2010;16:9398-412.

[111] Kamitakahara M. Preparation of highly functional artificial bones using the properties of calcium phosphates. *Journal of the Ceramic Society of Japan.* 2011;119:266-70.

[112] Chirila TV, Zainuddin. Calcification of synthetic polymers functionalized with negatively ionizable groups: A critical review. *Reactive & Functional Polymers.* 2007;67:165-72.

[113] Huang J, Lin YW, Fu XW, Best SM, Brooks RA, Rushton N, et al. Development of nano-sized hydroxyapatite reinforced composites for tissue engineering scaffolds. *Journal of Materials Science-Materials in Medicine.* 2007;18:2151-7.

[114] Stevanovic M, Maksin T, Petkovic J, Filipic M, Uskokovic D. An innovative, quick and convenient labeling method for the investigation of pharmacological behavior and the metabolism of poly(DL-lactide-co-glycolide) nanospheres. Nanotechnology. 2009;20.

[115] Stancu IC, Filmon R, Cincu C, Marculescua B, Zaharia C, Tourmen Y, et al. Synthesis of methacryloyloxyethyl phosphate copolymers and in vitro calcification capacity. *Biomaterials.* 2004;25:205-13.

[116] Liu F, Wang R, Cheng Y, Jiang X, Zhang Q, Zhu M. Polymer grafted hydroxyapatite whisker as a filler for dental composite resin with enhanced physical and mechanical properties. *Materials Science & Engineering C-Materials for Biological Applications.* 2013;33:4994-5000.

[117] Kishimoto T, Morihara Y, Osanai M, Ogata S, Kamitakahara M, Ohtsuki C, et al. Synthesis of poly(Pro-Hyp-Gly)(n) by direct polycondensation of (Pro-Hyp-Gly)(n), where n = 1, 5, and 10, and stability of the triple-helical structure. *Biopolymers.* 2005;79:163-72.

[118] Hosoya K, Ohtsuki C, Kawai T, Kamitakahara M, Ogata S, Miyazaki T, et al. A novel covalently crosslinked gel of alginate and silane with the ability to form bone-like apatite. *Journal of Biomedical Materials Research Part A.* 2004;71A:596-601.

[119] Dalas E, Kallitsis JK, Koutsoukos PG. Crystallization of Hydroxyapatite on Polymers. *Langmuir.* 1991;7:1822-6.

[120] Hossainy SFA, Hubbell JA. Molecular-Weight Dependence of Calcification of Polyethylene-Glycol Hydrogels. *Biomaterials.* 1994;15:921-5.

[121] Tamai H, Yasuda H. Preparation of polymer particles coated with hydroxyapatite. *Journal of Colloid and Interface Science.* 1999;212:585-8.

[122] Musyanovych A, Schmitz-Wienke J, Mailaender V, Walther P, Landfester K. Preparation of biodegradable polymer nanoparticles by miniemulsion technique and their cell interactions. *Macromolecular Bioscience.* 2008;8:127-39.

[123] Singh RK, Kim H-W. Inorganic nanobiomaterial drug carriers for medicine. *Tissue Engineering and Regenerative Medicine.* 2013;10:296-309.

[124] Guo Y-J, Long T, Chen W, Ning C-Q, Zhu Z-A, Guo Y-P. Bactericidal property and biocompatibility of gentamicin-loaded mesoporous carbonated hydroxyapatite microspheres. *Materials Science & Engineering C-Materials for Biological Applications.* 2013;33:3583-91.

[125] Pourbaghi-Masouleh M, Hesaraki S, Zamanian A, Khanlarkhani A. Gentamicin-PEGylated poor crystalline carbonated hydroxyapatite submicron particles. *Materials Letters.* 2013;113:130-3.

[126] Shinto Y, Uchida A, Korkusuz F, Araki N, Ono K. Calcium Hydroxyapatite Ceramic Used As a Delivery System for Antibiotics. *Journal of Bone and Joint Surgery-British Volume.* 1992;74:600-4.

[127] Lian X, Liu H, Wang X, Xu S, Cui F, Bai X. Antibacterial and biocompatible properties of vancomycin-loaded nano-hydroxyapatite/collagen/poly (lactic acid) bone substitute. *Progress in Natural Science-Materials International.* 2013;23:549-56.

[128] Josse S, Faucheux C, Soueidan A, Grimandi G, Massiot D, Alonso B, et al. Chemically modified calcium phosphates as novel materials for bisphosphonate delivery. *Advanced Materials.* 2004;16:1423-+.

[129] Faraji AH, Wipf P. Nanoparticles in cellular drug delivery. *Bioorganic & Medicinal Chemistry.* 2009;17:2950-62.

[130] Kester M, Heakal Y, Fox T, Sharma A, Robertson GP, Morgan TT, et al. Calcium Phosphate Nanocomposite Particles for In Vitro Imaging and Encapsulated Chemotherapeutic Drug Delivery to Cancer Cells. *Nano Letters.* 2008;8:4116-21.

[131] Barna JC, Williams DH. The structure and mode of action of glycopeptide antibiotics of the vancomycin group. *Annual Review of Microbiology.* 1984;38:339-57.

[132] Lin K, Zhou Y, Zhou Y, Qu H, Chen F, Zhu Y, et al. Biomimetic hydroxyapatite porous microspheres with co-substituted essential trace elements: Surfactant-free hydrothermal synthesis, enhanced degradation and drug release. *Journal of Materials Chemistry.* 2011;21:16558-65.

[133] Gu J, Xia F, Wu Y, Qu X, Yang Z, Jiang L. Programmable delivery of hydrophilic drug using dually responsive hydrogel cages. *Journal of Controlled Release.* 2007;117:396-402.

[134] Melville AJ, Rodriguez-Lorenzo LM, Forsythe JS. Effects of calcination temperature on the drug delivery behaviour of Ibuprofen from hydroxyapatite powders. *Journal of Materials Science-Materials in Medicine.* 2008;19:1187-95.

[135] Grossmann G, Grossman A, Ohms G, Breuer E, Chen R, Golomb G, et al. Solid-state NMR of bisphosphonates adsorbed on hydroxyapatite. *Magnetic Resonance in Chemistry.* 2000;38:11-6.

[136] Ezra A, Golomg G. Administration routes and delivery systems of bisphosphonates for the treatment of bone resorption. *Advances drug delivery reviews.* 2000;42:175-95.

[137] Rodriguez-Lorenzo LM, Fernández M, Parra J, Vazquez B, López-Bravo A, San Roman J. Targeting and controlled release of novel bisphosphonates by injectable, self-curing acrylic formulations. *J Biomed Mater Res B.* 2007;83:596–608.

[138] Rodriguez-Lorenzo LM, Vazquez B, San Roman J. Incorporation of 2nd and 3rd generation bisphosphonates on hydroxyfluorapatite. *key Eng Mater.* 2006;309-311:899-902.

[139] Rodriguez-Lorenzo LM, Vazquez B, San Román J. Surface modification of calcium hydroxyfluor carbonate apatites by bisphosphonates. *key Eng Mater.* 2005;284-286:357-60.

[140] Diez-Perez. Bisphosphonates. *Maturitas.* 2002;43:s19-s26.

[141] Kellinsalmi M, Monkkonen H, Monkkonen J, Leskela HV, Parikka V, Hamalainen M, et al. In vitro comparison of clodronate, pamidronate and zolendronic acid effects on rat osteoclasts and human stem cell-derived osteoblasts. *Basic and clinical pharmacology and toxicology.* 2005;97:382-91.

[142] Vargas N., Patiño C, Rodriguez-Lorenzo L.M., L T. Synthesis of hybrids compounds apatite-alendronate by reactive milling and effects on the structure and morphology of the apatite phase. *Ceramics International.* 2013;39:3921-9.

[143] Russell RGG, Watts NB, Ebetino FH, Rogers MJ. Mechanisms of action of bisphosphonates: similarities and differences and their potential influence on clinical efficacy. *Osteoporosis international.* 2008;19:733-59.

[144] Josse S, Faucheux C, Soueldan A, Grimandi G, Massiot D, Alonso B, et al. Novel biomaterials for bisphosphonate delivery. *Biomaterials.* 2005;26:2073-80.
[145] Tanaka T, Hirose M, Kotobuki N, Ohgushi H, Furuzono T, Sato J. Nano-scaled hydroxyapatite/silk fibroin sheets support osteogenic differentiation of rat bone marrow mesenchymal cells. *Materials Science and Engineering C.* 2007;27:817-23.
[146] Iafisco M, Palazzo B, Falini G, Di Foggia M, Bonora S, Nicolis S, et al. Adsorption and conformational change of myoglobin on biomimetic hydroxyapatite nanocrystals functionalized with alendronate. *Langmuir.* 2008;24:4924-30.
[147] Oyane A, Ootsuka T, Hayama K, Sogo Y, Ito A. Enhanced immobilization of acidic proteins in the apatite layer via electrostatic interactions in a supersaturated calcium phosphate solution. *Acta Biomaterialia.* 2011;7:2969-76.
[148] Krishnan A, Siedlecki CA, Vogler EA. Mixology of protein solutions and the Vroman effect. *Langmuir.* 2004;20:5071-8.
[149] Clare TL, Clare BH, Nichols BM, Abbott NL, Hamers RJ. Functional monolayers for improved resistance to protein adsorption: Oligo(ethylene glycol)-modified silicon and diamond surfaces. *Langmuir.* 2005;21:6344-55.
[150] Ribeiro N, Sousa SR, Monteiro FJ. Influence of crystallite size of nanophased hydroxyapatite on fibronectin and osteonectin adsorption and on MC3T3-E1 osteoblast adhesion and morphology. *Journal of Colloid and Interface Science.* 2010;351:398-406.
[151] Vogler EA. Protein adsorption in three dimensions. *Biomaterials.* 2011;33:1201-37.
[152] Iafisco M, Sabatino P, Lesci IG, Prat M, Rimondini L, Roveri N. Conformational modifications of serum albumins adsorbed on different kinds of biomimetic hydroxyapatite nanocrystals. *Colloids and Surfaces B: Biointerfaces.* 2010;81:274-84.
[153] Yang Q, Zhang Y, Liu M, Zhang Y, Yao S. Study of fibrinogen adsorption on hydroxyapatite and TiO2 surfaces by electrochemical piezoelectric quartz crystal impedance and FTIR-ATR spectroscopy. *Analytica Chimica Acta.* 2007;597:58-66.
[154] Israelachvili J, Wennerstrom H. Role of hydration and water structure in biological and colloidal interactions. *Nature.* 1996;379:219-25.
[155] Noh H, Vogler EA. Volumetric interpretation of protein adsorption: Mass and energy balance for albumin adsorption to particulate adsorbents with incrementally increasing hydrophilicity. *Biomaterials.* 2006;27:5801-12.
[156] Gagnon P. Improved antibody aggregate removal by hydroxyapatite chromatography in the presence of polyethylene glycol. *Journal of Immunological Methods.* 2008;336:222-8.
[157] Kandori K, Oda S, Fukusumi M, Morisada Y. Synthesis of positively charged calcium hydroxyapatite nano-crystals and their adsorption behavior of proteins. *Colloids and Surfaces B-Biointerfaces.* 2009;73:140-5.
[158] Mistry AS, Mikos AG. Tissue engineering strategies for bone regeneration. In: Yannas IV, editor. *Regenerative Medicine II: Clinical and Preclinical Applications* 2005. p. 1-22.
[159] Mrksich M. Using self-assembled monolayers to model the extracellular matrix. *Acta Biomaterialia.* 2009;5:832-41.
[160] Graham FL, Vandereb AJ. Transformation of Rat Cells by DNA of Human Adenovirus-5. *Virology.* 1973;54:536-9.

[161] Chowdhury EH. Self-assembly of DNA and cell-adhesive proteins onto pH-sensitive inorganic crystals for precise and efficient transgene delivery. *Current Pharmaceutical Design.* 2008;14:2212-28.

[162] Chowdhury EH, Akaike T. Fibronectin-coated nano-precipitates of calcium-magnesium phosphate for integrin-targeted gene delivery. *Journal of Controlled Release.* 2006;116:E68-E9.

[163] Uskokovic V. Surface charge effects involved in the control of stability of sols comprising uniform cholesterol particles. *Materials and Manufacturing Processes.* 2008;23:620-3.

In: Apatite
Editors: M. Iafisco and J. M. Delgado-López
ISBN: 978-1-63321-500-9

Chapter 12

APATITE AS BIOMATERIAL IN HARD TISSUE ENGINEERING: AN OVERVIEW

Chieko Yamagata[1*], Andrea D. Rodas[1], Olga Z. Higa[1], Agatha M. Misso[1], Fernando S. Silva[1] and Signo T. Reis[2]
[1]Nuclear and Energy Research Institute, Sao Paulo, SP, Brazil
[2]Missouri University of Science and Technology, Rolla, MO, US

ABSTRACT

The demand for synthetic biomaterials to replace and repair bone tissue lost from injury or disease has significantly increased in the last decade. Therefore, the development of bioactive materials is extremely important. Similar to the bone crystalline apatite phase, hydroxyapatite (HA) has garnered widespread consideration in advanced hard tissue engineering. The main reason is its osteoconduction by biologically active bone-like apatite formation through chemical reaction of its surface with surrounding simulated body fluid (SBF). Consequently, HA synthetic development as a biomaterial has been extensively investigated. Various methods, such as precipitation technique, sol-gel approach, hydrothermal technique, multiple emulsion technique, biomimetic deposition technique, electro deposition technique etc. synthesize hydroxyapatite nanopowder, with precise control of microstructure, particle shape and size. The precipitation technique, is widely used for HA powder synthesis, followed by the hydrothermal technique. The sol–gel technique has recently drawn researcher's attention due to its advantages, such as molecular-level mixing of reactants, low processing temperatures, and obtaining of HA monophasic. The biomimetic deposition technique is a supplementary method for HA synthesis. It consists of the nucleation and growth of nanosized carbonated and 'bone-mimic' HA at physiological pH and temperature, promoted by metastable SBF. The biomimetic process of calcium phosphate (Ca-P) nucleation and growth on titanium and titanium alloys have been extensively studied because of their relevance in orthopedic applications. The biomimetic deposition technique transforms glasses and glass-ceramics to bioactive materials, mainly for some silicate bio ceramics, because of their excellent apatite mineralization in SBF. Osteoconduction and bioactivity of HA justify its strong affinity to the host hard tissues,

[*] Corresponding Author Email: yamagata@ipen.br.

which is an important advantage in clinical applications. They include bone repair, bone augmentation, implants coating and filling in bone or tooth. The combination of apatite and wollastonite on W–A (Wollastonite-Apatite) sintered composite promotes an improvement on its mechanical properties, bioactivity and biodegradability.

Keywords: Apatite, biomaterials, hydroxyapatite, synthesis, bio ceramics

1. INTRODUCTION

During the past few decades, researchers have been focusing on the synthesis of various materials for biomedical applications as so-called biomaterials. The global market of biomaterials was estimated in 150–200 U$ billion in 2012 including all diagnostic therapeutic equipment [1]. The ten largest markets are the US, Japan, Germany, France, Italy, the UK, Brazil, China, Canada and Spain. The US market share leads the world, followed by Europe, and Japan. The largest market for biomaterial-based products is orthopedic biomaterials followed by cardiovascular and drug delivery materials [1].

The function of biomaterials is to replace diseased or damaged tissues. During the last four decades, a great revolution has occurred in the use of ceramics to improve the quality of human life, due to innovative use of ceramics known as bioceramics. Those ceramics have potential application for replacement or repairing of lost, injured or diseased bone [2]. Bone is the second most replaced organ after blood [3]. Bone defects are among the most significant worldwide health problem. Data of 2005 indicated more than 2 million bone implant procedures are performed worldwide each year [4]. The estimated cost of those procedures is about US$2.5 billion [5]. Almost all of the procedures involve obtaining bone from either the patient or another donor, and thus, the implants require two surgical procedures, even if the replacement tissue is available. The use of bio ceramics reduces the number of those invasive procedures. Reducing the number of procedures, the possibility of infection or other complications also decrease, as well as costs. This has been one of the main reasons for the development of bone substitute materials, including bio ceramic materials. More than 50 percent of the available bone implants alternatives consist of ceramics (including glasses) either alone or in combination with another material [6]. Currently, ceramics have become a diverse class of biomaterials. A long list of properties had evolved for successful biomaterials. They are selected to be highly biocompatible and anti thrombogenic, and should not be toxic, non-allergenic, non-carcinogenic, non-pyrogenic, and non-inflammatory [3]. Based on the reaction of the tissue to the biomaterial, those ceramics can be classified into three distinct categories: bioinert, bioactive, and bioresorbable [7]. Bio inert refers to a material that retains its structure in the body after implantation and does not induce any immunologic host reactions. Bioactive materials have the property of establishing chemical bonds with bone tissue, known as osseointegration. Degradable or resorbable materials are incorporated into the surrounding tissue, or they can even dissolve completely over a period by the bio system of the organism replacement without toxicity and rejection [8]. Among different classes of biomaterials, hydroxyapatite $Ca_{10}(PO_4)_6(OH)_2$ (HA) is a bio ceramic, widely used in various biomedical applications. Its applications include orthopedics and dentistry, as bone substitute and implants coating due to the high similarity to the inorganic material of bone and teeth. It has excellent biocompatibility and exceptional bioactivity [9,

10]. Based on crystallographic and chemical studies [11], synthetic HA has been recognized to be similar to naturally occurring apatite. The durability or failure of the implant-tissue integration depends on its interactions with the biological environment, especially if it is strongly connected with the surface of the implant [12]. The physical, chemical and biochemical properties of the implant surface control the performance of important processes such as protein adsorption, cell-surface interaction and cell/tissue development at the interface between the body and biomaterials [12]. The material for orthopedics and dentistry applications is required to have the ability to form, or to induce, the formation of apatite upon its surface after coming into contact with bone in the body. This intermediate apatite coating works for bonding the implant of the bioactive material to the bone [13]. There are two main routes for the manufacturing of osteoconductive scaffolds for tissue engineering: The first is incorporating bio ceramic particles in the scaffold through a variety of techniques [14] and second is coating the scaffold with a thin layer of apatite through biomimetic processes [15, 16]. These two routes show advantages and disadvantages. The biomimetic processes make the non-bioactive scaffolds readily fabricated to be osteoconductive.

2. CERAMICS FOR HARD TISSUE ENGINEERING

The incorporation of biomaterials into body tissue depends on aspects such as biocompatibility, design, and material properties. The use of ceramics such as calcium phosphate, metallic oxides, and glass ceramics has been widely reported [17-23]. Bio ceramics are classified according to the tissue reaction on the biomaterial as bioinert, bioactive, and bioresorbable [24]. Alumina (Al_2O_3), carbon, zirconia (ZrO_2), and single oxide ceramic are examples of bioinert ceramics [25, 26]. They are employed as bone plates, artificial valves, femoral-head component, and artificial joints [27]. They do not form a bond with bone but they allow the development of thin fibrous capsules in the interface. The commonly bioactive ceramics used in orthopedic operation are Ceravital®, bioglass, apatite-wollastonite (A-W) ceramic and dense HA. Bioactive ceramic undergoes surface chemical reactions [28], the response of ion exchange between the bioactive implant and the body fluids results in the development of a biologically active carbonated apatite layer on the implant. Because the bioactive ceramics have poor properties compared with bioinert ceramics, they are most commonly used as bone defect fillers provided in block structures, porous material or granules [29]. A-W glass ceramic constitutes an exception; it has good mechanical properties for example bending strength, fracture toughness and young's modulus, better than cortical bone, enough to be used in load bearing applications such as vertebral prostheses etc. [30-32]. Biodegradable or resorbable ceramics, as the name implies, degrade constantly and natural tissues upon implantation in the host [33, 34] substitute them. Resorbable ceramics include HA, calcium carbonate, calcium sulfate, carbonate HA, and β-tricalcium phosphate [27]. For bone implantation purposes, calcium phosphate and bioglass are ideal because they have the capability to produce a connection with bone tissue, different of metallic oxides that are bio inert.

Close to the bone crystalline apatite (A) mineral phase, HA is commonly used as substitutes or filling materials in bone surgery. Therefore, development of synthetic HA as hard tissue replacement materials has been extensively investigated.

3. Development of Hydroxyapatite (HA) as Biomaterial

The purity and particle characteristics of the final synthesized powder can affect the bioactivity, mechanical and biological interaction properties. As a result, these characteristics determine the application of the material and subsequently the selection of synthesis method that enables to control the exact demand of crystal morphology, chemical composition, crystallinity, particle size distribution, and agglomeration. Several methods that have been utilized for the synthesis of HA include precipitation technique, sol-gel approach, hydrothermal technique, multiple emulsion technique, biomimetic deposition technique, electro deposition technique etc. Some commonly used techniques are discussed below.

3.1. Precipitation Method

Precipitation technique, also known as wet chemical precipitation is one of the most widely used method for the synthesis of HA powder, due to mainly to the use of relatively low-cost raw materials, and low reaction temperatures. Careful control of the solution condition is critical in the wet chemical precipitation. A controlled pH and temperature of the solution is mandatory. The precipitated powder is usually calcined at 400-600°C or even at a higher temperature in order to obtain a stoichiometric apatite structure. In some cases, a well crystallized HA phase was only developed while approaching a sintering temperature of 1200°C. However, fast precipitation during phosphate solution titration (to calcium solution) leads to chemical non-homogeneity in the final product. Slow titration and diluted solutions are imperative to improve chemical homogeneity and stoichiometry of the resulting HA. In this method, HA can be prepared by adding orthophosphoric acid (H_3PO_4) [35] or ammonium hydrogen phosphate ($(NH_4)_2HPO_4$) [36] to calcium hydroxide. The following reactions take place:

$$10\ Ca(OH)_2 + 6\ H_3PO_4 \rightarrow Ca_{10}(PO_4)_6(OH)_2 + 18\ H_2O \quad (1)$$

$$10\ Ca(OH)_2 + 6\ (NH_4)_2HPO_4 \rightarrow Ca_{10}(PO_4)_6(OH)_2 + 6\ H_2O + 12\ NH_4OH \quad (2)$$

The characteristics, such as the size, shape, surface area and crystallinity of HA particles, obtained from orthophosphoric acid, are strongly dependent on the acid addition rate and the reaction temperature [35]. Santos et al. [36] produced HA by three different aqueous precipitation routes. Calcium hydroxide, calcium phosphate, ammonium phosphate and phosphoric acid were used as reagents. The synthesis parameters, such as pH and thermal treatment have strong effect on the Ca/P ratio crystalline phase obtained. Otherwise, a decrease of solution pH below about 9 could lead to the formation of Ca-deficient HA structure [37]. Gallo [38] has investigated the crystallographic properties of F^- substituted, with general formula $Ca_{10}(PO_4)6OH_{2\text{-}}Fx$, and Sr-F co-substituted hydroxyapatite nanopowders for potential biomedical applications. Nanopowder of hydroxyapatite was synthesized by a simple co-precipitation method. Rietveld refinement XRD characterization has been performed to understand the influence of the F^- addition on structural properties such as lattice parameters, crystal size, phase composition, and crystallinity degree. Fluoridated hydroxyapatite series with nominal Ca/P ratio equal to 1.67, 1.50 and 1.28, with nominal fluorine atomic substitution ranging from 0% to 100% have been synthesized by a precipitation method in aqueous solution, followed by a heat treatment up to 900°C.

Wang et al. [39] synthesized HA nanoparticles via wet chemical precipitation method, under normal environment pressure, varying a series of process conditions, which were supposed to influence the size and morphology of the particles. HA powders with several morphologies, such as sphere, rod, needle, wire and bamboo-leaf-like were obtained by controlling the synthesis parameters. The investigation revealed that the solvent systems, dispersant species and drying methods have effects on the particle size and dispensability. The initial pH values and reaction temperatures both played important roles in the morphology of HA, as well as the phase formation and degree of crystallinity. The attaining of the crystalline phases of synthetic HA ($Ca_{10}(PO_4)6(OH)_2$), such as pure hydroxyapatite and CaO, biphasic calcium phosphate (BCP, HA + β-TCP) or β- tri calcium phosphate ($Ca_3(PO_4)_2$, β-TCP), depends on the Ca/P ratio in the feed of the precipitation process [40, 41]. Some possible calcium phosphates, as by-products of the synthesis of HA and whose Ca/P ratios are quite close to one another, are presented in Tab 1.

The dispersion and morphology of obtained precipitation products can be modified by the presence of surfactants. Cationic [44-46], anionic [47, 48] or mixture of cationic and anionic surfactants [49] are largely used.

Table 1. Intermediate compounds in the formation of HA [42, 43]

Name	Abbreviation	Chemical formula	Ca/P ratio
Hydroxyapatite	HA	$Ca_{10}(PO_4)_6OH_2$	1.67
Carbonate hydroxyapatite	CHA	$Ca_{10}(PO_4)_6CO_3$	1.67
Fluor apatite	FHA	$Ca_{10}(PO_4)_6F_2$	1.67
α,β tri calcium phosphate	α, β -TCP	$Ca_3(PO_4)_2$	1.55
Amorphous calcium phosphate	ACP	$Ca_9(PO_4)_6$	1.55
Octa calcium phosphate	OCP	$Ca_8H_2(PO_4)_6$	1.33

3.2. Hydrothermal Method

In the last century, the hydrothermal technique was considered an important technology for synthesis of various ceramic materials, including HA [46]. Hydrothermal and solvo-thermal processing consists on the thermal treatment, in a sealed vessel reactor, of the solvent, in which the precursor ions is present. In the case of hydrothermal synthesis, water is the used solvent. In the hydrothermal synthesis, ceramic materials are directly crystallized from single or heterogeneous phase reactions in aqueous solutions at elevated temperature ($T > 25°C$) and pressure ($P > 100$ kPa) [46]. A specific hydrothermal method, for the HA synthesis, is characterized by the reaction between aqueous solution of Ca^{2+} and $PO4^{3-}$, in closed recipients under controlled temperature and/or pressure. The temperature can be elevated above the boiling point of the water, reaching the pressure of vapor saturation. Due to the conditions of temperature and pressure, the HA prepared by hydrothermal method is homogeneous with high degree of crystallinity and high sinterability, and controlled shape and composition that can be achieved even at low temperatures [50-52].

Zhang et al. [53] prepared HA rods by hydrothermal method from anhydrous dicalcium phosphate ($CaHPO_4$, DCPA) and $CaCO_3$ as starting materials. The temperature of reaction was in the range of 120 to 180°C. Through the reaction temperature at 140 °C for 24 hours, the maim product was HA and a small amount of β-TCP as a by-product was presented. Manafi et al. [54] have successfully synthesized nanostructured hydroxyapatite with uniform morphology; controllable size, and narrow-size distribution in diameter of dispersed particles. $CaHPO_4.2H_2O$, NaOH and distilled water were mixed, followed by adding 2-3 g of cetyltrimethyl ammonium bromide (CTAB). The hydrothermal synthesis was performed at 150°C for 2 hours in an electric oven.

Liu et al. [55] investigated the influence of pH and temperature on the structure and morphology of HA. The synthesis was conducted by different pH values ranging from 6 to 14, and the different temperatures in the range from 60 to 140 °C. They indicated that pH has strong effect on the morphology, crystallite size and crystallinity of HA. They also suggested that the single phase of HA was only obtained at 120°C, when pH of the reaction was kept at 9. This HA single phase exhibited nanosized whiskers with the aspect ratio of 8-20, and had the median diameter and length was 40 nm and 600 nm.

Xin [56] studied the synthesis of HA, dicalcium phosphate dehydrate (DCPD) and dicalcium phosphate anhydrous (DCPA) phases in simple EDTA assisted hydrothermal method. Pure phase DCPD, DCPA and HA nanocrystals were obtained by controlling the solution pH and temperature.

3.3. Sol-Gel Approach

Sol–gel technique has attracted much attention recently due to its recognized advantages to synthesize glass, glass–ceramic and ceramics powders. This technique allows molecular-level mixing of reactants, low processing temperatures, the control of particles size, and the obtaining of chemically homogeneous nanopowders, bulk amorphous monolithic solids and thin films [57-60]. The sol-gel approach is an effective route for the synthesis of monophasic HA [61] because the technique permits an exacting control of the process parameters.

Coating of different substrates seems to be the main contribution to the sol–gel processing of HA [62-65]. Sol-gel process derived HA materials are efficient to improve the interaction and stability at the artificial/natural bone interfaces in both *in vitro* and *in vivo* environment [66]. Studies revealed that, the HA synthesized by sol-gel is close to biological apatite and its bioresorbability is higher than conventional prepared powders. Depending on the processing parameters, a powder obtained from a typical sol–gel method usually shows a stoichiometric structure with a high specific surface area and a small cluster size [67-69]. The generation of secondary phase (usually calcium oxide, CaO) in the process [70, 71], constitutes the major disadvantage of the technique, since this phase has demonstrated to be harmful to the HA biocompatibility. The high cost of the starting materials, especially alkoxide-based precursors, is the other inconvenience of the method. A variety of calcium and phosphorous precursor combinations were employed for HA sol-gel synthesis.

Sanosh et al. [72] reported the synthesis of high purity nanosized β-Tri calcium phosphate (β –TCP) powder via sol–gel method using calcium and phosphorous as precursors. Particle distribution analysis showed a skewed distribution plot where particle size was between 70 and 80 nm. Sranya [73] prepared HA nanopowders by the sol–gel emulsion technique. The water medium was emulsified, in a support solvent (oil medium) in the presence of a surfactant. Gelation of the sol droplets was achieved by adding gelling agent and maintaining the desired pH. Encinas-Romero [74] reported the preparation of hydroxyapatite-wollastonite biocomposite by an alternative sol-gel route. Calcium nitrate and ammonium phosphate were used as precursors of hydroxyapatite, and high purity natural wollastonite was added to the precursors in the ratios of 20, 50 and 80 percent by weight immersed in aqueous medium. Hydroxyapatite was crystallized at relatively low temperatures of about 350°C, while the wollastonite remained unreacted. The final biocomposites, which were sintered at 1200°C for 5 h, showed the presence of hydroxyapatite and wollastonite as unique phases.

3.4. Biomimetic Deposition Method

A supplementary method for the synthesis of HA is the biomimetic deposition technique. This technique consists of the nucleation and growth of nanosized carbonated and 'bone-mimic' HA at physiological pH and temperature, promoted by metastable simulated body fluid (SBF). The SBF is prepared according to the procedure described by Kokubo [75], obeying the chemical composition of human body fluid, with various ions concentrations nearly similar to the inorganic constituent of human blood plasma. Table 2 shows the composition of prepared SBF solution and blood plasma.

Table 2. Ion concentrations of the SBF and blood plasma (mM)

	Na^+	K^+	Mg^{2+}	Ca^{2+}	Cl-	HCO_3^-	HPO_4^{2-}	SO_4^{2-}
SBF	142.0	5.0	1.5	2.5	147.8	4.2	1.0	0.5
Blood plasma	142	5	1.5	2.5	103	27	1.0	0.5

The biomimetic method is an important technique for the reconstruction of bone defects: after bone loss or when the amount of damaged bone is too large for self-reconstruction. In such cases, the damaged bones are reconstructed using alternative materials. Normally the autograph technique transfers other parts of the bones of the same patient. Another technique used is allograft, which is transferred bone from other donors. The first procedure has a problem of limited amount of tissue being available. The second has problems related to overcoming the foreign body reaction and infections, when it is implanted into the recipient. Therefore, artificial materials are required to repair bone defects. Artificial ceramic or metallic bones can be implanted for reconstruction. Hydroxyapatite ceramics are used as important materials for bone reconstruction due to their bone-bonding ability. They are also compatible with various tissues [76]. Although the static mechanical strength of sintered hydroxyapatite is comparable to that of cortical bone, this material is predisposed to fatigue failure under conditions of high tensile loading. Consequently, it is unsuitable for applications in load-bearing situations. These problems may be solved by hydroxyapatite coating on the metallic implants. The hydroxyapatite-coated metallic implants show high tensile strength and ductility of the metal, and bioactivity of hydroxyapatite. Various methods have been used to prepare hydroxyapatite coatings, such as thermal spraying [77, 78], sputter coating [79], pulsed laser ablation [80], dip coating [81], sol–gel [82], electrophoretic deposition [83], and hot isostatic pressing [84], etc. Biomimetic approaches for bioactive HA coatings have emerged as highly promising techniques that enable in vitro HA formation in SBF. Different levels according to the composition, structure, morphology, bulk and surface chemical-physical properties, the biomimetic of synthetic materials process takes place. Recently, biomaterials research focuses overcoming the limitations of calcium phosphate ceramics as well as improving their bio reactivity exploring the single advantages of nanotechnology [85]. The application of biomimetic HA synthetic crystals, in orthopedics, are very promising, mainly due to its nanodimensional and morphological similarity with bone crystals.

There are some advantages of the biomimetic method, such as, low processing temperatures; developing bonelike apatite; complex shape coatings and incorporating bone growth stimulating factors [86]. The hydroxyapatite layer derived from biomimetic processes improves cell differentiation in chondrocyte cells [87] and forms the subsequent bone matrix deposition that helps strong bonding to bone [88].

3.4.1. Biomimetic HA Coating on Metals

The nucleation and growth of calcium phosphates (Ca-P) on titanium and its alloys have been extensively studied due to their relevance in orthopedic applications [89]. This nucleation and growth of calcium phosphates can be explained by the surface calcification of the implant, resulted from the calcium and phosphate ions supersaturating, which are present in the body fluids, and the surface properties of titanium (Ti) [90, 91]. In aqueous medium, a hydrated layer of oxide naturally covers titanium. The hydroxyl group on the surface of titanium favors the chemical bond between the calcium and phosphate ions [89, 92-95]. The calcification of Ti and titanium oxides depends on degree of supersaturating in the solution [91]. The substrate roughness affects the nucleation, and growth of Ca-P film on the titanium implants [90]. Because of these physic-chemical properties, a biomimetic coating process has been developed [94-98]. By this technique, bone-like apatite coatings could be deposited on titanium implants within 7 days [9] by heterogeneous nucleation of Ca-P from simulated body fluids (SBFs), which are supersaturated towards hydroxyapatite at pH=7.25 and at 37 °C [94, 99]. Barrére [97] reported the kinetics of the process in detail. The formation of the calcium phosphate layer on the metal surface is controlled by the concentrations of calcium and phosphate ions [100]. Magnesium and carbonate ions favor heterogeneous nucleation rather than crystal growth [101, 102].

3.4.2. Biomimetic HA Coating on Bioactive Glasses and Glass-Ceramics

More recently, the biomimetic deposition technique has been adapted to new bioactive glasses and glass-ceramics materials, which are largely used in bone regeneration [103]. The surface reactivity of those materials with biological fluids has been extensively studied. Bioactive glasses and glass-ceramics [104, 105] are important materials used for repair and reconstruction of diseased bone tissues. A main feature of those materials is their ability (bioactivity) to form an apatite-like layer on their surfaces in vivo and in vitro [106]. Numerous bioactive materials showing good surface hydroxyapatite (HA) formation ability have already been synthesized [107]. The simultaneous dissolution of silicates and ensuing formation of Si–OH groups is necessary [108-111] to create nucleation site for HA development. Most common bioactive materials comprising CaO and SiO_2 follow this mechanism for biomimetic HA formation on their surfaces. Although the mechanism of apatite formation has not been completely clarified, studies showed that bioglass and glass-ceramics containing Ca and Si, might induce apatite formation in simulated body fluid (SBF) [112]. Studies revealed the excellent bioactivity of some calcium silicate ceramics, such as wollastonite (triclinic $CaSiO_3$) [112-113], pseud wollastonite (monoclinic $CaSiO_3$) [114, 115], diopside ($CaMgSi_2O_6$) [116], tricalcium silicate (Ca_3SiO_5) [117], akermanite ($Ca_2MgSi_2O_7$) [45-47] and combeite ($Na_2Ca_2Si_3O_9$) [118]. Bioactive ceramics of CaO-MgO-SiO_2 system have been studied due to their mechanical properties and the fact that their degradation rate can be controlled [119]. An investigation of bioactive glass powders by sol-gel processing showed that powders produced by this technique are more bioactive than the melt-derived glasses of the same composition [120]. A novel modified sol-gel process was proposed to synthesize CaO-MgO-SiO_2 bioactive glass-ceramic powder [121]. Non-aggregated SiO_2 aerogel particles were initially prepared by surfactant template sol-gel technique, under HCl acid-catalyzed hydrolysis from Na_2SiO_3, as described previously [122]. A HNO_3 solution of Ca and Mg ions was embedded on prepared silica particles. A glass-ceramic powder was obtained by thermal treatment of the embedded material. The ceramic body, obtained from sintering of the powder, at 1320 ^{0}C for 3h, presented the bioactive crystalline phases: wollastonite ($CaSiO_3$), akermanite ($Ca_2MgSi_2O_7$) and dicalcium silicate (Ca_2SiO_4). In Figure 1 [121], XRD patterns peaks of those phases can be observed.

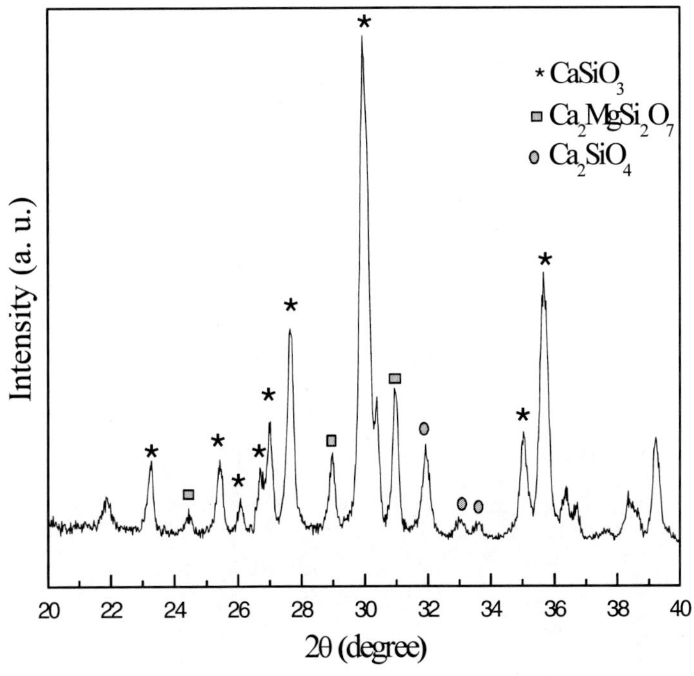

Figure 1. XRD pattern of the CaO-MgO-SiO_2 ceramic obtained after sintering at 1320 ^{0}C for 3h.

FTIR spectra of this ceramic body submitted to the biomimetic processes for 7, 14 and 24 days by soaking in SBF, are showed in Figure 2 [121]. Phosphate absorption band at about 1047 cm^{-1} is observed on spectra of the samples soaked for 7, 14 and 21 days, which indicates the formation of apatite layer on the surface of the immersed samples of the ceramic [123]. This absorption band increases from the sample of 7days to the sample of 14 days, but it virtually decreases for the sample of 21 days. This might indicates that apatite layer has formed on the surface of the sample of 7 days and settled on the 14 days sample. For the sample of 21 days, the formed apatite layer on its surface, perhaps, dissolved over the time. Figure 3 [121] shows SEM micrographs of the sintered ceramic before (Figure 3 (a)) and after 14 days (Figure 3b)) soaking in SBF. In Figure 3(a), a continuously connected crystalline grain with open pores microstructure of sintered body is observed. In Figure 3(b), particles of hydroxyapatite, as confirmed by FTIR spectra in Figure 2, with nanomicrospheres morphology [124] uniformly deposited on the surface are observed.

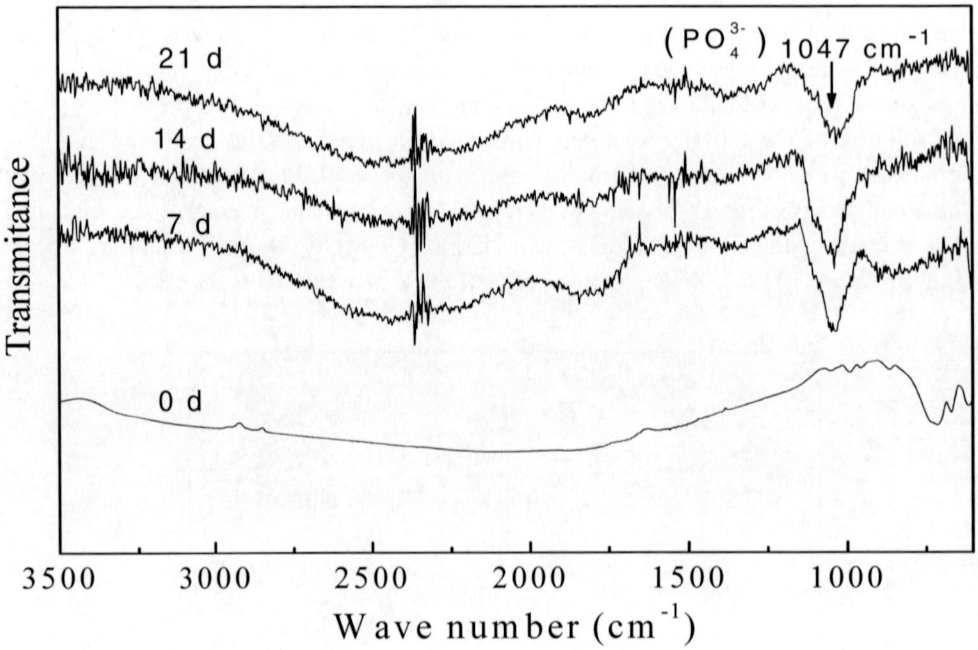

Figure 2. FTIR spectra of sintered ceramic surface, before (0d) and after 7 (7d), 14 (14d) and 21 (21d) days soaking in SBF.

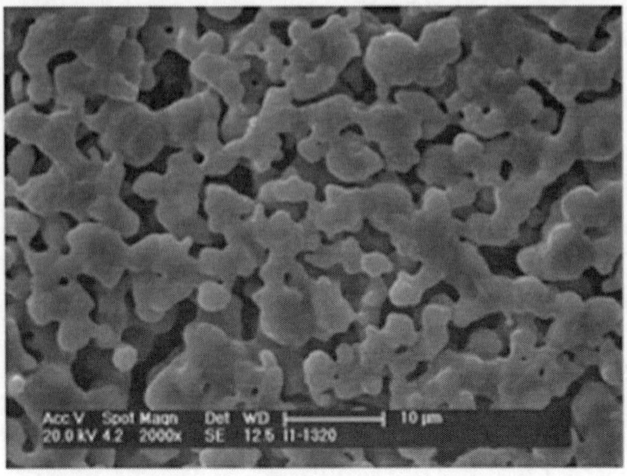

(a)

Figure 3. (Continued).

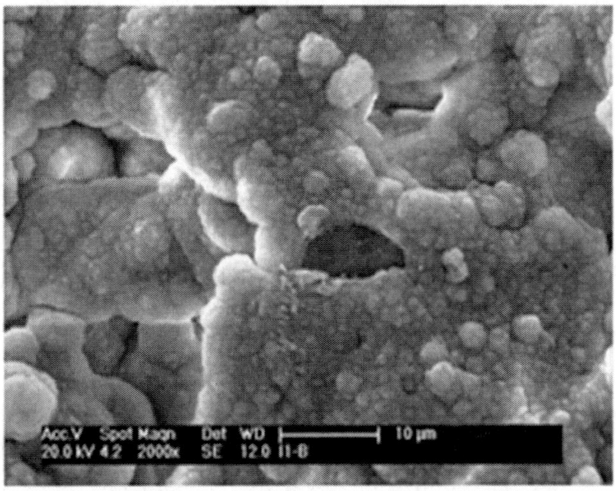

(b)

Figure 3. SEM micrographs of sintered CaO-MgO-SiO_2 glass-ceramic body surface, before (a) and after 14 days (b) soaking in SBF.

The mechanism of the bioactivity, as reported in the literature, it starts with rapid ion exchange between the alkaline ions from the glass surface and the hydrogen ions from SBF solution, to form of silanol groups. By polycondensation of silanol groups, the developing of a silica gel layer takes place. This layer promotes the adsorption of Ca^{2+} and $PO_4{}^{3-}$ ions from SBF solution and subsequently those ions react, forming the hydroxyl carbonate apatite (HCA) layer. From the literature, this mechanism has been observed in MgO-containing glasses as well as in MgO-free glasses. There is a strong divergence in the literature concerning the effect of magnesium on bioactive glasses. A review on magnesium containing bioactive glasses for biomedical applications has been recently presented [125]. A study [126] concluded that the mechanical properties are improved with increasing the MgO contents, whereas the apatite-formation ability, in SBF, and cell proliferation are decreased.

4. Aplication of Synthetic HA in Hard Tissue Engineering

The close chemical similarity of HA to natural bone apatite has led to wide research efforts to use synthetic HA as a bone substitute and/or replacement in biomedical application [127, 128]. As introduced earlier, HA is the main mineral element of human hard tissues. It can be found in teeth and bone. Application of HA can mainly to fast bone formation and solid biological fixation to bony tissue due its excellent osteoconductivity and bioactivity. Synthetic HA, as a biomaterial, has garnered extensive consideration in advanced hard tissue engineering. Tissue engineering applies the principles of engineering and life sciences to develop biological substitutes to seed living cells into biocompatible and eventually biodegradable environment (the scaffold), and then cultivates the system in a bioreactor so that the initial cellular population can develop into a tissue [129, 130]. The strong affinity of HA to host hard tissues is a greater advantage in clinical applications compared to most other bone substitutes such as allografts or metallic implants [131]. Synthetic HA has been widely used to repair hard tissues. Common uses include bone repair, bone augmentation, as well as coating of implants or acting as fillers in bone or teeth [132- 135].

The use of HA without additives for load bearing bone regeneration application is inefficient due to its low strength. There have been many attempts to improve the mechanical properties of HA scaffold by fiber and whisker reinforcement, coating with bioactive glass, bioactive polymers and nanocomposites [136, 137]. To overcoming mechanical limitations of HA, one approach is to combine it with a tougher phase, producing a composite by reinforcing it with suitable materials [138]. Recently, calcium silicate based bioceramics have been studied as potential substitutes for hard tissue like bone, because of their superior bioactivity compared to HA [139]. Wollastonite $CaSiO_3$ (W), a calcium silicate ceramic, has been widely used as a filler to fabricate apatite composites for improving mechanical properties. The acicular structure of W has been reported to strengthen the apatite crystals [140]. The addition of different amounts of wollastonite in W–A (Wollastonite-Apatite) sintered composite promotes an improvement on its mechanical properties, bioactivity and biodegradability [141, 142].

References

[1] Bergmann C. P., Stumpf A., 2013, Dental Ceramics, Microstructure Properties and Degradation, Series: Topics in Mining, Metallurgy and Materials Engineering, VII, 9-13.

[2] Hench L.L., 1991, Bioceramics: From Concept to Clinic, *J. Am. Ceram. Soc.* Vol.74, Issue 7, 1487- 1510.

[3] Cormack A. N., Tilocca D. A., 2012, Structure and biological activity of glasses and ceramics, *Phil. Trans. R. Soc. A* . 370, 1271–1280.

[4] Giannoudis P. V., Dinopoulos H. & Tsiridis E., 2005, Bone substitutes: an update. *Injury,* 36, S20–S27. (doi:10.1016/j.injury.2005.07.029).

[5] Desai B. M., 2007, Osteobiologics, *Am. J. Orthop.* 36, 8–11.

[6] Damien C. J., ParsonsJ. R., 1991, Bone graft and bone graft substitutes: a review of current technology and applications, *J. Appl. Biomater.* 2, 187–208.

[7] Dubok,V. A., 2000, Bioceramics - Yesterday, Today, Tomorrow, *Powder Metallurgy and Metal Ceramics* 39(7-8), 381-394.

[8] Patel, N. R., Gohil, P.P., 2012. A Review on Biomaterials: Scope, Applications& Human Anatomy Significance, *International Journal of Emerging Technology and Advanced Engineering* Volume 2, Issue 4, 91-101.

[9] Chen Q. Z., Wong, C. T., Lu W. W., Cheung K. M. C., Leong J. C. Y., Luk K. D. K., 2004, Strengthening mechanisms to crystalline hydroxyapatite in vivo, *Biomaterials,* 25, 4243-4254.

[10] Wang, M., 2003, Developing bioactive composite materials for tissue replacement, *Biomaterials*, 23, 2133-2151.

[11] Posner A. S., 1985, The structure of bone apatite surfaces. *J. Biomed. Mater. Res.*, 19, 241-250.

[12] Liu X., Zhao X., Fu Ricky K.Y, Joan P.Y. Ho, Chuanxian D., Paul K.C., 2005, Plasma-treated nano-structured TiO_2 surface supporting biomimetic growth of apatite, *Biomaterials*, 26, 6143-6150.

[13] Giavaresi G., Branda F., Causa F., Luciani, G., Fini, M., Aldini, N.N., Rimondini L., Ambrosio L., Giardino R., 2004, Poly(2-hydroxyethyl methacrylate) biomimetic coating to improve osseointegration of a PMMA/HA/Glass composite implant: *In vivo* mechanical and histomorphometric assessments. *Int. J. Artif. Organs 27*, 674–680.

[14] Sachlos E., Czernuszka J.T., 2003, Making tissue engineering scaffolds work. Review on the application of solid freeform fabrication technology to the production of tissue engineering scaffolds, *Eur. Cell Mater.*, *5*, 29–40.

[15] Yang F., Wolke J.G.C., Jansen J.A., 2008, Biomimetic calcium phosphate coating on electrospun poly(å-caprolactone) scaffolds for bone tissue engineering, *Chem. Eng. J. 137*, 154–161.

[16] Dean-Mo L., Troczynski T., Wenjea J., Seng T., 2001, Water-based sol-gel synthesis of hydroxyapatite: process development, *Biomaterials,* Vol. 22, pp. 1721-1730.

[17] Al-Sanabani J. S., Madfa A. A., Al-Sanabani F. A., 2013, Application of Calcium Phosphate Materials in Dentistry, *International Journal of Biomaterials,* Article ID 876132,12 pages, http://dx.doi.org/10.1155/2013/876132.

[18] Ivan F. D., Andreea M., Tanase C. E., Butnaru M., Vereştiuc L., 2013, Biomimetic Composites Based on Calcium Phosphates and Chitosan - Hyaluronic Acid with Potential Application in Bone Tissue Engineering, *Key Engineering Materials,* 587, 191-196.

[19] Neel E. A. A., Chrzanowski W., Knowles J. C., 2014, Biological performance of titania containing phosphate-based glasses for bone tissue engineering applications, *Mat. Sci. Eng.: C***,** 35, 307–313.

[20] Boutin P., 2014, Total arthroplasty of the hip by fritted alumina prosthesis. Experimental study and 1st clinical applications, *Orthopaedics & Traumatology: Surgery & Research* 100, 15–21.

[21] Saadaldin S. A., Rizkalla A. S., 2014, Synthesis and characterization of wollastonite glass–ceramics for dental implant applications, *Dental Materials,* 30, 3, 364–371.

[22] Diba M., Goudouri O. M., Tapia F., Boccaccini A. R., march 2014, Magnesium-containing bioactive polycrystalline silicate-based ceramics and glass-ceramics for biomedical applications, in press.

[23] Baino F., Vitale-Brovarone C., 2014, Mechanical properties and reliability of glass–ceramic foam scaffolds for bone repair, *Mat. Let.*, 118, 27–30.

[24] Thamaraiselvi T. V. and Rajeswari S., 2004, Biological Evaluation of Bioceramic Materials - A Review, *Trends Biomater. Artif. Organs*, Vol 18 (1), 9-17.

[25] Maccauro G., Iommetti P. R., Raffaelli L.and Manicone P. F., 2011, Alumina and Zirconia Ceramic for Orthopaedic and Dental Devices, in Biomaterials Applications for Nanomedicine, Edited by Prof. Rosario Pignatello, ISBN 978-953-307-661-4, 299-308.

[26] Adeel A., 2014, Implantable zirconia bioceramics for bone repair and replacement: A chronological review, *Materials Express,* 4, 1, 1-12.

[27] Billotte W., 2002, Ceramic Biomaterials, *in Biomaterials, CRC Press,* 21–53,

[28] Rokusek D., Davitt C., Bandyopadhyay A., Bose S., Hosick H.L., 2005, Interaction of human osteoblasts with bioinert and bioactive ceramic substrates, *J. Biomed. Mater. Res.* A, 1,75, 3, 588-94.

[29] Hench L., and Kokubo, T., 1998, Properties of Bioactive Glasses and Glass-Ceramics. Handbook of Biomaterial Properties. London: Chapman-Hall, pp. 355–356.

[30] Kokubo T., 1991, Bioactive Glass Ceramics: Properties and Applications. *Biomaterials*, 12, 2, 155-63.

[31] Kokubo T., Shigematsu M., Nagashisma Y. et al., 1982, Apatite and wollastonite containing glass ceramics for prosthetic application, *Bull. Inst. Chem. Res. Kyoto Univ.*, 60, 260-268.

[32] Shyu J. J., Wu J. M., 1990, Crystallization of MgO-CaO-SiO_2 -P_2O_5 glass, *J. Am. Ceram. Soc.*, 73, 1062-1068.

[33] Gisep, A., 2002, Research on Ceramic Bone Substitutes: Current Status", *Injury,* 33, 2, 88–92.

[34] Hamadouche, M. and Sedel, L., 2000, "Ceramics in Orthopaedics", *J. Bone Joint. Surg. Br.,* 82, 8, 1095–1099.

[35] Tuzlakoglu K., Reis R.L., 2007, Formation of bone-like apatite layer on chitosan fiber mesh scaffolds by a biomimetic spraying process, *J. Mater. Sci. Mater. Med.,* 18, 1279–1286.

[36] Ferraz M.P., Monteiro F.J., Manuel C.M., 2004, Hydroxyapatite nanoparticles: A review of preparation methodologies, *J. Appl. Biomater. Biomech.*, 2, 74-80.

[37] Santos M.H., Oliveira M., Freitas S. P., Mansur H. S., Vasconcelos W. L., 2004, Synthesis control and characterization of hydroxyapatite prepared by wet precipitation process. *Mater. Res.,* 7, 4, 625-630.

[38] Gallo R., 2011, Synthesis and characterization of substituted apatites for biomedical aplications, thesis, Università degli Studi di Padova Facoltá di Ingegneria Corso di Laurea Magistrali in Ingegneria dei Materiali,

[39] Peipei W., Caihong L., Haiyan G., Xuerong J., Hongqiang W., Kaixing L., 2010, Effects of synthesis conditions on the morphology of hydroxyapatite nanoparticles produced by wet chemical process, *Powder Technology,* 203, 315–321.

[40] Salma K., Berzina L., Borodajenko N., 2010, Calcium phosphate bioceramics prepared from wet chemically precipitated powders, *Processing and Application of Ceramics*, 4, 45– 51.

[41] Afshar A., Ghorbani M., Ehsani N., Saeri M.R., Sorrell C.C., 2003, Some importante factors in the wet precipitation process of hydroxyapatite, *Materials and Design*, 24, 197–202.

[42] Narasara, T. S. B., Phebe, D. E., 1996, Some physico-chemical aspects of hydroxyapatite, Narasaraju, *J. Mater. Sci. 31*, 1-21.

[43] Wang L. and Nancollas G. H., 2008, Calcium Orthophosphates: Crystallization and Dissolution, *Chem. Rev.* 108, 11, 4628–4669.

[44] Wang Y.J., Chen J.D., Wei K., Zhang N.G. S.H., Wang X. D., 2006, Surfactant-assisted synthesis of hydroxyapatite particles, *Materials Letters*, 60, 3227–3231.

[45] Shanti P. M. S. L., Ashok M., Balasubramanian T., Riyasdeen A., Akbarsha M.A., 2009, Synthesis and characterization of nan-hydroxyapatite at ambient temperature using cationic surfactant, *Materials Letters,* 63, 2123–2125.

[46] Coelho J. M., Moreira J. A., Almeida A., Monteiro F. J., 2010, Synthesis and characterization of HAp nanorods from a cationic surfactant template method, *J. of Mat. Sci.: Mat. in Med.*, 21, 2543–2549.

[47] Michael P., Shantini S. L., Ashok M., Mangalaranja R.V., Balasubramaniam T., 2012, Synthesis and characterization of crystalline hydroxyapatite spheroids using template for bioapplications, *Key Eng. Mat.,* v. 493-494, 723-727.

[48] Kolodziejczak-Radzimska A., Samuel M., Paukszta D., PiaseckiA., Jesionowski T., 2014, Synthesis of hydroxyapatite in presence of anionic surfactant, *Physicochem. Probl. Miner. Process.* 50, 1, 225−236.

[49] Sun. Q., Wang, H., The surface surfactants on the topography of HA, *Adv. Mat. Res.,* 2013, v.690-693, 466-469.

[50] Suchanek W. L., Riman R. E., 2006, Hydrothermal synthesis of advanced ceramic powders, *Adv. Sci. Technol.,* 45, 184-193.

[51] Yoshimura M., Sujaridworakun P., Koh F., 2004, Hydrothermal conversion of calcite crystals to hydroxyapatite, *Mat. Sci. Eng.* - C, 24, 4, 521-525.

[52] Riman R. E., Suchanek W. L., Bytappa K., Chen C. W., Shuk P. and C. S. Oakes, 2002, Solution synthesis of hydroxyapatite designer particulates, *Solid State Ionics*, 151, 393-402.

[53] Zhang X. and Vecchio K. S., 2007, Hydrothermal synthesis of hydroxyapatite rods, *J. Crystal Growth,* 308, 133-140.

[54] Manafi S. A., Joughehdoust S., 2009, Synthesis of Hydroxyapatite Nanostructure by Hydrothermal Condition for Biomedical Application, *Iranian Journal of Pharmaceutical Sciences,* 5(2), 89-94.

[55] Liu J., Ye X., Wang H., Zhu M., Wang B., and Yan H., 2003, The influence of pH and temperature on the morphology of ydroxyapatite synthesized by hydrothermal method, *Ceram. International,* 29, 629-633.

[56] Renlong X., Fuzeng R., Yang L., 2010, Synthesis and characterization of nano-crystalline calcium phosphates with EDTA-assisted hydrothermal method, *Materials and Design*, 31, 1691–1694.

[57] Sanosh K., Chu M.C., Balakrishnan A., Lee Y. J., Kim T., Cho S. J., 2009, Synthesis of nano hydroxyapatite powder that simulate teeth particle morphology and composition. *Curr. Appl. Phys*., 9, 1459–62.

[58] Ramanan S. R., Venkatesh R., 2004A study of hydroxyapatite fibers prepared via sol–gel route. *Mater. Lett*., 58, 3320–3333.

[59] Anee T., Ashok M., Palanichamy M., Narayana K. S., 2003, A novel technique to synthesize hydroxyapatite at low temperature, *Mater. Chem. Phys*., 80, 725–730.

[60] Montero M., Saenz A., Castano V., 2009, Synthesis of nano-hydroxyapatite from silica suspensions through chemical compensation. *J. Exp. Nanosci*., 4, 193–202.

[61] Balamurugan A., Michel J., Faure J., Benhayoune H., Wortham L., Sockalingum G., Banchet V., Bouthors S., Laurent-Maquin D., Balossier G., 2006, Synthesis and structural analysis of sol gel derived stoicheometric monophasic hydroxyapatite, *Ceramics-Silikaty*, 50, 1, 27-31.

[62] W. Weng, S. Zhang, K. Cheng, H. Qu, P. Du, G. Shen, J. Yuan, G. Han, 2003, "Sol–gel preparation of bioactive apatite films", *Surface and Coatings Technology*, 167, 292-296.

[63] W. Weng, G. Han, P. Du, G. Shen, J. Yang, 2002, "The effect of citric acid addition on sol–gel preparation of apatite films", *Mater. Chem. Phys.,* 77, 578-582.

[64] Cheng K., Weng W., Han G., Du P., Shen G., Yang J., Ferreira J.M.F., 2003, Sol–gel derived fluoridated hydroxyapatite films, *Mat. Res. Bul.*, 38, 89-97.

[65] Choi D., Marra K., Kumta P. N., 2004, Chemical synthesis of hydroxyapatite/poly (caprolactone) composite, *Mat. Res. Bul.*, 39, 417-432.

[66] Li P., Groot K., 1994, Better bioactive ceramics through sol-gel process. *J. Sol-gel Sci. Technol.,* 2, 797-801.

[67] Fathi M., Hanifi A., Mortazavi V., 2008, Preparation and bioactivity evaluation of bone-like hydroxyapatite nanopowder, *J. Mater. Process. Tech.,* 202, 536–42.

[68] Kuriakose T.A. Kalkura S.N., Palanichamy M., Arivuoli D., Dierks K., Bocelli G., et al. 2004, Synthesis of stoichiometric nano crystalline hydroxyapatite by ethanolbased sol–gel technique at low temperature. *J. Cryst. Growth*, 263, 517–23.

[69] Sanosh K., Chu M.C., Balakrishnan A., Kim T., Cho S.J., 2009, Preparation and characterization of nano-hydroxyapatite powder using sol–gel technique, *Bull. Mater. Sci.*, 32, 465–70.

[70] Bigi A., Boanini E., Rubini K., 2004, Hydroxyapatite gels and nanocrystals prepared through a sol–gel process. *J. Solid State Chem.*, 177, 3092–3098.

[71] Eshtiagh-Hosseini H., Housaindokht M.R., Chahkandi M., 2007, Effects of parameters of sol–gel process on the phase evolution of sol–gel-derived hydroxyapatite, *Mater. Chem. Phys.*, 106, 310–316.

[72] Sanosh K.P., Min-Cheol C., Balakrishnan A., Kim T.N., Seong-Jai C., 2010, Sol–gel synthesis of pure nano sized β-tricalcium phosphate crystalline powders, *Current Applied Physics*, 10, 68–71.

[73] Saranya K., Kowshik M. and Ramanan S. R., 2011, Synthesis of hydroxyapatite nanopowders by sol–gel emulsion technique, *Bull. Mater. Sci.,* 34, 7, 1749–1753.

[74] Encinas-Romero M.A., Peralta-Haley J., Valenzuela-García J. L., Castillón-Barraza F.F., 2013, Synthesis and Structural Characterization of Hydroxyapatite-Wollastonite Biocomposites, Produced by an Alternative Sol-Gel Route, *J. Biomat. & Nanobiotechnology*, 4, 4, 327-333.

[75] Kokubo T., Kushitani H., Sakka S., Kitsugi T. and Yamamuro T., 1990, Solutions able to reproduce in vivo surface-structure changes in bioactive glass-ceramic A-W, *J. Biomed. Mater. Res.,* 24, 721-734.

[76] LeGeros R.Z., 2002, Properties of osteoconductive biomaterials: calcium phosphates, *Clin. Orthop. Relat. Res.*, 395, 81-98.

[77] Mangal R., Amit B., Susmita B., 2011, Induction plasma sprayed nano hydroxyapatite coatings on titanium for orthopaedic and dental implants, *Surface and Coatings Technology*, 2785–2792.

[78] Nakamura M., Kobayashi A., Nozaki K., Horiuchi N., Nagai A., Yamashita K., 2014, Improvement of Osteoblast Adhesion Through Polarization of Plasma-Sprayed Hydroxyapatite Coatings on Metal, *J. Med. Biol. Eng.*, 34, 1, 44-48.

[79] Mohseni E., Zalnezhad E., and Bushroa A. R., 2014, Comparative investigation on the adhesion of hydroxyapatite coating on Ti–6Al–4V implant: A review paper, *Inter. J. Adhesion and Adhesives,* 48, 238-257.

[80] Cleries L., Martinez E., Fernandez-Pradas J. M., Sardin G., Esteve J., Morenza J. L., 2000, Mechanical properties of calcium phosphate coatings deposited by laser ablation, *Biomaterials,* 21, 967–971.

[81] Choi J., Bogdanski D., Koller M., Esenwein S. A., Muller D., Muhr G., Epple M., 2003, Calcium phosphate coating of nickel-titanium shape-memoryalloys. Coating procedure and adherence of leukocytes and platelets, *Biomaterials*, 24, 3689–3696.

[82] Liu D., Yang Q., Troczynski T., 2002, Sol-gel hydroxyapatite coatings on stainless steel substrates. *Biomaterials,* 23, 691–698.

[83] Ma J., Wang C., Peng K. W., 20003, Electrophoretic deposition of porous hydroxyapatite scaffold, *Biomaterials,* 24, 3505–3510.

[84] Wie H., Hero H., Solheim T., 1998, Hot isostatic pressing-processed hydroxyapatite-coated titanium implants: light microscopic and scanning electron microscopyinvestigations, *Int. J. Oral Maxillofac Implants,* 13, 837–844.

[85] Kalita S. J., Bhardwaj A., Bhatt H. A., 2007, Nanocrystalline calcium phosphate ceramics in biomedical engineering, *Material Science Engeneering* C, 27, 441-449.

[86] Habibovic P., Barrere F., van Bliterswijk C. A., de Groot K., Layrolle P., 2002, Biomimetic hydroxyapatite coating on metal implants. *J. Am. Ceramic. Soc.,* 85, 517–522.

[87] Loty C., Loty S., Kokubo T., Forest N., and Sautier J.M., 1997, Prefabricated Biological Apatite Formation on a Bioactive Glass-Ceramic Promotes In Vitro Differentiation of Fetal Rat Chondrocytes, Bioceramics. Paris: Elsevier Science Publishers, 219–222.

[88] Loty C., Sautier J.M., Boulekbache H., Kokubo T., Kim H.M., and Forest N.J., 2000, In vitro bone formation on a bonelike apatite layer prepared by a biomimetic process on a bioactive glass-ceramic, *Biomed. Mater. Res*., 49, 423-434.

[89] Tengvall P., Lundstrom I., 1992, Physico-chemical considerations of titanium as a biomaterial, *Clin. Mater*., 2, 115-134.

[90] Zhu P., Masuda Y., Koumoto K., 2004, The effect of surface charge on hydroxyapatite nucleation. *Biomaterials,* 25, 3915-3921.

[91] Wu W., Nancollas G. H., 1997, Kinetics of nucleation and crystal growth of hydroxyapatite and fluoroapatite on titanium oxide surfaces. *Colloids Surf* B: Biointerfaces, 10, 87–94.

[92] Combes C., Rey C., Freche M., 1998, XPS and IR study of dicalcium phosphate dehydrate. *Colloids Surf* B: Biointerfaces, 11, 15–27.

[93] Li P., Ohtsuki C., Kokubo T., Nakanishi K., Soga N., de Groot K., 1994, Role of hydrated silica, titania and alumina in forming biologically active bone-like apatite on implants, *J. Biomed. Mater. Res.,* 28, 7–15.

[94] Healy K. E., Ducheyne P., 1992, Hydration and preferential molecular adsorption on titanium in vitro. *Biomaterials*, 13, 8, 553–561.

[95] Kokubo T., Kushitani H., Abe Y., Yamamuro T., 1989, Apatite coating on various substrates in simulated body fluids, *Bioceramics,* 2, 235–2342.

[96] Barrére F., Layrolle P., van Blitterswijk K., 2000, Fast formation of biomimetic Ca-P coating on Ti6Al4V. *Mater. Res. Soc. Symp. Proc*. 2000, 599, 135–140.

[97] Habibovic P., Barrere F., van Blitterswijk C.A., de Groot K., Layrolle P., 2002, Biomimetic hydroxyapatite coating on metal implants, *J. Am. Ceram. Soc*., 85, 3, 517–22.

[98] Barrere F., van Blitterswijk C. A., de Groot K., Layrolle P., 2002, Nucleation of biomimetic Ca-P coatings on Ti6A14V from a SBF_5 solution: influence of magnesium. *Biomaterials*, 23, 10, 2211–2220.

[99] Pattanayak D. K, Yamaguchi S., Matsushita T., Nakamura T., Kokubo T., 2012, Apatiteforming ability of titanium in terms of pH of the exposed solution, J. R. Soc. Interface, 74, 2145-2155.

[100] Newesely H., 1961, Changes in crystal types of low solubility calcium phosphates in presence of accompanying ions. *Arch. Oral Biol.*, *6*, S174–S180.
[101] Eanes E.D., Rattner S.L., 1980, The effect of magnesium on apatite formation in seeded supersaturated solutions at pH 7.4., *J. Dent. Res.*, *60*, 1719–1723.
[102] Tomazic B., Tomson M., Nancollas G.H., 1975, Growth of calcium phosphates on hydroxyapatite crystals: The effect of magnesium, *Arch. Oral Biol.*, *20*, 803–808.
[103] Cao W., Hench L.L., Bioactive materials, *Ceram. Int.*, *22*, 493–523.
[104] Verné E., Bretcanu O., Balagna C., Bianchi C. L., Cannas M., Gatti S., Vitale-Brovarone C., 2009, Early stage reactivity and in vitro behavior of silica-based bioactive glasses and glass-ceramics, *J. Mater. Sci. Mater. Med.*, *20*, 75–87.
[105] Burg J. L., Porter S., Kellam J. F., 2000, Biomaterial developments for bone tissue engineering, *J. Biomater,* 21, 2347–2359.
[106] Kokubo T., 1991, Bioactive glass ceramics: properties and applications, *Biomaterials*, 155-163.
[107] Hench L.L., 1998, Bioceramics, *J. Am. Ceram. Soc.*, 81, 1705–1728.
[108] Karlsson K. H., 1998, Bioactivity of glass and its relation to glass structure, *Glass Phys. Chem.* 24, 280–284.
[109] Karlsson K. H., Backman R., Hupa M., 2002, An equilibrium study of phosphate precipitation on bioactive glass, *Key Eng. Mater.* 218–220, 103–108.
[110] Karlsson K. H., 2004, Bioactivity of glass and bioactive glasses for bone repair, *Glass Technol.,* 45,157–161.
[111] Agathopoulosa S., Tulyaganova D. U., Ventura J. M. G., Kannana S., Karakassides M.A., Ferreira J. M. F., 2006, Formation of hydroxyapatite onto glasses of the CaO–MgO–SiO_2 system with B_2O_3, Na_2O, CaF_2 and P_2O_5 additives, *Biomaterials* 27, 9, 1832–1840.
[112] Liu X.Y., Ding C.X., Chu P.K., 2004, Mechanism of apatite formation on wollastonite coatings in simulated body fluids, *Biomaterials* 25, 1755–1761.
[113] Liu X. Y., Ding C. X., Wang Z., 2001, Apatite formed on the surface of plasma sprayed wollastonite coating immersed in simulated body fluid, *Biomaterials, 22*, 2007-2012.
[114] Sarmento C., Luklinska Z.B., Brown L., Anseau M., De Aza P.N., De Aza S., et al., 2004. In vitro behavior of osteoblastic cells cultured in the presence of pseudowollastonite ceramic, *J. Biomed. Mater. Res.* A, 69, 351–358.
[115] De Aza P.N., Luklinska Z.B., Martinez A., Anseau M.R., Guitian F., 2000, Morphological and structural study of pseudowollastonite implants in bone, *J. Microsc.,* 197, 60–67.
[116] Iwata N. Y., Lee G. H., Tokuoka Y., Kawashima N., 2004, Sintering behavior and apatite formation of diopside prepared by coprecipitation process*, Colloids Surf.* B: *Biointerfaces,* 34, 239–245.
[117] Zhao W. Y., Chang J., 2005, Preparation and characterization of novel tricalcium silicate bioceramics, *J. Biomed. Mater. Res.* A 73, 86–89.
[118] Du R. L., Chang J., 2004. Preparation and characterization of bioactive sol–gel derived $Na_2Ca_2Si_3O_9$, *J. Mat. Sci.: Materials in Medicine,* 15, 12, 1285-1289.
[119] Wu C., Chang J., 2007, Degradation, bioactivity, and cytocompatibility of diopside, akermanite, and bredigite ceramics, *J.Biomed. Mater. Res. B Appl. Biomater.,* 83, 1, 153-160.

[120] Li R., Clark A. E., Hench L. L., 1991, An investigation of bioactive glass powders by sol-gel processing, *J. Appl. Biomater.*, 2, 231– 239.

[121] Yamagata C., Paiva M. R. S, Higa O. Z., Rodas A. D., Silveira A. C. F., Reis S. T, Sol gel derived CaO-MgO-SiO_2 ceramic glass system: preparation and in vitro characterization, http://confsys.encs.concordia.ca/ICCM19/AllPapers/FinalVersion/YAM80662.pdf

[122] Yamagata C., Elias D. R., Paiva M. R. S., Misso A. M., Castanho S. R. H. M., 2012, Influence of the precursor concentration on the characteristics of silica powder obtained from Na_2SiO_3 by a facile low temperature synthesis process, *J. Mat. Sci. Eng.* B2 (8), 429-436.

[123] Cerruti M. G., 2004, *Characterization of bioactive glasses. Effect of the immersion in solutions, that simulate body fluids,* Ph.D. thesis, University of Turin, Department of Chemistry IFM.

[124] Ostomel A. T., Shi Q., Tsung C. K., Liang H., Stucky G. D., 2006, Spherical bioactive glass with enhaced rates of hidroxyapatite deposition and hemostatic activity, *Small*, 2, 11, 1261-1265.

[125] Diba M., Tapia F., Boccaccini A. R. and Strobel L. A., 2012, Magnesium-Containing Bioactive Glasses for Biomedical Applications, *International Journal of Applied Glass Science*, 3, 221–253.

[126] Souza M. T., Crovace M. C., Schröder C., Eckert H., Peitl O., Zanotto E. D., 2013, Effect of magnesium ion incorporation on the thermal stability,dissolution behavior and bioactivity in Bioglass-derived glasses, *J. Non-Crystalline Solids*, 382, 7–65.

[127] Hutmacher D. W., Schantz J. T., Lam C. X. F., Tan K. C., Lim T. C., 2007, State of the art and future directions of scaffold based bone engineering from a biomaterials perspective, *J. Tissue Eng. Regenerat. Med.*, 1, 245–60.

[128] Habraken W., Wolke J. G. C., Jansen J. A., 2007, Ceramic composites as matrices and scaffolds for drug delivery in tissue engineering, *Adv. Drug. Deliv. Rev.*, 59, 234–48.

[129] Langer R., Vacanti J. P., 1993, *Tissue Engineering Science*, 260, 920–926.

[130] Khademhosseini A., Langer R., Borenstein J.,Vacanti J.P., 2006, Microscale technologies for tissue engineering and biology. *Proc. Natl. Acad. Sci. USA*, 103, 2480-2487.

[131] Bauer T. W., Geesink R. C., Zimmerman R., McMahon J. T., 1991, Hydroxyapatite-coated femoral stems. Histological analysis of components retrieved at autopsy, *J. Bone Joint Surg.*73, 1439-1452.

[132] Johnson A. J. W., Herschler B. A., 2011, A review of the mechanical behavior of CaP and CaP/polymer composites for applications in bone replacement and repair. *Acta Biomater.*,7, 16–30.

[133] Bohner M., 2010, Design of ceramic-based cements and putties for bone graft substitution. *Eur. Cells Mater.*, 20, 1–12.

[134] Ginebra M. P., Espanol M., Montufar E.B., Perez R. A., Mestres G., 2010, New processing approaches in calcium phosphate cements and their applications in regenerative medicine. *Acta Biomater.*, 2863–2873.

[135] Dorozhkin S.V., 2009, Calcium orthophosphate-based biocomposites and hybrid biomaterials, *J. Mater. Sci.*, 44, 2343–2387.

[136] Jun I., Song J., Choi W., Koh Y., Kim H., 2007, Porous hydroxyapatite scaffolds coated with bioactive apatite–wollastonite glass-ceramics, *J. Am. Ceram. Soc.*, 90, 2703–2708.

[137] Freyman M., Yannas I.V., Gibson L.J., 2011, Cellular materials as porous scaffolds for tissue engineering, *Progress in Mater. Sci.,* 46, 273–282.

[138] Suwanprateeb J., Sanngam R., Suvannapruk W., Panyathanmaporn T., 2009, Mechanical and in vitro performance of apatite-wollastonite glass ceramic reinforced hydroxyapatite composite fabricated by 3D-printing, *J. Mater. Sci. Mater Med*., 20, 6, 1281-1289.

[139] Ni S., Chang J., Chou L., 2006, A novel bioactive porous $CaSiO_3$ scaffold for bone tissue engineering, J. Biom. Mat. Res., 76, 196–205.

[140] Kokubo T., A/W glass-ceramic: processing and properties, in: L.L. Hench, J. Wilson (Eds.), An Introduction to Bioceramics, World Scientific, Singapore, 1993, pp. 75–88.

[141] Lin K., Zhang M., Zhai W., Qu H., Chang J., 2011, Fabrication and characterization of hydroxyapatite/wollastonite composite biocera- mics with controllable properties for hard tissue repair, *J. Am. Ceram. Soc.,* 94, 99–105.

[142] Martin A., Romero E., Salinas S. A., Jesus L., Garcia V., Payan S.R., Castillo′n-Barraza F.F., 2010, Mechanical and bioactive behavior of hydroxyapatite–wollastonite sintered composites, *Inter. J. Appl. Ceram. Techn.*, 7, 164–177.

INDEX

C

D

G

H

I

J

K

L

M

N

O

P

Q

R

T

U

V

W

X

Y

Z